Sustainability of Transport Infrastructures

Sustainability of Transport Infrastructures

Guest Editors

Joel R. M. Oliveira
Hugo Silva
R. Christopher Williams
Zejiao Dong

MDPI

Basel • Beijing • Wuhan • Barcelona • Belgrade • Novi Sad • Cluj • Manchester

Guest Editors
Joel R. M. Oliveira
Department of Civil Engineering
University of Minho
Guimaraes
Portugal

Hugo Silva
Department of Civil Engineering
University of Minho
Guimaraes
Portugal

R. Christopher Williams
Department of Civil, Construction, and Environmental Engineering
Iowa State University
Ames
United States

Zejiao Dong
School of Transportation Science and Engineering
Harbin Institute of Technology
Harbin
China

Editorial Office
MDPI AG
Grosspeteranlage 5
4052 Basel, Switzerland

This is a reprint of the Special Issue, published open access by the journal *Sustainability* (ISSN 2071-1050), freely accessible at: www.mdpi.com/journal/sustainability/special_issues/infrastructures.

For citation purposes, cite each article independently as indicated on the article page online and using the guide below:

Lastname, A.A.; Lastname, B.B. Article Title. *Journal Name* **Year**, *Volume Number*, Page Range.

ISBN 978-3-7258-2902-6 (Hbk)
ISBN 978-3-7258-2901-9 (PDF)
https://doi.org/10.3390/books978-3-7258-2901-9

Contents

About the Editors

Joel R. M. Oliveira

Dr. Joel R. M. Oliveira is an Assistant Professor in the Department of Civil Engineering at the University of Minho and a member of ISISE. He completed his Ph.D. at the University of Nottingham, UK, in 2006. His research focuses on sustainable road pavements, particularly life-cycle assessment, pavement preservation, and recycling. Dr. Oliveira is actively involved in academic leadership and international collaborations.

Hugo Silva

Dr. Hugo M. R. D. Silva is an Assistant Professor in the Department of Civil Engineering at the University of Minho and a member of ISISE. His research focuses on sustainable asphalt materials, including waste incorporation, recycling, bio-asphalt binders, and self-healing technologies. Dr. Silva also directs the Master's Degree in Civil Engineering and participates in international research collaborations and editorial activities.

R. Christopher Williams

Dr. R. Christopher Williams is the Gerald and Audrey Olson Professor of Civil Engineering at Iowa State University and the Director of the Asphalt Materials and Pavements Program at the Institute for Transportation. His research expertise includes asphalt materials, sustainable pavement design, and biopolymer applications. Dr. Williams has extensive experience in research, education, and leadership in the field of transportation infrastructure.

Zejiao Dong

Dr. Zejiao Dong is a Full Professor at the School of Transportation Science and Engineering at Harbin Institute of Technology, China. His research interests focus on pavement materials and the sustainability of transportation infrastructure. Dr. Dong has contributed significantly through his research and academic collaborations.

Preface

This reprint compiles a selection of research articles from the Special Issue, titled "Sustainability of Transport Infrastructures", addressing critical challenges and innovative solutions in the field. This collection of articles highlights the growing importance of sustainable practices in transportation systems, which play a vital role in reducing environmental impacts, promoting economic efficiency, and enhancing social equity.

This reprint begins with a comprehensive overview, discussing holistic approaches to sustainable road maintenance policies and their assessment. Following this, it explores crucial challenges, including the decarbonization of public transport and the sustainable management of resources in light-duty vehicles, emphasizing the pressing need for climate action and resource conservation.

A significant portion of this collection explores sustainable materials in road construction, showcasing the potential of steel slag, recycled concrete aggregates, and bio-based materials to replace traditional, resource-intensive components. Urban mobility solutions, such as bike-sharing systems and permeable pavements, are also examined, offering practical applications to enhance city sustainability.

The final section focuses on advanced pavement assessment and maintenance methodologies, demonstrating how technologies like life cycle assessment and polymer-modified asphalts can contribute to long-term environmental and economic benefits.

By addressing topics ranging from innovative materials to systemic approaches, this Book represents a comprehensive resource for researchers, practitioners, and policymakers. By providing actionable insights, this Book facilitates the creation of sustainable transport infrastructure capable of meeting current and future challenges.

The editors hope this reprint inspires collaboration and innovation in creating resilient, eco-friendly transport systems supporting global sustainability goals.

Joel R. M. Oliveira, Hugo Silva, R. Christopher Williams, and Zejiao Dong
Guest Editors

sustainability

MDPI

Editorial

Editorial of "Sustainability of Transport Infrastructures"

Joel R. M. Oliveira [1,*], Hugo M. R. D. Silva [1], R. Christopher Williams [2] and Zejiao Dong [3]

1 Department of Civil Engineering, Institute for Sustainability and Innovation in Structural Engineering, University of Minho, Campus de Azurem, 4800-058 Guimarães, Portugal; hugo@civil.uminho.pt
2 Department of Civil, Construction, and Environmental Engineering, Iowa State University, Ames, IA 50011, USA; rwilliam@iastate.edu
3 School of Transportation Science and Engineering, Harbin Institute of Technology, Harbin 150090, China; hitdzj@hit.edu.cn
* Correspondence: joliveira@civil.uminho.pt

Sustainability in transport infrastructures has become a key priority for governments, industries, and researchers worldwide. Alongside their substantial environmental repercussions, efficient transportation systems contribute significantly to societies' economic and social progress. Recent international goals, such as the United Nations Sustainable Development Goals [1], have underscored the need to reduce the carbon footprint of transportation. Organisations like the European Union [2] have also set ambitious targets for lowering greenhouse gas emissions. These objectives have pushed the industry to adopt greener technologies, use materials more efficiently, and develop innovative strategies to enhance the resilience and longevity of transport networks [3].

In the context of pavement construction, sustainability efforts focus on reducing the consumption of natural resources, minimising energy use, and lowering the environmental impact of infrastructure projects. These goals have been achieved through the increased integration of bio-based binders [4] and advanced design techniques, along with the adoption of circular economy principles, which promote the reuse and recycling of materials to reduce waste and extend the life cycle of resources used in construction [5]. Moreover, self-healing materials have shown promise as an innovative approach to enhance the durability of pavements by allowing minor damage to repair itself, potentially reducing maintenance needs and extending infrastructure lifespans [6]. Sustainable transport infrastructures are also essential for enhancing social equity, as they improve accessibility and connectivity, particularly in underserved communities [7]. Resilience to climate change is another critical aspect [8]. With the increasing frequency of extreme weather events, it is crucial to design transport infrastructures that can withstand such conditions, ensuring long-term performance [9,10].

However, despite these advancements, several challenges remain. One of the primary issues is the durability and long-term performance of sustainable materials under varying environmental conditions, such as temperature fluctuations and moisture exposure [11]. Despite the increasing adoption of sustainable practices, the ability of these materials to endure different environmental stresses is still not fully understood. Additionally, although life cycle assessment (LCA) is widely recognised as a valuable tool for evaluating environmental impact [12], its practical application in projects remains limited [13]. More robust policies are needed to address these gaps and ensure the consistent implementation of LCA and other sustainable practices across different regions and transportation networks [14].

This collection of research papers on "Sustainability of Transport Infrastructures" brings together studies addressing these critical challenges. The articles span a range of topics, including decarbonising transport systems, utilising recycled and bio-materials in pavement construction, and conducting life cycle analyses of infrastructure solutions. Taken together, these contributions provide a solid foundation for understanding the

Citation: Oliveira, J.R.M.; Silva, H.M.R.D.; Williams, R.C.; Dong, Z. Editorial of "Sustainability of Transport Infrastructures". *Sustainability* **2024**, *16*, 10158. https://doi.org/10.3390/su162310158

Received: 3 October 2024
Accepted: 14 November 2024
Published: 21 November 2024

current state of sustainable transport infrastructure and offer pathways to address existing knowledge gaps.

A significant focus of this volume is the increasing use of recycled materials in pavement construction in alignment with the principles of the circular economy. Research shows that incorporating recycled asphalt pavement, reclaimed concrete aggregates, and other industrial by-products, such as steel slag, can significantly reduce the environmental footprint of road construction. In addition to maintaining or improving the mechanical properties of pavements, these materials help reduce reliance on virgin resources and thus lower construction costs [15]. In summary, using recycled materials supports both environmental sustainability and economic efficiency by promoting the circular use of resources.

Decarbonisation in transportation is another central theme in this Special Issue. Several articles examine how technologies such as warm mix asphalt (WMA) and alternative binders contribute to reducing carbon emissions. For instance, WMA enables lower production temperatures, thus cutting fuel consumption and greenhouse gas emissions during asphalt production. These studies highlight practical solutions already implemented in real-world scenarios, demonstrating how they can significantly reduce the carbon footprint of transport infrastructure.

Life cycle assessment is also a crucial research theme presented here. LCA enables a comprehensive evaluation of environmental impacts throughout the entire lifespan of a transport system, including stages such as material extraction, production, construction, maintenance, and eventual disposal or recycling. Several contributions in this issue illustrate how LCA can effectively compare the environmental performance of different materials and maintenance strategies. Consequently, these insights hold significant value for policymakers, enabling them to make well-informed decisions about which infrastructure solutions are the most sustainable.

Looking ahead, several promising directions for research and development stand out. For example, bio-based materials, such as bio-binders derived from waste products, offer great potential for reducing the environmental impact of pavement construction [16]. However, further research is needed to evaluate the long-term performance of these materials under real-world conditions. Additionally, the emergence of smart infrastructure systems presents another exciting avenue for exploration. Integrating sensors and monitoring technologies into pavements makes it possible to collect real-time data on road performance [17]. These data could then optimise maintenance schedules, extend infrastructure lifespans, and reduce the need for resource-intensive repairs [18]. In combination with sustainable construction practices, these technologies could transform how transport networks are built and maintained.

Moreover, policy frameworks must also evolve to support the adoption of sustainable materials and practices. Technological advancements are essential, but strong policy measures are equally necessary to encourage the use of greener technologies and ensure compliance with sustainability standards [19]. Policymakers can find valuable guidance in the research presented in this volume. Indeed, it demonstrates how sustainable practices can be effectively implemented and highlights the benefits of adopting a life cycle approach to decision-making.

The contributions to this Special Issue offer a comprehensive view of the current state of sustainable transport infrastructure. By addressing key challenges, such as using recycled materials, reducing carbon emissions, and implementing life cycle assessments, the research offers theoretical insights and practical applications to guide future infrastructure projects. As the transportation sector continues to evolve in response to environmental, social, and economic pressures, collaboration between researchers, engineers, and policymakers is essential. Over time, these endeavours will pave the way for the emergence of technically viable and more sustainable transport solutions.

Future research will be crucial in advancing these initiatives, particularly in biomaterial development [20], smart infrastructure, policy integration, circular economy principles, and digitalisation of construction processes [21]. The integration of artificial intelligence (AI), data mining, and building information modelling (BIM) also offers promising pathways in sustainability assessments [18]. For instance, AI and data mining can enhance performance monitoring and predictive maintenance [22], allowing for optimised resource use and more sustainable pavement management. Additionally, combining BIM with life cycle assessment (LCA) enables more precise sustainability analyses by providing detailed insights into material use, energy consumption, and environmental impacts across the infrastructure life cycle [23]. By continuing to explore these avenues, society will move closer to achieving truly sustainable transport infrastructure that meets the needs of both present and future generations.

Conflicts of Interest: The authors declare no conflicts of interest.

References

1. United Nations. *Transforming Our World: The 2030 Agenda for Sustainable Development*; United Nations: New York, NY, USA, 25 September 2015. Available online: https://sdgs.un.org/2030agenda (accessed on 25 September 2024).
2. European Commission. *A New Circular Economy Action Plan For a Cleaner and More Competitive Europe, COM(2020) 98*; European Commission: Brussels, Belgium, 11 March 2020. Available online: https://eur-lex.europa.eu/legal-content/EN/TXT/?uri=COM:2020:98:FIN (accessed on 1 October 2024).
3. Patil, G.; Pode, G.; Diouf, B.; Pode, R. Sustainable Decarbonization of Road Transport: Policies, Current Status, and Challenges of Electric Vehicles. *Sustainability* **2024**, *16*, 8058. [CrossRef]
4. Kazemi, M.; Wang, H.; Fini, E. Bio-based and nature inspired solutions: A step toward carbon-neutral economy. *J. Road. Eng.* **2022**, *2*, 221–242. [CrossRef]
5. Rahman, M.T.; Mohajerani, A.; Giustozzi, F. Recycling of waste materials for asphalt concrete and bitumen: A review. *Materials* **2020**, *13*, 1495. [CrossRef] [PubMed]
6. Anupam, B.R.; Sahoo, U.C.; Chandrappa, A.K. A methodological review on self-healing asphalt pavements. *Constr. Build Mater.* **2022**, *321*, 126395. [CrossRef]
7. Xu, X.; Zhu, G.; Zhang, C.; Zhao, X.; Li, Y. Research Progress of the Impacts of Comprehensive Transportation Network on Territorial Spatial Development and Protection. *Land* **2024**, *13*, 479. [CrossRef]
8. Alhjouj, A.; Bonoli, A.; Zamorano, M. A Critical Perspective and Inclusive Analysis of Sustainable Road Infrastructure Literature. *Appl. Sci.* **2022**, *12*, 12996. [CrossRef]
9. Saleh, M.; Hashemian, L. Addressing Climate Change Resilience in Pavements: Major Vulnerability Issues and Adaptation Measures. *Sustainability* **2022**, *14*, 2410. [CrossRef]
10. Asres, E.; Ghebrab, T.; Ekwaro-Osire, S. Framework for Design of Sustainable Flexible Pavement. *Infrastructures* **2022**, *7*, 6. [CrossRef]
11. Shi, B.; Dong, Q.; Chen, X.; Gu, X.; Wang, X.; Yan, S. A comprehensive review on the fatigue resistance of recycled asphalt materials: Influential factors, correlations and improvements. *Constr. Build. Mater.* **2023**, *384*, 131435. [CrossRef]
12. Aryan, Y.; Dikshit, A.K.; Shinde, A.M. A critical review of the life cycle assessment studies on road pavements and road infrastructures. *J. Environ. Manag.* **2023**, *336*, 117697. [CrossRef] [PubMed]
13. Alaloul, W.S.; Altaf, M.; Musarat, M.A.; Javed, M.F.; Mosavi, A. Systematic review of life cycle assessment and life cycle cost analysis for pavement and a case study. *Sustainability* **2021**, *13*, 4377. [CrossRef]
14. Moins, B.; France, C.; Van den bergh, W.; Audenaert, A. Implementing life cycle cost analysis in road engineering: A critical review on methodological framework choices. *Renew. Sustain. Energy Rev* **2020**, *133*, 110284. [CrossRef]
15. Loureiro, C.D.A.; Moura, C.F.N.; Rodrigues, M.; Martinho, F.C.G.; Silva, H.M.R.D.; Oliveira, J.R.M. Steel Slag and Recycled Concrete Aggregates: Replacing Quarries to Supply Sustainable Materials for the Asphalt Paving Industry. *Sustainability* **2022**, *14*, 5022. [CrossRef]
16. Zhang, Y.; Ding, P.; Zhang, L.; Luo, X.; Cheng, X.; Zhang, H. Green roads ahead: A critical examination of bio-bitumen for sustainable infrastructure. *Front. Mater.* **2024**, *11*, 1382014. [CrossRef]
17. Wang, X.; Zhang, Y.; Li, H.; Wang, C.; Feng, P. Applications and challenges of digital twin intelligent sensing technologies for asphalt pavements. *Autom. Constr.* **2024**, *164*, 105480. [CrossRef]
18. Sakr, M.; Sadhu, A. Recent progress and future outlook of digital twins in structural health monitoring of civil infrastructure. *Smart Mater. Struct.* **2024**, *33*, 033001. [CrossRef]
19. Ruiz, A.; Vinke-De Kruijf, J.; Santos, J.; Keijzer, E.; Volker, L.; Dorée, A. The slow implementation of sustainable innovations in the asphalt paving sector: The role of actors and their interactions. *Constr. Manag. Econ.* **2024**, *42*, 902–925. [CrossRef]
20. Kalama, D.M.; Pipintakos, G.; Van den bergh, W. Time Travel Through Asphalt Bio-Binder Innovations. In *RILEM Bookseries*; Springer Science and Business Media B.V.: New York, NY, USA, 2024; Volume 58, pp. 35–43.
21. Gao, C.; Wang, J.; Dong, S.; Liu, Z.; Cui, Z.; Ma, N.; Zhao, X. Application of Digital Twins and Building Information Modeling in the Digitization of Transportation: A Bibliometric Review. *Appl. Sci.* **2022**, *12*, 11203. [CrossRef]

22. Ranyal, E.; Sadhu, A.; Jain, K. Road Condition Monitoring Using Smart Sensing and Artificial Intelligence: A Review. *Sensors* **2022**, *22*, 3044. [CrossRef] [PubMed]
23. Oreto, C.; Biancardo, S.A.; Abbondati, F.; Veropalumbo, R. Leveraging Infrastructure BIM for Life-Cycle-Based Sustainable Road Pavement Management. *Materials* **2023**, *16*, 1047. [CrossRef] [PubMed]

sustainability

MDPI

Article

Towards Road Sustainability—Part I: Principles and Holistic Assessment Method for Pavement Maintenance Policies

Anne de Bortoli [1,*], Adélaïde Féraille [2] and Fabien Leurent [3]

[1] LVMT, Ecole des Ponts, Université Gustave Eiffel, 4-20 Boulevard Newton, Cité Descartes, Champs-sur-Marne, CEDEX 2, F-77447 Marne-la-Vallée, France

[2] Navier Laboratory, Ecole des Ponts, Université Gustave Eiffel, 6-8 Avenue Blaise-Pascal, Cité Descartes, CEDEX 2, F-77455 Marne-la-Vallée, France; adelaide.feraille@enpc.fr

[3] CIRED, UMR 8568, Campus du Jardin Tropical, 45 bis Avenue de la Belle Gabrielle, CEDEX 2, F-94736 Nogent-sur-Marne, France; fabien.leurent@enpc.fr

* Correspondence: anne.de-bortoli@enpc.fr

Abstract: Assessing the holistic sustainability of public policies remains a challenge rarely taken up due to a lack of adequate assessing methods. Frequently, only environmental and/or financial aspects are addressed, rather than the three pillars, including macro- and micro-economic as well as social performance. This paper presents an assessment method to fully compare the performance of pavement resurfacing policies for all its stakeholders and considering pavement–vehicle interactions. First, an analytical and then systemic approach to road maintenance highlights all its stakeholders, and a complete set of sustainability indicators is proposed to quantify the various impacts of maintenance programs: tax revenues, road operator's and users' savings, domestic production and employment, net present value, users' time savings and noise reduction health benefits, as well as protection of natural resources, biodiversity and human health. Second, specific physical models of road condition (International Roughness Index) and its role in pavement–vehicle interaction in terms of vehicle consumption and wear as well as traffic noise are introduced. Then, equations to calculate these indicators are presented based on a comparison of existing assessment methods. The final transdisciplinary method pulls from road engineering, industrial ecology, acoustics and economics. It especially combines environmental and economic life cycle assessments and economic input–output analysis, as well as financial and socioeconomic appraisals. Finally, this article takes up the interdisciplinary challenge of building a fully holistic assessment method to help decision makers properly address sustainability, and its general algorithm can be adapted to assess a variety of transportation policies.

Keywords: holistic sustainability; road maintenance policies; multicriteria decision making support; life cycle assessment; macroeconomics; socioeconomic appraisal; public investments; pavement–vehicle interaction

Citation: de Bortoli, A.; Féraille, A.; Leurent, F. Towards Road Sustainability—Part I: Principles and Holistic Assessment Method for Pavement Maintenance Policies. *Sustainability* **2022**, *14*, 1513. https://doi.org/10.3390/su14031513

Academic Editors: Joel R.M. Oliveira, Hugo Silva, R. Christopher Williams and Zejiao Dong

Received: 11 December 2021

Accepted: 19 January 2022

Published: 28 January 2022

1. Introduction

The sustainability of road transportation strongly depends on infrastructure maintenance policies, but the impacts of these policies have never been quantified with a comprehensive triple bottom line framework. As an example, in France, road transportation accounts for 8% of jobs and 13% of the gross domestic product, and it generates annual tax revenues of EUR 45 billion [1]. The average household spends more than 10% of its budget on cars [2]), and the average individual spends a cumulative period of 4 years on the road in their life (calculated from the French households transportation survey [3]). These socioeconomic figures have concomitant environmental implications, as 29% of primary energy consumption [4] and 28% of greenhouse gas (GHG) emissions [5] are attributable to road transportation in the country. Road vehicles are also responsible for local pollution, particularly particulate matter emissions, including PM2.5, that alone would cause almost

10,000 annual deaths in France (see Supplementary Material). Moreover, the impact of road noise affects 16 million people, around 1 million of them at critical health thresholds [6].

Road conditions—and thus pavement maintenance policies—influence this performance. For instance, road surface deterioration increases consumption and wear for the vehicles traveling over them due to pavement–vehicle interactions (PVI) [7], thereby affecting the overall efficiency of the road and especially its environmental performance [8]. Nevertheless, the holistic sustainability of pavement maintenance policies has never been assessed comprehensively. While some methods were developed to make triple bottom line assessments of specific activities or products, such as buildings in Turkey [9], small and medium manufacturing companies [10] or urban food system governance [11], methods are still needed for transportation [12] and its infrastructure maintenance plans. Numerous methods claim a sustainability approach while being in fact mono-pillar, i.e., considering only one of the three dimensions: social, environmental or economic.

The main objective of this article is to build a comprehensive method to assess the holistic sustainability of road maintenance policies, with a focus on pavement resurfacing and PVI. Indeed, besides major refurbishments, pavement resurfacing is the most common and important kind of road maintenance operation, consisting of (re)building the top layer of a road surface, by spreading road materials over a thickness of 1 to less than 10 cm. This paper's subobjectives are as follows: first, to define road maintenance stakeholders and the categories of holistic sustainable impacts affecting them; second, to introduce specific physical models of road condition (international roughness index (IRI)) and its role on the pavement–vehicle interaction in terms of vehicle consumption and wear as well as traffic noise and third, to put forward a set of equations to fully specify the assessment model and make it computable. A companion article "Part II" will present a case study on maintenance policies for an interurban highway in France and show the practicability of the method and its concrete usefulness.

2. Literature Review

2.1. Mono-Pillar Sustainability Methods and Hybridization

Fairly extensive literature exists on the environmental evaluation of roads through life cycle assessment (LCA) [13], sometimes used specifically for the question of resurfacing [14–21]. Life cycle cost analysis (LCCA) can also be used to financially optimize roads [22–25]. Next, the input–output analysis (IOA) approach ([26] is an operational method for grasping the macroeconomic impacts of an economic policy on domestic production and employment [27]. It has been used in the transportation sector to assess the impact of a new road in the U.S. [28] or in France [29], of rolling out electric vehicles in France [30] or of the Paris–Bordeaux high-speed rail line in France [31], but never for road resurfacing. Finally, hybrid methods are used to assess the impact of roads excluding road maintenance. For example, a socioeconomic appraisal method is used by the French Ministry of Transportation to assess the public utility of a new transportation infrastructure, which combines macroeconomic, financial, social (time saved) and environmental dimensions [29]. This method has been hybridized with LCA in the case of an urban road rehabilitation operation [32]. However, it is not suitable for analyzing maintenance and there are deficiencies in its scope. Moreover, whole life cost analysis—hybridizing LCA and LCCA—was recently used in a pavement management study [33].

2.2. PVI Models and Road Maintenance Sustainability

Many models of vehicle operating costs have been developed since the 1970s [7]: incremental models linking the IRI with vehicle consumption and wear were developed for the World Bank between 1971 and 2005 by the Massachusetts Institute of Technology (MIT) [34], the Transportation Research Lab [35–37] and Birmingham University [38]. Originally developed for roads and vehicles in developing countries, these models were first calibrated for a developed country in the U.S. by Michigan State University [7]. This calibration reveals consumption differences of around 3%, 8% and 130% in fuel, tires and suspension systems, respectively, between vehicles traveling on a very smooth surface (IRI = 1 m/km) and vehicles running on a road in fairly poor condition (IRI = 6 m/km), these percentages depending on the vehicle type and its roll speed.

Finally, several studies have assessed some sustainability consequences of vehicle consumption due to road surface conditions. We summarize the characteristics of these studies in Table 1 and describe them further.

Table 1. Characteristics of road assessments considering PVI.

Lead Author and Year	Indicators				Consumption of Goods			
	Costs	GHG	Energy	Others	Roadworks	Fuel	Tires	Suspension Parts
Wang 2012b		X	X		X	X		
Chong 2017		X	X		X	X		
Wang 2014	X	X			X	X		
Bryce 2014			X		X	X		
Yu 2015		X	X	X (LCA)	X	X		
Yang 2015	X	X	X		X	X		
Santos 2015		X	X	X (LCA)	X	X		
Santos 2017	X	X	X	X (LCA)	X	X		
Pellecuer 2014		X	X	X (LCA)	X	X		
Guevara 2017	X			Time	X	X	X	X

The models of vehicle consumption sensitivities to the IRI established by Chatti and Zaabar [7] have been used several times to assess the effectiveness of road maintenance practices. Wang et al. [8] compared the improvements in GHG emissions and energy consumption obtained from a resurfacing program leading to a lower than usual IRI, including in the scope of their assessment both the impact of roadwork and the fuel consumption of the vehicles traveling on the roads. Chong and Wang [39] conducted a fairly similar study, extending the latter approach with an optimization program considering the initial dimensions of the road. Wang, Harvey and Kendall [40] assessed the optimum potential gain in GHG emissions in relation to road maintenance policy in the State of California and calculated the cost of mitigation in dollars per ton of CO_2 equivalent saved. Bryce et al. [41] presented a five-year optimization model that minimizes total energy consumed in both road maintenance and fuel consumption, constrained by the road operator's costs and road conditions. Yu et al. [42] developed an optimization program of the same type, extending the environmental scope to GHG emissions, acidification potential and respiratory effects, using a weighting method. Yang et al. [43] compared the impact of resurfacing programs with differences in the proportion of reclaimed asphalt pavement included in the new resurfacing material from an environmental perspective—GHG emissions and primary energy consumption—and from the perspective of the financial impact on the road operator. The only environmental effect from PVI considered was fuel consumption. Similarly, Santos et al. [44] compared the environmental performance of several resurfacing techniques but also left out the impact of differences in tire and suspension wear arising from PVI. This performance was assessed using the environmentally extended input–output method but not LCA in a second comparative study considering tires and suspension parts [33]. The latter study also includes an economic performance indicator, in the form of net present value (NPV), including the economic flows associated with road maintenance as well as vehicle consumption, calculated with a single discount rate of 2.3%. A decision support method for road surface management, developed by Pellečuer et al. [45], calculates the monetized health impacts of road noise and atmospheric pollution, as well as the cost of carbon dioxide emissions associated with traffic, under different road surface conditions. It also includes the variation in fuel consumption but not wear on tires and suspension. This study considers the monetary cost of noise pollution.

Finally, the marginal effect of vehicle consumptions according to road condition on the time spent by the users to operate their vehicle is considered by Guevara [46]. Nevertheless, no studies have compared the excess time spent because of roadwork with the time saved on the logistics of vehicle consumptions (journeys to gas stations to fill up and to service stations for tire and suspension maintenance).

2.3. Article's Scientific Contributions

As shown in the literature review, the methods developed to assess the sustainability of road maintenance policies are numerous but fragmentary with respect to the range of impacts on sustainable development and on stakeholders, as well as in consideration of the consequences of PVI. We thus propose a method that completes the existing approaches by adding the social, macroeconomic and financial aspects missing in the literature and by considering all the consequences of PVI—in particular the previously neglected environmental contributions of tire and suspension system wear. A novel method is also integrated to quantify the impact of road pavement resurfacing on traffic noise over time and ultimately damage to road local residents' health.

3. Method Overview

3.1. Identifying Road Maintenance Stakeholders and a Comprehensive Classification of Impacts

We built the method based on a reflection on which fundamental goals a public policy should follow towards a sustainable pathway, considering all four thinking schemes of socioeconomic ethics [47], main findings of the economics of welfare [48] and happiness [49] and the analysis of the original French texts theoretically driving the national social contract [50–52]. After a thorough analysis, the socioeconomic factors of happiness which are related to the strategies of road maintenance are: health, safety and security, employment, leisure and free time, resources and natural environment (details are available in the Supplementary Material and a Ph.D. thesis manuscript [53]). These factors affect the following stakeholders: the asset manager (i.e., road operator), the users, the residents, the State (or Government) and the environment.

Then, from a road maintenance stakeholder analysis and the multiple interests underlined in the literature review, we selected a set of indicators to holistically assess the multiattribute performance of a road resurfacing policy. The number of indicators was arbitrarily restricted to around 10 to avoid decision making cognitive saturation [54]. Non-quantitative impacts such as user comfort were not considered, and road safety was not accounted for due to its inherent multifactorial aspect, only partially depending on the infrastructure itself [55], thus difficult to separate from the other factors. Finally, the set of eleven indicators is composed of: users' time savings, health protection related to road noise, protection of resources, biodiversity and human health, road operator's and users' savings, domestic production and employment, tax revenue and national economic savings.

The stakeholders, impact causality chains and selected indicators are illustrated in the Figure 1. It shows how PVI are at the center of the stakeholders' impacts.

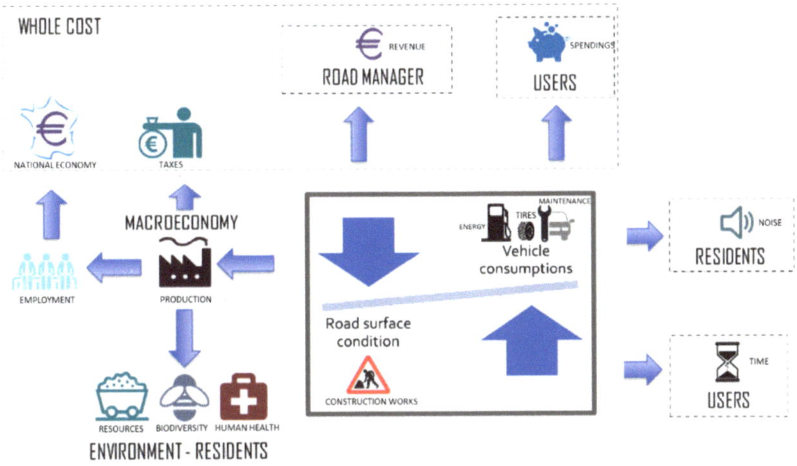

Figure 1. Systemics of road maintenance impacts: stakeholders' analysis, causality chains and impact categories per stakeholder.

3.2. Overview of the Method's Architecture

The architecture of the integrated method that we developed to assess the different indicators selected in Figure 1 is represented in Figure 2. As an input, it uses IRI data of the road studied at the time t_0. Depending on the pavement resurfacing scheme, it simulates the evolution of the IRI over time and calculates the associated inventories of consumptions and emissions on the assessment period: road works and vehicle consumptions, as well as noise emissions. It then combines assessment methods covering all aspects relating to road maintenance sustainability for the stakeholders: whole cost analysis, financial and socioeconomic assessments, LCA and Leontief's IOA. Combining these methods allows for calculating the main impacts caused by a maintenance policy as a result of the speed and intensity of road surface deterioration throughout its lifespan and on the different stakeholders highlighted previously.

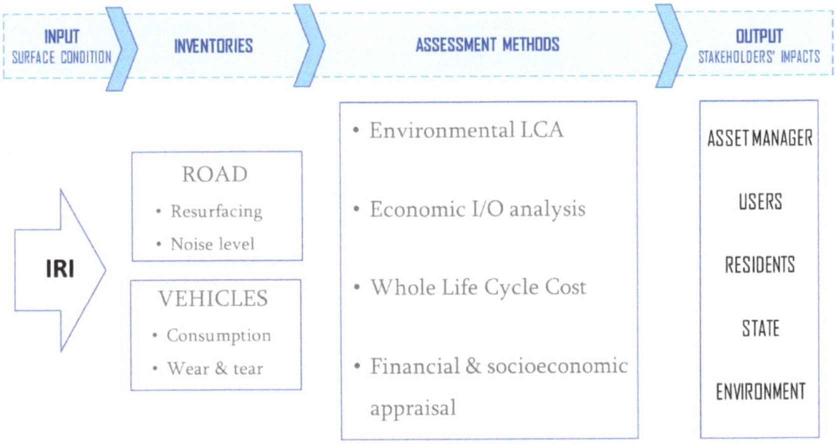

Figure 2. Architecture of the integrated method to assess the sustainability of pavement resurfacing programs (IRI: international roughness index; LCA: life cycle assessment; I/O: input–output).

3.3. *Practical Use of the Method*

The method was developed to be applied at the scale of the road section. The section must be considered over an observation period that is long enough to contain several resurfacing cycles and where the physical condition of the section is identical at the end of the analysis period for all the maintenance plans considered. A maintenance plan consists of a temporal sequence of resurfacing operations. Several resurfacing techniques exist, differing in terms of the thickness of the materials used, the quality of these materials and the construction technique of the surface layer. Each technique presents a specific financial, macroeconomic and environmental performance profile from the point of view of the maintenance operation. On the other hand, each maintenance technique has a potentially different impact on road surface condition and therefore the consumption and wear of the vehicles it supports. The vehicle population carried by the road section is separated into different categories of vehicles. The different consequences of PVI on these categories in terms of consumption and wear are accounted for. The impacts of this consumption and wear, as well as those of road works, are considered over their entire life cycles, as macroeconomic, financial and environmental consequences happen over their entire supply chain. The environmental aspects quantified are therefore calculated on a global scale. On the other hand, the health impact of road noise for residents relates to a local and not a global indicator. Finally, macroeconomic indicators focus on a national approach.

4. Specific Physical Models

4.1. *Capturing the Evolution of Road Surface Condition*

Considering the literature, the surface condition is expected to be a key parameter of road maintenance sustainability. From the IRI data for the road section to be assessed from time t_0, we calculate the evolution of the IRI over the entire lifespan of the road surface, between two resurfacing operations and before/after resurfacing operations.

4.1.1. IRI Evolution between Two Resurfacing Operations

The studies published on changes in the IRI between two resurfacings [8,23,43,56–60] propose different formulas in the mathematical form described in Equation (1)—where a, b and d are calibration parameters, and t is time.

$$IRI(t) = a(t+b)^d \quad (1)$$

In previous studies, this ultimately corresponds to deterioration speeds ranging from 0.019 m/km·year [60] to 0.23 m/km·year [59]. We recommend calibrating the equation for every road section or network assessed using field measures.

4.1.2. Effect of Resurfacing Operations on the IRI

The effect of resurfacing works on the IRI can be estimated by calibrating the American equation established by Wang et al. [59] using the IRI data for the road assessed as recalled in Equation (2), where e and g are calibration coefficients depending on the resurfacing technique, t_{R+} the time just after resurfacing and t_{R-} the time just before.

$$IRI(t_{R+}) = e \times IRI(t_{R-}) + g \quad (2)$$

By default, for a resurfacing technique with a thickness between 3 and 7.5 cm, one can take the parameter values of the study referred to, recalculated here in international system units: e = 0.40 and g = 0.67 m/km [59].

4.2. *Consumptions and Emissions over the Pavement Surface's Lifetime*

Vehicle consumptions and emissions partly depend on the IRI, which partly depends on the resurfacing program and thus on maintenance operations. To compare the sustainability of different road maintenance programs, we assess for each option the resurfacing works demand, the vehicle consumptions—in fuel, tires and shock absorbers—as well as the road noise emissions. PVI intensity and consequences depend on the kind of vehicle considered. To model the traffic, we choose a typology with 4 categories of vehicles like Chatti and Zaabar's method [7]: passenger car (PC), light commercial vehicle (LCV), small heavy vehicle (2 axles) (SHV) and large heavy vehicles (3 axles and more) (LHV).

4.2.1. Resurfacing Works Demand

Based on the resurfacing programs to be tested, with Equation (3), we calculate the resurfacing works demand RWD_k (in square meters) using a resurfacing technique k depending on the width ω of the pavement to be resurfaced in meters, its length λ in meters and N_k, the number of resurfacing operations of type k over the assessment period.

$$RWD_k = \omega \times \lambda \times N_k \tag{3}$$

4.2.2. Excess Vehicle Consumptions

Based on the data for pavement condition and dynamic traffic over time, we calculate the excess consumptions due to pavement surface irregularities over the pavement's lifetime for each vehicle category and consumable type—fuel, tires and suspensions. To do this, we consider the French average consumptions and select, after a model comparison, the IRI consumption sensitivity factor from HDM-4 [38] calibrated for the U.S. [61], which we adapt to the speeds actually observed on the different types of French road networks by statistical regression (see Supplementary Material). French speed limits are very similar to those in other European countries, but the equation adaptations can be conducted using the same method for each country's real conditions.

- Excess fuel consumption

The equations to quantify excess fuel consumption (EFC), depending on IRI, vehicle type and road type, are calculated in the Supplementary Material and presented in Table 2 for each kind of intercity network in France considering the average speeds on these networks in km/h for light vehicles (LV)—PC and LCV—and heavy vehicles (HV)—SHV and LHV. Again, regressions can be calculated using the same method for each country's real conditions.

Table 2. Equations to calculate EFC depending on IRI and type of vehicle and road.

		EFC = f (IRI)	**R^2**
Highways (118/88)	PC	EFC = 0.0233 IRI + 0.975	0.9927
	LCV	EFC = 0.00710 IRI + 0.996	0.9317
	Small HV	EFC = 0.00870 IRI + 0.992	0.9939
	Large HV	EFC = 0.0170 IRI + 0.981	0.9799
Express ways (101/84)	PC	EFC = 0.0239 IRI + 0.975	0.9966
	LCV	EFC = 0.0079 IRI + 0.994	0.9694
	Small HV	EFC = 0.0092 IRI + 0.991	0.9975
	Large HV	EFC = 0.0177 IRI + 0.980	0.9823
National/rural roads (82/79)	PC	EFC = 0.0245 IRI + 0.976	0.9983
	LCV	EFC = 0.0088 IRI + 0.993	0.9914
	Small HV	EFC = 0.0097 IRI + 0.991	0.9996
	Large HV	EFC = 0.0185 IRI + 0.980	0.9848

Fuel consumption due to the deterioration of the surface of the road over time can be calculated using Equation (4):

$$FC_{j,k} = \lambda \times \sum_i \int_{t_0}^{t_f} EFC_{j,i}(t) \times AFC_{i,j}(t) \times Q_j(t) \tag{4}$$

with $FC_{j,k}$ the fuel volume of type k overconsumed on a road of type i over a period from t_0 to t_f in hundredths of liters, λ the length of the road section evaluated in kilometers, $EFC_{j,i}$ the EFC of a vehicle of type j on a road of type i in percentage, $AFC_{i,j}$ the dynamic average fuel consumption of a vehicle of type j on a road of type i in liters per 100 km and Q_j the traffic flow of type j vehicles on the road section considered in number of vehicles.

- Excess tire wear

The equations to calculate excess tire wear (ETW), depending on IRI, vehicle type and road type, are calculated in the Supplementary Material and presented in Table 3. for each kind of intercity network in France and LV and HV average speeds, respectively. Specific national regressions can be calculated using the same method for any country.

Table 3. Equations to calculate ETW depending on IRI and type of vehicle and road.

	Vehicle	ETW = f (IRI)	R^2
Highways (118/88)	PC	ETW = 0.0168 IRI + 0.9812	0.9854
	LCV	ETW = 0.0102 IRI + 0.9927	0.9687
	Small HV	ETW = 0.0122 IRI + 0.9866	0.9907
	Large HV	ETW = 0.0089 IRI + 0.9917	0.9983
Express ways (101/84)	PC	ETW = 0.0136 IRI + 0.9856	0.9884
	LCV	ETW = 0.0088 IRI + 0.9929	0.9776
	Small HV	ETW = 0.0012 IRI + 0.987	0.9921
	Large HV	ETW = 0.0085 IRI + 0.9923	0.9972
National/rural roads (82/79)	PC	ETW = 0.0100 IRI + 0.9905	0.9865
	LCV	ETW = 0.0073 IRI + 0.9931	0.9839
	Small HV	ETW = 0.0117 IRI + 0.9874	0.9938
	Large HV	ETW = 0.008 IRI + 0.9931	0.9949

Tire consumption due to the deterioration of the surface of the road over time can be calculated using Equation (5):

$$TC_{i,k} = \lambda \times \sum_i \int_{t_0}^{t_f} ETW_{i,j}(t) \times TWR_{i,j}(t) \times Q_j(t) \tag{5}$$

with $TC_{i,k}$ the number of tires of type k overconsumed on the road of type i, λ the length of the road section evaluated (in kilometers), $ETW_{i,j}$ the ETW for a vehicle of type j on a road of type i, $TWR_{i,j}$ the kilometric tire wear ratio for a vehicle of type j on a road of type i and Q_j the traffic flow of type i vehicles.

- Excess suspension wear

According to Chatti and Zaabar's model, vehicles age faster on roads only with IRIs over 3 m/km. In our model, we will consider that only the suspension systems age faster under high IRIs. Over the 3 m/km threshold, the equations to calculate the shock absorber lifespan, depending on IRI, vehicle type and road type, are calculated in the Supplementary Material and presented in Table 4, for each kind of intercity network in France and LV and HV average speeds, respectively. These equations can be recalculated for any network in any country using the same process presented in the Supplementary Material.

Table 4. Equations to calculate the shock absorber lifespan (SAL) depending on IRI and type of vehicle and road.

		SAL = f (IRI)	R^2
Highways (118/88)	PC	SAL = 139 643.exp(−0.183 * IRI)	0.9750
	LCV	SAL = 139 643.exp(−0.183 * IRI)	0.975
	Small HV	SAL = 218 765.exp(−0.271 * IRI)	0.9874
	Large HV	SAL = 181 842.exp(−0.207 * IRI)	0.9651
Express ways (101/84)	PC	SAL = 155 221.exp(−0.183 * IRI)	0.9750
	LCV	SAL = 155 221.exp(−0.183 * IRI)	0.975
	Small HV	SAL = 226 487.exp(−0.271 * IRI)	0.9874
	Large HV	SAL = 188 260.exp(−0.207 *IRI)	0.9651
National/rural roads (82/79)	PC	SAL = 177 331.exp(−0.183 * IRI)	0.9750
	LCV	SAL = 177 331.exp(−0.183 * IRI)	0.975
	Small HV	SAL = 236 940.exp(−0.271 * IRI)	0.9874
	Large HV	SAL = 196 949.exp(−0.207 * IRI)	0.9651

Shock absorber consumption for IRIs above 3 m/km can be calculated using Equation (6):

$$SAC_{i,k} = \lambda \times \sum_i \int_{t_0}^{t_f} \frac{1}{SAL_{i,j}(t)} \times Q_j(t) \tag{6}$$

with $SAC_{i,k}$ the number of shock absorbers of type k consumed on the road type i, λ the length of the road section evaluated (in kilometers), $SAL_{i,j}$ the SAL for a vehicle of type j on a road of type i and Q_j the traffic flow of type j vehicles.

Note that this time, we directly calculate a consumption rather than an overconsumption: this does not change the final calculation, as it will be conducted on a differential approach between resurfacing programs.

4.2.3. Noise Emission

- Unitary noise linear power level over time

Noise characterization can be calculated through the unitary noise linear power level (Lw/m) corresponding to the noise power emitted by one vehicle on one meter. To understand its evolution over time, the statistical models of tire–road noise evolution from the European benchmark method [62] are updated using the most recent version of the French road noise database established by the CETE of East, a public technical center that reports to the French Ministry of Transportation (see Supplementary Material). These new models, different for heavy vehicles (HVs) and light vehicles (LVs), are presented in Table 5. Rx corresponds to acoustic categories of pavement rolling course, these categories being detailed in the Supplementary Material. Although the model is updated using a French database, physical acoustic phenomena is the same everywhere; thus, it must be usable in other countries with similar climate conditions, where pavement ageing is equivalent.

Table 5. Unitary noise linear power level per size of vehicle, depending on the pavement age t (in year), for t > 2 years and for three different categories of pavement surface Rx.

Rx	Light Vehicles	Heavy Vehicles
R1	2.2ln(t − 1) + Lw/m (t = 2 years)	1.3ln(t − 1) + Lw/m (t = 2 years)
R2	2.7ln(t − 1) + Lw/m (t = 2 year)	1.6ln(t − 1) + Lw/m (t = 2 years)
R3	$\begin{cases} 0.2(t-2) + \frac{Lw}{m(t=2\ years)}\ if\ t \in [2;10] \\ \frac{Lw}{m(t=10\ years)}\ if\ t > 10 \end{cases}$	$\begin{cases} 0.125(t-2) + \frac{Lw}{m(t=2\ years)}\ if\ t \in [2;10] \\ \frac{Lw}{m(t=10\ years)}\ if\ t > 10 \end{cases}$

- Total linear power level per vehicle

The noise of one motorized vehicle rolling on pavement is calculated by adding two noise components: the noise emitted by the motor and the noise emitted by the tire–road interaction. To calculate the unitary noise linear power level at t = 2 years, Lw/m (t = 2 years) per component and type of vehicle, we use formulas from Sétra [62] detailed in the Supplementary Material in the conditions of French interurban roads. We obtain Table 6.

Table 6. Average noise power level—motor and tire–road components—per meter of source line for a two-year-old rolling course.

	Lw/m—Motor Component (dB(A))			Lw/m—Tire–Road Component (dB(A))								
				R1			R2			R3		
	DR/NR	ER	HR	DR/NR	ER	HR	DR/NR	ER	HR	DR/NR	ER	HR
1 LV	42	43	43	49	50	52	53	54	56	55	57	58
1 HV	50	50	51	59	60	60	62	63	63	63	64	64

DR = department road; NR = national road; ER = express road; HR = highway road.

We can then calculate Lw/m (t = 2 years) per type of vehicle (LV or HV) by adding the noise source u—motor and tire–road noise—using the generic noise addition Equation (7).

$$L_{tot} = L(\sum_{u} Source_u) = 10\log\left[\sum_{u} 10^{\frac{L_u}{10}}\right] \quad (7)$$

- Total linear power level per road lane

From the calculation of these linear power levels (emitted by an HL or an LV) for a pavement surface of category Rx of age t (tire–road noise component only), the total linear power level per road lane can be calculated by adding the noise of all the vehicles by period of time—day or night. To do so, Equation (8) is the formula to calculate the linear density ρ_j in vehicles of type j per meter, with q_j the traffic in number of vehicles j per hour and Vj the average speed for vehicles of type j.

$$\rho_j = \frac{q_j}{Vj \times 1000} \quad (8)$$

Then, Equation (9) is the formula to calculate L, the noise level of a road per meter depending on the linear density of heavy and light traffic d_{LV} and d_{HV} during the day (6 a.m. to 10 p.m.) or the night (10 p.m. to 6 a.m.) and the noise level of one HV and LV L_{LV} and L_{HV} per meter.

$$L\left(\sum_{LV+HV} j\right) = 10\log\left[\rho_{LV}10^{\frac{L_{LV}}{10}} + \rho_{HV}10^{\frac{L_{HV}}{10}}\right] \quad (9)$$

- Additivity of the line sources' noises and temporal weighting

In the case of multiple road lanes, the noise of each line source must be added to obtain the linear power level of the road. Several methods can be considered to add the line sources' noise levels, depending on the road and noise receptor configuration. The method must be selected depending on the configuration studied.

- Calculation of the sound power

With Equation (10), we then calculate the sound power $W_{/m}$ emitted by the traffic from the linear power levels of the road, using Equation (9), with $W_{0,linear}$ = 1 pW/m = 1×10^{-12} W/m [63].

$$W_{/m} = W_{0,linear} \times 10^{\frac{L_{w/m}}{10}} \quad (10)$$

The sound level is then integrated over the length λ of the road section studied to obtain the average sound powers during the day (Wj) or the night (Wn) according to the following Equation (11):

$$W_{tot} = \int_0^l W_{/m} dl = W_{/m} \times \lambda \tag{11}$$

This result is then converted into acoustic energy emitted by multiplying by the time period considered. Our sound energy varying with the aging of the road surface and the latter being calculated over a time step of one year, we therefore multiply W_{tot} by 3600 s × 365 days × the number of hours of the period considered (16 h for the day and 8 h for the night) to obtain Ej and En, respectively, the daytime and nighttime acoustic energies emitted in $J(A)$.

5. Three Pillar Indicators' Algorithm

The classification of performance indicators between the three pillars of sustainable development is a matter of perspective when an indicator can be attached to two or three pillars. We propose the following classification, recognizing this potential intrinsic overlap. The method aims to assess the benefits of an alternative maintenance scheme (Alt) compared to a business-as-usual (BAU) program. Thus, performance indicators will always be calculated as the difference between the alternative scheme's performance and the BAU scheme, using the generic Equation (12):

$$\begin{aligned} \text{Benefit}(\text{Alt}) &= \text{NegativeImpact}(\text{BAU}) - \text{NegativeImpact}(\text{Alt}) \\ &= \text{PositiveImpact}(\text{Alt}) - \text{PositiveImpact}(\text{BAU}) \end{aligned} \tag{12}$$

5.1. Environmental Metrics

5.1.1. LCA: Characterization Methods and Indicators

We use LCA to evaluate the environmental impacts, following ISO standards 14040 and 14044 [64,65]. This method quantifies the environmental impact of a system—being a product or an activity—by inventorying input and output flows crossing the system over its life cycle and calculating their impact on the environment using characterization methods. The inputs and outputs are product, energy or material flows. Input flows come from the natural environment or the technosphere. Based on this inventory, characterization methods relate the flows to their potential impact on the environment, potentially considering their fate and the exposure of specific ecosystems. These methods make it possible to calculate many different environmental impacts, and the same type of indicator—e.g., impact on climate change, acidification or eutrophication—can be calculated in multiple ways depending on the characterization method chosen. The quality of a characterization method depends on the good representativeness of the physical realities modeled: the method must be scientifically up-to-date and adapted to the area of assessment. Two kinds of LCA indicators exist: midpoint and endpoint indicators. Midpoint indicators focus on single environmental problem—such as climate change or acidification—while endpoint indicators aggregate midpoint indicators in the three areas of environmental protection: natural resources, ecosystems and human health. A complete midpoint indicator set has a dozen indicators. As we want to restrict our indicator set to around ten, we choose to select endpoint indicators, which allow covering the total environmental damage concisely. Selecting LCA endpoint indicators rather than midpoint indicators is doubly beneficial: focusing on meaningful environmental indicators for the preservation of our planet and restricting the number of performance indicators to help decision makers while encompassing all the midpoint impact categories, including the most popular one: climate changecontribution. Further adaptations of this method do not exclude calculating a complete midpoint set to assess the environmental impacts of resurfacing policies and, for instance, a climate change contribution indicator.

5.1.2. Selection of Endpoint Indicators

We select the two operational environmental endpoint indicators from the method IMPACT World+ [66]: damage to ecosystems (renamed "biodiversity") and to human health. IMPACT World+ is the update of IMPACT 2002+, LUCAS and EDIP, and currently the most scientifically advanced characterization method in LCA [67]. The IMPACT World+ biodiversity damage indicator covers the effects on biodiversity of (short- and long-term) marine, freshwater and soil acidification; freshwater and marine eutrophication; land use; ecotoxicity (short- and long-term); climate change (short- and long-term); ionizing radiation and heat pollution affecting water quality and water availability (for terrestrial and aquatic freshwater ecosystems). The biodiversity loss calculated is expressed in PDF.m^2.yr: it corresponds to the potentially disappeared fraction of species over a surface area of one squared meter over one year. The IMPACT World+ indicator for damage to human health includes the following effects: climate change (short- and long-term), carcinogenic and non-carcinogenic toxicities (short- and long-term), ionizing radiation, ozone layer depletion, formation of fine particulates and photochemical oxidants and availability of water. This indicator is expressed in *DALY*, meaning disability-adjusted life years. One *DALY* is equivalent either to one year of potential life lost through premature death or one year of productivity lost as a result of incapacity. Note that the human health damage indicator includes the impact of air pollution over the entire supply chain of road maintenance consumption and road usage. Air pollution due to roads is a major public health issue, even more when highly trafficked and localized in densely populated areas. However, manufacturing is also a major source of air pollution: we chose an indicator that does not discriminate the damage to human health due to air pollution around roads and around facilities worldwide (where, for instance, spare parts or fuel are produced), whereas socioeconomic appraisals only consider local pollutions.

A third endpoint indicator focusing on damage to non-renewable resources and calculated with the ReCiPe method [68] comes to complete this environmental set of indicators. The latter, expressed in dollars, counts the economic impacts relating to the consumption of fossil and mineral resources, using a marginal cost approach. This approach is underpinned by the notion of resource scarcity: the scarcer a resource, the more its consumption will generate additional future extraction costs. ReCiPe offers the most advanced resource damage indicator, as it is an update of the Eco-Indicator and CML methods [67].

5.1.3. Calculation of the Metrics

Equation (13) is the formula to calculate the impact of a unitary consumption $I_{j,k,o}$ for each type of consumption k—a type of tire, suspension system or fuel needed to travel one kilometer—per type of vehicle j, for each type o of the three damage indicators. It sums the products of each flow of consumption or emissions $flow_c$ listed in the corresponding consumption life cycle inventories by characterization factor $CF_{c,o}$ obtained from the methods IMPACT World+ or ReCiPe, we.

$$I_{j,k,o} = \sum flow_c \times CF_{c,o} \tag{13}$$

The environmental impact of type o related to switching from a BAU resurfacing scenario to an alternative scenario is then calculated by multiplying the unitary impact of each type of consumption k by the number of units consumed over the assessment period for the total traffic. Consumption in one resurfacing scenario can be calculated with Equations (3)–(6).

5.2. Social Metrics

Two social indicators are considered in the method: users' time savings and the impact of road noise on residents' health. The latter could be considered an environmental indicator. However, as it involves the local population exposed to the road's traffic, it is considered a social indicator.

5.2.1. Road Noise Health Impact Indicator

The method includes an indicator of the impact of road noise on local residents' health $DALY_{\Delta t}$. It combines the road noise model developed in the noise inventory section of the article with a road noise LCA model developed by Meyer (2017). More specifically, we will use Meyer's characterization factors $CF_{\Delta t}$ that relate the noise energy $E_{\Delta t}$ emitted by the road to human health damage according to Equation (14), with $CF_{day} = 6.61 \times 10^{-7}\ DALY/J(A)$ and $CF_{night} = 1.25 \times 10^{-5}\ DALY/J(A)$.

$$DALY_{\Delta t} = E_{\Delta t} \times CF_{\Delta t} = W_{\Delta t} \times \Delta t \times CF_{\Delta t} = W_{day} \times \Delta t(day) \times CF_{day} + W_{night} \times \Delta t(night) \times CF_{night} \quad (14)$$

The residents' health damage due to traffic noise will be calculated over the assessment period according to Equation (15).

$$HealthDamage_{residents} = \int_{t_0}^{t_f} DALY_{\Delta t} dt \quad (15)$$

Contrary to air pollution impacts on human health that are considered over the entire supply chain, for this indicator, we calculate the impact of road noise on human health of people living around the road, due to a lack of database on noise emissions of different activities.

5.2.2. Users' Time Saving Indicator

This indicator considers, based on the resurfacing program as well as fuel consumption, tire and suspension wear, the time spent by users in roadwork zones but also in journeys to gas stations or garages for vehicle maintenance and operation.

We use Equation (16) to calculate the time loss for a vehicle of type j due to a work zone, with $\theta_{alternative,j}$ the time to cross the work zone during construction for the type of vehicle j and $\theta_{standard,j}$ the standard time to cross the section. $\theta_{alternative,j}$ can relate to time lost due to congestion, construction traffic lights, reduced speed limitations or detours.

$$TimeLoss_{works,j} = \theta_{works,j} - \theta_{standard,j} \quad (16)$$

Then, the time lost on the road section over the assessment horizon is calculated by summing the time lost for the four types of vehicles j and all the traffic during the n resurfacing operations, with $q_j(t)$ the hourly traffic of type j vehicles, Δt_m the duration of the resurfacing operation k and N_k the number of resurfacings of type k over the assessment period, as shown in Equation (17).

$$TimeLoss_{works} = \sum_k \sum_j Time_{loss,j} \times q_j(t) \times N_k \Delta t\_m_k \quad (17)$$

On the specific road network of type i, the total time lost due to vehicle consumption of type c (tires, shock absorbers or fuel) is calculated with Equation (18) from N_c, the vehicle spare part consumption in number of total replacements of tires and shock absorbers or the volumes of 100 L of fuel for the vehicle of type j, and the estimated duration of the various operating and maintenance activities $OD_{c,j}$, with $c = 1$ relating to tire consumption, $c = 2$ to shock absorber consumption and $c = 3$ to fuel consumption.

$$TimeLoss_{vehicle\ consumptions,i} = \sum_j \sum_c N_c \times OD_{c,j} = \sum_j TC_{i,j} \times OD_{1,j} + \sum_j SAC_{i,j} \times OD_{2,j} + \sum_j \frac{FC_{i,j}}{100} \times OD_{3,j} \quad (18)$$

The users' time savings generated by an alternative resurfacing scheme compared to a reference are calculated using Equation (12) based on Equation (19):

$$TimeLoss_{users} = TimeLoss_{vehicle \quad consumptions} + TimeLoss_{works} \tag{19}$$

5.3. Economic Metrics

Based on resurfacing operations and vehicle consumption, we calculate the economic impacts of resurfacing programs, i.e., their macroeconomic effects and their financial consequences for the stakeholders concerned: the road operator, users and government. This requires cost models for vehicles and resurfacing consumption.

5.3.1. Users' Costs

The excess cost to use their vehicles for the road users $VehicleExpenses_{users}$, discounted over time at a rate r, can be calculated over the assessment period by multiplying each type c of excess consumption EE_k—namely, FC_k the excess fuel consumption of type k, TC_k the number of tires of type k overconsumed and SAC_k the number of shock absorbers of type k consumed—by their costs $C_k(t)$ in constant currency, following Equation (20).

$$VehicleExcessExpenses_{users} = \int_{T=t_0}^{t_f} \sum_k \frac{1}{(1+r)^{T-t_0}} \times EE_k(t) \times C_k(t) \tag{20}$$

By default, we propose to consider a discount rate a equal to the government's 10-year borrowing rate, as it represents the household financial market reality.

5.3.2. Road Operator's Costs

The cost to maintain the road for the operator $MaintenanceExpenses_{operator}$, discounted over time at a rate r, can be calculated over the assessment period by multiplying the surfaces of the road maintained using the resurfacing technique k, $RWD_k(t)$, by $C_k(t)$, the cost of this technique per square meter over time in constant currency, following Equation (21).

$$MaintenanceExpenses_{operator} = \int_{T=t_0}^{t_f} \sum_k \frac{1}{(1+r)^{T-t_0}} \times RWD_k(t) \times C_k(t) \tag{21}$$

By default, we propose to consider a discount rate r equal to the rate of return of the operator if it is a private company or to the government's 10-year borrowing rate if the road is publicly operated.

5.3.3. Domestic Production and Employment

To calculate domestic production and employment content indicators, we use Leontief's IOA method [26,69]. This method is used by governments to analyze the national accounts and produce their macroeconomic projections. It uses input–output (I/O) tables representing production by economic sectors and their mutual dependencies to calculate the macroeconomic impact of an economic shock, i.e., a change in demand on certain economic sectors.

First, we convert the previously calculated physical consumption items from Equations (3)–(6) into monetary flows employing cost models. These models are specific to a country and a period. The calculation can be conducted on an annual basis to potentially consider dynamic cost models. Then we apply the Leontief method using the I/O tables—generally supplied by the national ministries for the economy—to calculate an indicator of domestic production. It includes the annual sum of direct and indirect domestic production resulting from all demands for roadwork, fuel and service station changes to tires and suspension. Two formulas are central. First is that of the technical coefficients in Equation (22) to calculate the share of intermediate consumption of a sector per unit of production of this sector, with α_{ij} the coefficient corresponding to the share of expenditure on the product i

in production of branch j, ter_{ij} the matrix of intermediate consumption and $Y = [y_j]$ the matrix of production.

$$\alpha_{ij} = \frac{ter_{ij}}{y_j} \tag{22}$$

Then, the inverse Leontief relation, noted in Equation (23), makes it possible, from the matrix $A = [\alpha_{ij}]$ of technical coefficients and the vector f of demand in a sector, to calculate the production sectorial effects P, with I the identity matrix.

$$P = (I - A)^{-1} \times f \tag{23}$$

We can calculate an annual vector f following Equation (24), where each element of the vector f corresponds to the differential consumption (in currency) of a maintenance scenario compared to the reference. Hence, only the elements corresponding to pavement construction, vehicle maintenance, tires, shock absorbers and fuel consumption (or larger activities depending on the country economic nomenclature), will be nonempty.

$$f_i = \int_{t_0}^{t_f} \Delta consumption_i(t) \times UnitaryPrice_i(t) \tag{24}$$

The direct and indirect domestic production related to the maintenance scheme assessed can be calculated by summing the elements of vector P.

Then, based on the production vector, Equation (25) uses a national job content vector for the different economic sectors $[JobContent]$, which is, most of the time, supplied by national ministries for one country economy to calculate domestic employment in Full-Time Equivalent (FTE) jobs, relating to these demands, on the supply chain within the country where road maintenance is performed.

$$\text{FTE} = [JobContent] \times f \tag{25}$$

5.3.4. Tax Revenues

The tax revenues collected on consumption for roadwork, fuel, tires and suspensions, discounted over time at a rate r, are calculated using Equation (26), with f_t the demand vector for the year T in constant currency and $\%tax_t$ the tax rate vector (or a matrix in the case of tax rate variations over time), indicating the tax rate applicable for each type of consumption for the year T.

$$TaxRevenue = \sum_{T=t_0}^{t_f} \frac{f_T \times \%tax_T}{(1+r)^{T-t_0}} \tag{26}$$

By default, we propose to consider a discount rate a relative to the government's 10-year borrowing rate, as it represents the State financial market reality.

5.3.5. Integrated National Economic Indicator

A government is often interested in the total cost of a policy such as a road maintenance program, combining the financial flows for the different stakeholders. The NPV indicator can be used to estimate the financial interest for an actor or group of actors in an operation: the larger it is, the more financially interesting the operation. It is used in particular by decision makers and road managers to justify their decisions. We use this concept to calculate the multistakeholder financial impact of a maintenance scheme—for the operator, users and tax authority. We recall in Equation (27) the generic formula for calculating an NPV, i.e., the sum of the financial flows—incomes and expenses that we detailed earlier for all the stakeholders—in constant currency, which is actualized each year over the entire duration of the assessment period, with E_T the expenses, I_T the incomes planned for the year t, and a the discount rate.

$$\mathrm{NPV} = \sum_{T=t_0}^{t_f} \frac{I_T - E_T}{(1+r)^{T-t_0}} \tag{27}$$

We suggest setting r at 2.5%, which is the risk-free discount rate recommended for the valuation of major investments in France. A risk-free rate choice allows us not to override the impact of future events [29] and, thus, to limit the burden of negative impacts for future generations.

6. Discussion

The geographical scope of the calculated impacts varies amongst the indicators: this scope goes from the close area around the road studied with the indicator of damage to human health due to road noise to a worldwide perimeter with the three LCA indicators, passing through a national scope with macroeconomic indicators. This heterogeneity of perimeters is partly justified by the scope of the impacts or the contribution of the various phenomena involved. For example, it is preferable to assess the environmental impacts at a global scale to limit the burden's geographic shifts. In addition, climate change, an important component of damage indicators, is based on global dynamics. On the contrary, it is likely that most of the noise changes linked to road maintenance policies are located around roads rather than around the manufacturing sites of road maintenance consumables because the impact of road maintenance policies on the production of these sites is marginal. Then, economic solidarity mostly occurs at a national level. However, the impacts of air pollution around roads and, more generally, in the country generate national socioeconomic impacts due to the cost of healthcare. Thus, comparing local or even national impacts to impacts in the rest of the world would advocate for or against further discretizing the two scales and potentially aggregate health damage due to noise and local air pollution into a resident human health damage indicator.

From a road management point of view, this method could benefit from future developments. It is for now restricted to the evaluation of resurfacing, but combined optimization of structural design and maintenance treatment scheduling over the lifespan of a road would allow for capturing the feedback effects between the mechanical behavior of the road structure and its surface layers. Such an approach will require advances in the field of road aging modeling. Next, the models linking road surface condition to vehicle fuel consumption and wear on tires and suspensions are central to the assessment mechanics in the method: these need to be developed and validated in regional conditions to enhance reliability of results. Uncertainty calculations could also be conducted to assess this reliability. Finally, a few indicators could be added. First is a user comfort indicator: riding comfort partly depends on the rolling course condition; nevertheless, we did not find any conclusive approach to propose an adequate quantitative indicator. Ideally, at the level of the road network, this user comfort indicator could be exploited in a more systemic way to calculate the impact of a resurfacing policy by considering the modification of the users' routes that may result from the evolution of the surface conditions. The selection of the route according to comfort can also imply differences in time, speed and consumption of travel, which are challenging to capture. Second is a pavement safety and reliability indicator quantifying the consequence of the road condition and, for example, cumulating the following costs: noninjury incidents, such as vehicle damage, goods breakage and damage for the freight, and monetized impact of injury accidents and fatalities. However, in practice, road safety depends on many factors, and, to our knowledge, the impact of the surface condition has not been decoupled from other factors [55]. Third is the tax revenue indicator, which could be refined to consider the financial impact of maintenance policies on social allowances from public authorities, e.g., on the unemployment allowances budget, depending on the employment performance of the policy.

7. Conclusions

In this article, we present the first method of holistic sustainability assessment developed to our knowledge in the field of transportation. Notably, it integrates the three pillars of sustainability and considers important macroeconomic and other state-of-the-art indicators compared to existing sustainability rating systems of pavement and transportation projects [70]. It especially completes the existing approaches by adding the social, macroeconomic and financial aspects missing in the literature and by considering all the consequences of PVI—in particular the previously neglected environmental contributions of tire and suspension system wear. A novel method is also integrated to quantify the impact of road pavement resurfacing on traffic noise over time and, ultimately, damage to local residents' health. A set of indicators that is both comprehensive and concise is needed to support decision makers for whom triple bottom line performance remains a theoretical concept that they struggle to grasp in its entirety. Indeed, holistic sustainability can hardly be defined "above ground", i.e., in a generic way without looking at a specific system or object. Therefore, we have developed this method on the specific example of road maintenance policies, considering sustainability under the French social contract. While the calculation of the inventories (Section 4) is specific to road maintenance, and the PVI equations are calibrated in French conditions as an example, the performance calculation algorithm (Section 5) can be used to assess all kinds of transportation policies in any country. The selection of the relevant quantitative indicators is based on an integrated vision of sustainability by including all stakeholders highlighted by the literature review. The benefits of an alternative road maintenance policy compared to standard practices for these stakeholders are assessed on all three pillars of sustainable development. Thus, this report offers a positive or upbeat consequential approach to triple bottom line decision making, in the manner of the environmental handprint concept [71], that reverses the vision of the classic environmental footprint calculated in attributional LCA. Finally, this method can be used by any road manager or road owner willing to fully understand and tackle its sustainability responsibility.

Supplementary Materials: Detailed information referred to in the main text as accessible in the supplementary material can be found in the supplementary material document following this article. The following supporting information can be downloaded at: https://www.mdpi.com/article/10.3390/su14031513/s1. The code for the method is available on the GitHub repository accessible at the following link: https://github.com/Anne2B/PhD.git (accessed on 15 January 2022).

Author Contributions: Conceptualization: A.d.B., A.F. and F.L.; methodology: A.d.B., A.F. and F.L.; software: A.d.B.; validation: A.d.B.; formal analysis: A.d.B.; investigation: A.d.B.; resources: F.L.; data curation: A.d.B.; writing—original draft preparation: A.d.B.; writing—review and editing: A.d.B. and F.L.; visualization: A.d.B.; supervision: A.F. and F.L.; project administration: F.L.; funding acquisition: F.L. All authors have read and agreed to the published version of the manuscript.

Funding: This research was funded by the industrial chair ParisTech-VINCI in "eco-design of buildings and infrastructure".

Acknowledgments: The authors would like to thank the industrial chair ParisTech-VINCI in "eco-design of buildings and infrastructure", which funded the Ph.D. thesis from which this work was derived.

Conflicts of Interest: The authors declare no conflict of interest.

Abbreviations

List of abbreviations. The abbreviations included in the text are reported alphabetically.

Abbreviation	Full Form
BAU	business-as-usual
DALY	disability-adjusted life years
DR	department road
EFC	excess fuel consumption
ER	express road
ETW	excess tire wear
FTE	full-time equivalent
GHG	greenhouse gas
HR	highway road
HV	heavy vehicle
I/O	input–output
IOA	input–output analysis
IRI	international roughness index
LCA	life cycle assessment
LCCA	life cycle cost analysis
LCV	light commercial vehicle
LHV	large heavy vehicle
LV	light vehicle
MIT	Massachusetts Institute of Technology
NPV	net present value
NR	national road
PC	passenger car
PDF	potentially disappeared fraction
PVI	pavement–vehicle interactions
SAL	shock absorber lifespan
SHV	small heavy vehicle

References

1. Union Routière de France. Faits & Chiffres 2018-Statistiques des Transports en France et en Europe. 2018. Available online: https://fr.calameo.com/read/0039965789c6b488b8029 (accessed on 5 October 2021).
2. Visse, P.-E. Évolution du Budget Automobile des Ménages Français Depuis 1990. *DGCCRF*. 14 April 2013. Available online: http://www.economie.gouv.fr/files/directions_services/dgccrf/documentation/dgccrf_eco/dgccrf_eco14.pdf (accessed on 8 December 2021).
3. Enquête Nationale Transport et Déplacements 2008. MEDDE. 2008. Available online: http://www.statistiques.developpement-durable.gouv.fr/transports/trv/deplacement-mobilite/mobilite-reguliere-locale.html (accessed on 28 October 2021).
4. CGDD. Chiffres Clés de L'énergie-Édition 2018. Commissariat Général au Développement Durable. 2018. Available online: https://www.statistiques.developpement-durable.gouv.fr/sites/default/files/2018-10/datalab-43-chiffres-cles-de-l-energie-edition-_2018-septembre2018.pdf (accessed on 7 November 2021).
5. CITEPA. Données D'émissions de Gaz à Effet de Serre Dans L'air en France Métropolitaine, Avril 2018, Format SECTEN. 2019. Available online: https://www.citepa.org/images/III-1_Rapports_Inventaires/SECTEN/CITEPA-chiffres-cles-2018-d.zip (accessed on 25 February 2019).
6. Bedeau, L.; Piquandet, J.; Duhautois, S.; Jonsson, E. Les Français et Les Nuisances Sonores. *TNS Sofres-MEEDM*. 2010. Available online: https://www.tns-sofres.com/sites/default/files/2010.06.29-nuisances-sonores.pdf (accessed on 18 November 2021).
7. Chatti, K.; Zaabar, I. *Estimating the Effects of Pavement Condition on Vehicle Operating Costs*; Transportation Research Board: Washington, DC, USA, 2012. Available online: http://onlinepubs.trb.org/onlinepubs/nchrp/nchrp_rpt_720.pdf (accessed on 7 November 2021).
8. Wang, T.; Lee, I.-S.; Kendall, A.; Harvey, J.; Lee, E.-B.; Kim, C. Life cycle energy consumption and GHG emission from pavement rehabilitation with different rolling resistance. *J. Clean. Prod.* **2012**, *33*, 86–96. [CrossRef]
9. Ulubeyli, S.; Kazanci, O. Holistic sustainability assessment of green building industry in Turkey. *J. Clean. Prod.* **2018**, *202*, 197–212. [CrossRef]
10. Chen, D.; Thiede, S.; Schudeleit, T.; Herrmann, C. A holistic and rapid sustainability assessment tool for manufacturing SMEs. *CIRP Ann.* **2014**, *63*, 437–440. [CrossRef]

11. Landert, J.; Schader, C.; Moschitz, H.; Stolze, M. A Holistic Sustainability Assessment Method for Urban Food System Governance. *Sustainability* **2017**, *9*, 490. [CrossRef]
12. Hüging, H.; Glensor, K.; Lah, O. Need for a Holistic Assessment of Urban Mobility Measures–Review of Existing Methods and Design of a Simplified Approach. *Transp. Res. Procedia* **2014**, *4*, 3–13. [CrossRef]
13. Santero, N.J.; Masanet, E.; Horvath, A. Life-cycle assessment of pavements. Part I: Critical review. *Resour. Conserv. Recycl.* **2011**, *55*, 801–809. [CrossRef]
14. Mladenovič, A.; Turk, J.; Kovač, J.; Mauko, A.; Cotič, Z. Environmental evaluation of two scenarios for the selection of materials for asphalt wearing courses. *J. Clean. Prod.* **2015**, *87*, 683–691. [CrossRef]
15. Kucukvar, M.; Noori, M.; Egilmez, G.; Tatari, O. Stochastic decision modeling for sustainable pavement designs. *Int. J. Life Cycle Assess.* **2014**, *19*, 1185–1199. [CrossRef]
16. Vidal, R.; Moliner, E.; Martínez, G.; Rubio, M.C. Life cycle assessment of hot mix asphalt and zeolite-based warm mix asphalt with reclaimed asphalt pavement. *Resour. Conserv. Recycl.* **2013**, *74*, 101–114. [CrossRef]
17. Tatari, O.; Nazzal, M.; Kucukvar, M. Comparative sustainability assessment of warm-mix asphalts: A thermodynamic based hybrid life cycle analysis. *Resour. Conserv. Recycl.* **2012**, *58*, 18–24. [CrossRef]
18. Yu, B.; Lu, Q. Life cycle assessment of pavement: Methodology and case study. *Transp. Res. Part D Transp. Environ.* **2012**, *17*, 380–388. [CrossRef]
19. Cuenoud, J.L. *Valorcol: Asphalt Mix Complying with Environment and Sustainable Development*; RGRA: Paris, France, 2011.
20. Huang, Y.; Bird, R.; Heidrich, O. Development of a life cycle assessment tool for construction and maintenance of asphalt pavements. *J. Clean. Prod.* **2009**, *17*, 283–296. [CrossRef]
21. Chiu, C.-T.; Hsu, T.-H.; Yang, W.-F. Life cycle assessment on using recycled materials for rehabilitating asphalt pavements. *Resour. Conserv. Recycl.* **2008**, *52*, 545–556. [CrossRef]
22. Babashamsi, P.; Yusoff, N.I.M.; Ceylan, H.; Nor, N.G.M.; Jenatabadi, H.S. Evaluation of pavement life cycle cost analysis: Review and analysis. *Int. J. Pavement Res. Technol.* **2016**, *9*, 241–254. [CrossRef]
23. Santos, J. A Comprehensive Life Cycle Approach for Managing Pavement Systems. Ph.D. Thesis, Universidade de Coimbra, Coimbra, Portugal, 2015. Available online: https://estudogeral.sib.uc.pt/bitstream/10316/30093/1/A%20Comprehensive%20Life%20Cycle%20Approach%20for%20Managing%20Pavement%20Systems.pdf (accessed on 22 November 2021).
24. Ferreira, A.; Santos, J. LCCA System for Pavement Management: Sensitivity Analysis to the Discount Rate. *Procedia-Soc. Behav. Sci.* **2012**, *53*, 1172–1181. [CrossRef]
25. FHWA. Life-Cycle Cost Analysis in Pavement Design—In Search of Better Investment Decisions. US Department of Transportation-Federal Highway Administration, Pavement Division Interim Technical Bulletin Publication No. FHWA-SA-98-079. 1998. Available online: https://www.fhwa.dot.gov/infrastructure/asstmgmt/013017.pdf (accessed on 22 November 2021).
26. Leontief, W.W. Quantitative Input and Output Relations in the Economic Systems of the United States. *Rev. Econ. Stat.* **1936**, *18*, 105. [CrossRef]
27. Quirion, P. L'effet net sur L'emploi de la Transition Énergétique en France: Une Analyse Input-Output du Scénario Négawatt. CIRED, Document de Travail No 46-2013. 2013. Available online: http://www2.centre-cired.fr/IMG/pdf/CIREDWP-201346.pdf (accessed on 2 October 2021).
28. Wubneh, M. *US Highway 17 and Its Impact on the Economy of Eastern North Carolina*; Urban & Regional Planning Program Department of Geography East Carolina University: Greenville, NC, USA, 2008.
29. Quinet, E. L'Évaluation Socioéconomique des Investissements Publics. Rapport Final. Tome 1, 2013. Report. Commissariat général à la stratégie et à la Prospective. Available online: https://www.strategie.gouv.fr/sites/strategie.gouv.fr/files/atoms/files/cgsp_evaluation_socioeconomique_29072014.pdf (accessed on 2 October 2021).
30. Leurent, F.; Windisch, E. Electric vs. Gasoline-Powered Vehicles: The Effects on a Nation's Economic Production and Public Finances. Routes/Roads, N° 357. 2013. Available online: https://routesroadsmag.piarc.org/en/Routes-Roads-Magazine-Issue-357-Climate-Change/1685,Routes-Roads-Magazine-Electric-Vs-Gasoline-Effects-Economy-Finances#c3e63u7JL70 (accessed on 12 October 2021).
31. Fouqueray, E. Evaluation de L'impact Économique de Court Terme et de Moyen Terme des Chantiers de Grandes Infrastructures de Transport-Le cas de la LGV SEA Tours-Bordeaux. Ph.D. Thesis, Université de Poitiers-Faculté de Sciences Économiques, Poitiers, France, 2016.
32. de Bortoli, A. Consequential environmental Life Cycle Assessment and socio-economic analysis-hybridization test on a Parisian project of Bus Rapid Transit. In Proceedings of the AVNIR International Conference, Lille, France, 8–9 November 2016. [CrossRef]
33. Santos, J.; Flintsch, G.; Ferreira, A. Environmental and economic assessment of pavement construction and management practices for enhancing pavement sustainability. *Resour. Conserv. Recycl.* **2017**, *116*, 15–31. [CrossRef]
34. Moavenzadeh, F.; Berger, F.; Brademeyer, B.; Wyatt, R. *The Highway Cost Model: General Framework*; Massachusetts Institute of Technology Department of Civil Engineering Research Report No 75-4; Massachusetts Institute of Technology: Cambridge, MA, USA, 1975.
35. Abaynayaka, S.W.; Hide, H.; Robinson, R.; Rolt, J. Prediction of road construction and vehicle operating costs in developing countries. *Proc. Inst. Civ. Eng.* **1977**, *62*, 419–446.

36. Kerali, R.; Parsley, L.; Robinson, R.; Snaith, M. Development of a Microcomputer based Model for Road Investment in Developing Countries. In Proceedings of the Second International Conference on Civil and Structural Engineering Computing, London, UK, 3–5 December 1985; pp. 83–86. [CrossRef]
37. Parsley, L.L.; Robinson, R. *The TRRL, Road Investment Model for Developing Countries (RTIM2)*; TRLRL Laboratory Report 1057; Transport and Road Research Laboratory: Crowthorne, UK, 1982.
38. Bennett, C.R.; Greenwood, I.D. Volume 7: Modeling Road User and Environmental Effects in HDM-4, Version 3.0, International Study of Highway Development and Management Tools (ISOHDM), World Road Association (PIARC). 2003. Available online: http://www.lpcb.org/index.php/documents/papers-and-reports/reports-and-books/194-2003-modelling-road-user-and-environmental-effects-in-hdm-4/file (accessed on 17 October 2021).
39. Chong, D.; Wang, Y. Impacts of flexible pavement design and management decisions on life cycle energy consumption and carbon footprint. *Int. J. Life Cycle Assess.* **2016**, *22*, 952–971. [CrossRef]
40. Wang, T.; Harvey, J.T.; Kendall, A. Reducing greenhouse gas emissions through strategic management of highway pavement roughness. *Environ. Res. Lett.* **2014**, *9*, 034007. [CrossRef]
41. Bryce, J.; Katicha, S.; Flintsch, G.; Sivaneswaran, N.; Santos, J. Probabilistic Life-Cycle Assessment as Network-Level Evaluation Tool for Use and Maintenance Phases of Pavements. *Transp. Res. Rec. J. Transp. Res. Board* **2014**, *2455*, 44–53. [CrossRef]
42. Yu, B.; Gu, X.; Ni, F.; Guo, R. Multi-objective optimization for asphalt pavement maintenance plans at project level: Integrating performance, cost and environment. *Transp. Res. Part D Transp. Environ.* **2015**, *41*, 64–74. [CrossRef]
43. Yang, R.; Kang, S.; Ozer, H.; Al-Qadi, I.L. Environmental and economic analyses of recycled asphalt concrete mixtures based on material production and potential performance. *Resour. Conserv. Recycl.* **2015**, *104*, 141–151. [CrossRef]
44. Santos, J.; Bryce, J.; Flintsch, G.; Ferreira, A.; Diefenderfer, B. A life cycle assessment of in-place recycling and conventional pavement construction and maintenance practices. *Struct. Infrastruct. Eng.* **2014**, *11*, 1199–1217. [CrossRef]
45. Pellecuer, L.; Assaf, G.J.; St-Jacques, M. Influence of Pavement Condition on Environmental Costs. *J. Transp. Eng.* **2014**, *140*, 04014050. [CrossRef]
46. Guevara, C.A. Mode-valued differences of in-vehicle travel time Savings. *Transportation* **2016**, *44*, 977–997. [CrossRef]
47. Arnsperger, C.; van Parijs, P. *Éthique Économique et Sociale*; La Découverte: Paris, France, 2007.
48. Pigou, A. The Economics of Welfare, Macmillan and Co. London. 1932. Available online: http://www.econlib.org/library/NPDBooks/Pigou/pgEW.html (accessed on 2 October 2021).
49. White, A. A Global Projection of Subjective Well-being: A Challenge to Positive Psychology? *Psychtalk* **2007**, *56*, 17–20.
50. Charte de L'environnement de 2004. 2005. Available online: https://www.legifrance.gouv.fr/Droit-francais/Constitution/Charte-de-l-environnement-de-2004 (accessed on 17 October 2021).
51. Déclaration des Droits de l'Homme et du Citoyen de 1789. Available online: https://www.legifrance.gouv.fr/Droit-francais/Constitution/Declaration-des-Droits-de-l-Homme-et-du-Citoyen-de-1789 (accessed on 2 December 2021).
52. Préambule de la Constitution du 27 Octobre 1946. Available online: https://www.legifrance.gouv.fr/Droit-francais/Constitution/Preambule-de-la-Constitution-du-27-octobre-1946 (accessed on 17 October 2021).
53. de Bortoli, A. Pour un Entretien Routier Durable: Prise en Compte des Conséquences de L'interaction Chaussée-Véhicule Dans L'aide à la Décision des Politiques de Resurfaçage-Illustration par un cas Autoroutier Français [Toward Sustainable Road Maintenance: Taking into Account Vehicle-Pavement Interactions into the Decision-Making Process-Illustration by a French Highway Case Study]. Université Paris Est-Ecole des Ponts ParisTech. 2018. Available online: https://www.researchgate.net/publication/333965224_Pour_un_entretien_routier_durable_prise_en_compte_des_consequences_de_l\T1\textquoterightinteraction_chaussee-vehicule_dans_l\T1\textquoterightaide_a_la_decision_des_politiques_de_resurfacage_-_illustration_par_un_cas_autoroutier_fran (accessed on 28 November 2021).
54. Sapir, J. Théorie de la régulation, conventions, institutions et approches hétérodoxes de l'interdépendance des niveaux de decision. In *Décisions Économiques*; Economica: Paris, France, 1998; pp. 169–215.
55. Cerezo, V.; Gothié, M. Pavement Prediction Performance Models and Relation with Traffic Fatalities and Injuries. France. October 2008. Available online: https://www.researchgate.net/profile/Veronique_Cerezo/publication/281328742_Pavement_prediction_performance_models_and_relation_with_traffic_fatalities_and_injuries/links/55faa08a08ae07629e0427d3/Pavement-prediction-performance-models-and-relation-with-traffic-fatalities-and-injuries.pdf (accessed on 22 October 2021).
56. Bryce, J.; Santos, J.; Flintsch, G.; Katicha, S.; McGhee, K.; Ferreira, A. Analysis of rolling resistance models to analyse vehicle fuel consumption as a function of pavement properties. In *Asphalt Pavements*; CRC Press: Boca Raton, FL, USA, 2014; pp. 263–273. [CrossRef]
57. Yang, R. *Development of a Pavement Life Cycle Assessment Tool Utilizing Regional Data and Introducing an Asphalt Binder Model*; University of Illinois at Urbana-Champaign: Urbana, IL, USA, 2014. Available online: https://www.ideals.illinois.edu/bitstream/handle/2142/50651/Rebekah_Yang.pdf?sequence=1 (accessed on 2 December 2021).
58. Wang, T.; Harvey, J.; Kendall, A. *Network-Level Life-Cycle Energy Consumption and Greenhouse Gas from CAPM Treatments*; University of California Pavement Research Center Research Report UCPRC-RR-2014-05; UC Davis: Davis, CA, USA, 2013.

59. Tseng, E. Construction of Pavement Performance Models for Flexible Pavement Wheelpath Cracking and IRI for the California Department of Transportation New Pavement Management System. Master's Thesis, University of California Davis, Davis, CA, USA, 2012. Available online: https://search.proquest.com/openview/c0d1baa8afc851c4962b1399cc5d5b3b/1?pq-origsite=gscholar&cbl=18750&diss=y (accessed on 24 November 2021).
60. McGhee, K.; Gillespie, J. Impact of a Smoothness Incentive/Disincentive on Hot-Mix Asphalt Maintenance Resurfacing Costs. FHWA/VTRC 06-R28, VTRC 06-R28. 2006. Available online: https://ntl.bts.gov/lib/37000/37300/37317/06-r28.pdf (accessed on 2 September 2021).
61. Zaabar, I.; Chatti, K. Calibration of HDM-4 Models for Estimating the Effect of Pavement Roughness on Fuel Consumption for U.S. Conditions. *Transp. Res. Rec. J. Transp. Res. Board* **2010**, *2155*, 105–116. [CrossRef]
62. Sétra. Road noise prediction—1-Calculating sound emissions from road traffic. Report Q-SETRA–11-ED13–FR+ENG, 120p, Reference SKU1916164304, Paris, France. 2009. Available online: https://www.cerema.fr/fr/centre-ressources/boutique/road-noise-prediction (accessed on 2 September 2021).
63. Sétra. Prévision du Bruit Routier 1-Calcul des Émissions Sonores Dues au Trafic Routier. Sétra. 2009. Available online: https://www.cerema.fr/fr/centre-ressources/boutique/prevision-du-bruit-routier-calcul-emissions-sonores-dues-au (accessed on 2 October 2021).
64. International Organization for Standardization. *ISO 14040:2006-Environmental Management—Life Cycle Assessment—Principles and Framework*; ISO: Geneva, Switzerland, 2006.
65. International Organization for Standardization. *ISO 14044:2006-Environmental Management—Life Cycle Assessment—Requirements and Guidelines*; ISO: Geneva, Switzerland, 2006.
66. Bulle, C.; Margni, M.; Patouillard, L.; Boulay, A.-M.; Bourgault, G.; De Bruille, V.; Cao, V.; Hauschild, M.; Henderson, A.; Humbert, S.; et al. IMPACT World+: A globally regionalized life cycle impact assessment method. *Int. J. Life Cycle Assess.* **2019**, *24*, 1653–1674. [CrossRef]
67. Curran, M.A. *Life Cycle Assessment Student Handbook*; Wiley: Hoboken, NJ, USA, 2015.
68. Huijbregts, M.A.J.; Steinmann, Z.J.N.; Elshout, P.M.F.; Stam, G.; Verones, F.; Vieira, M.; Zijp, M.; Hollander, A.; van Zelm, R. ReCiPe2016: A harmonised life cycle impact assessment method at midpoint and endpoint level. *Int. J. Life Cycle Assess.* **2017**, *22*, 138–147. [CrossRef]
69. Leontief, W. Environmental Repercussions and the Economic Structure: An Input-Output Approach. *Rev. Econ. Stat.* **1970**, *52*, 262–271. [CrossRef]
70. FHWA. *State of the Practice on Sustainability Rating Systems*; Federal Highway Administration, Federal Highway Administration: Washington, DC, USA, 2019.
71. Norris, G. *Handprint-Based NetPositive Assessment*; Center for Health and the Global Environment, Harvard T.H. Chan School of Public Health: Cambridge, MA, USA, 2015. Available online: https://hwpi.harvard.edu/files/chge/files/handprint-based_netpositive_assessment.pdf (accessed on 10 October 2021).

sustainability

MDPI

Article

Towards Road Sustainability—Part II: Applied Holistic Assessment and Lessons Learned from French Highway Resurfacing Strategies

Anne de Bortoli [1,*], Adélaïde Féraille [2] and Fabien Leurent [1,3]

[1] Laboratoire Ville Mobilité Transport, Ecole des Ponts ParisTech, University Gustave Eiffel, 4-20 Boulevard Newton, Cité Descartes, CEDEX 2, 77447 Marne-la-Vallée, France; fabien.leurent@enpc.fr

[2] Laboratoire Navier, Ecole des Ponts ParisTech, University Gustave Eiffel, 6-8 Avenue Blaise-Pascal, Cité Descartes, 77420 Champs-sur-Marne, France; adelaide.feraille@enpc.fr

[3] Centre International de Recherche sur l'Environnement et le Développement, UMR 8568, Campus du Jardin Tropical, 45 bis Avenue de la Belle Gabrielle, 94736 Nogent-sur-Marne, France

* Correspondence: anne.de-bortoli@enpc.fr or anne.debortoli@polymtl.ca

Abstract: Roads are major transportation infrastructure whose sustainability of maintenance practices has never been holistically assessed due to a lack of a proper method. This paper applies a newly developed assessment method (see article part I) on a 10-km-long section of French highway to fully compare the performance of various types of pavement resurfacing policies, for all the maintenance stakeholders, and considering pavement–vehicle interaction (PVI). After presenting the highway section and the parametrization of the model, four alternative resurfacing frequencies are compared to the French standard maintenance scenario over the pavement lifespan. Results show that increasing resurfacing frequency generates gains in terms of domestic production and employment, environmental damage (health, biodiversity, resources), user budgets, and local residents' health damage created by traffic noise. Conversely, it entails financial losses for the road operator and government (tax revenues and net present value), as well as time losses for users. On the contrary, the consequences of a decrease in this frequency are the opposite. Excess fuel consumption due to PVI governs the scale of the environmental and financial gains or losses of highway maintenance policies. Optima in terms of health returns on investment and user savings appear to be around a 50% increase in maintenance funding: for each additional euro spent by the operator, there is a user gain of 3.5 euros and a human health gain of 710 euros. Sensitivity analyses indicate that the marginal gains are highly sensitive to the thickness of the resurfacing technique for macroeconomic indicators, global Net Present Value, and operator savings, while the gains are proportional to the traffic and International Roughness Indicator deterioration speed for tax revenue, users' savings, time savings, noise, and environmental metrics. The other indicators are either slightly or not sensitive to these parameters. To conclude, the entire road maintenance system must be redesigned, from the tax system and funding schemes to the prioritization of road "green practices", to align all the stakeholders' interests towards a globally more sustainable road system.

Keywords: road maintenance; sustainability; key performance indicators; pavement asset management; public investment policies; life cycle

check for updates

Citation: de Bortoli, A.; Féraille, A.; Leurent, F. Towards Road Sustainability—Part II: Applied Holistic Assessment and Lessons Learned from French Highway Resurfacing Strategies. *Sustainability* **2022**, *14*, 7336. https://doi.org/10.3390/su14127336

Academic Editors: Hugo Silva, Joel R.M. Oliveira, R. Christopher Williams and Zejiao Dong

Received: 17 April 2022
Accepted: 10 June 2022
Published: 15 June 2022

1. Introduction

Roads are key infrastructures that support most of the world's transportation activity [1,2], explaining the wide corpus of research on their sustainability [3]. However, the studies carried out around sustainability solutions are fragmentary, often approached through the prism of a single discipline, as shown in the literature review of part I of this double article on pavement sustainability assessment [4]. Examining these solutions quantitatively through a holistic assessment to quantify the levers of sustainability is needed to develop sound public policy recommendations. This is the challenge that we address in this article on the issue of road maintenance, by applying the holistic method developed in part I of this double article [4].

As road networks mature, new infrastructure construction becomes scarce [5–7]; this mechanically concentrates growing economic, environmental, and social issues around network maintenance, especially since a road never actually reaches the end of its life [8], and is therefore never entirely rebuilt, but more or less heavily maintained. Thus, studying the sustainability of road maintenance policies seems to us to be the heart of the challenge in moving towards sustainable roads, for instance by using greener materials or construction practices [9]. Moreover, from a functional point of view, the road infrastructure aims to support vehicles. Its function is therefore to allow the movement of goods and people carried by these vehicles. The sustainability performance of a road also encompasses the performance of the vehicles circulating on it, in particular because of the Pavement–Vehicle Interactions (PVI) described in part I of this double article [4], which can generate substantial excess fuel consumption and accelerated deterioration of vehicle suspensions and tires, even under a limited surface condition deterioration [4,10].

The maintenance of a pavement consists of improving the surface or structure characteristics of a road through construction operations [11]. This improvement is achieved through more or less heavy constructive operations, ranging from simple crack filling—which consists of injecting a hydrocarbon binder into road cracks—to major rehabilitation, which consist in rebuilding the pavement over a thickness of twenty centimeters [11–13]. In our article, we focus on asphalt resurfacing operations; in France, it consists of the construction of a 0.5 to 9 cm-thick layer (0.2 to 3.5 inches) of materials on the surface of the pavement, potentially after a planning operation, i.e., a purge of deteriorated road surface materials. According to French road constructors, resurfacing activities represent a major part of their activity and sales. Around ten major resurfacing techniques are used around the world, varying in the materials used—bitumen or emulsion, type of aggregate —, the material manufacturing method— manufacturing temperature, in a plant or on-site— —, the construction process, building machine used, resurfacing type —, and the new rolling course thickness [11]. A maintenance program consists of a succession of resurfacing operations spaced out over time. Today, French road managers mainly have a financial and technical approach to maintaining pavements [12,14–16], and rarely include environmental criteria on carbon footprint, primary energy consumption, or water consumption [17]. Neither the three pillars of sustainability nor the consequences of PVI are considered in French resurfacing targets, due to a lack of knowledge on these two aspects and how to assess them. However, each maintenance operation has direct sustainability consequences linked to the consumption of materials and energy for the construction, and indirect consequences linked to the effect of the works on the subsequent road surface condition, and ultimately on vehicle consumptions.

The first objective of this article is to quantify the sustainability impacts of the maintenance program of a highway from a systemic point of view, i.e., including the consequences of the pavement condition on the traffic which circulates over it. In this article, we assume that road maintenance policies are the key to sustainable roads, a hypothesis that we test for highways through a French case study and multiple sensitivity analyses. We apply the new holistic assessment method developed in part I of this double article to perform a quantitative assessment of the sustainability impacts generated by altering resurfacing periods in a case study, in order to learn lessons on good maintenance practices for highway resurfacing in France (final objective). Sensitivity analyses are also conducted to extrapolate the conclusions from a specific case study to a high variety of representative conditions in France and Europe. The article also aims at demonstrating the practicability of the holistic assessment method presented in part I of this double article [4] on concrete cases. Parameters to customize part I's equations and databases—life cycle inventories, microeconomic and macroeconomic datasets—are developed for that purpose.

2. Case Study Presentation

2.1. A Typical French Highway

2.1.1. Presentation

We chose to study a highway on the French network in open country, which is typical in terms of climate, structure, traffic composition, geometry, and topography. This highway has been anonymized, as requested by its concessionaire. It is a four-lane highway, with two lanes in each direction. This geometry represents around two-thirds of the French highway network [18]. We recovered data records for a 100 km-long section of this highway from the pavement management system of its operator, Cofiroute, updated in January 2017.

2.1.2. Past Maintenance Strategy

We conducted a statistical analysis relating to the age of the road surface materials on the selected section of the highway. Most of the rolling courses of this highway are very thin asphalt concrete overlay (VTACO) or, more rarely, semi-coarse asphalt concrete overlay (SCAO). Figure 1 shows the statistical distribution of the rolling course ages. The average age calculated for the rolling courses is 6.87 years, thus the highway is resurfaced every 13.7 years in average. The analysis also shows that 20% of the rolling course linear is older than 15 years old.

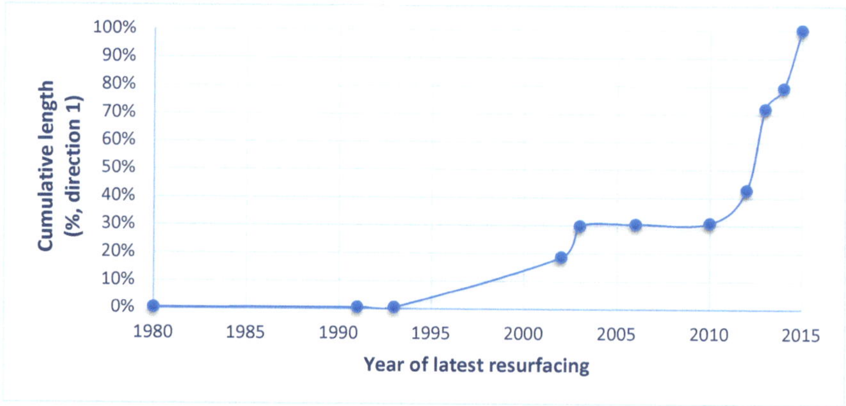

Figure 1. Cumulative percentage of the surfaces of the highway studied, by year of resurfacing, on 1 January 2017.

2.2. Selection and Presentation of the Section Studied

Within the 100 km section of this highway, we selected a 10 km-long section based on the criteria of national representativeness and standard dimensions for a highway maintenance operation.

2.2.1. Structure and Condition of the Section

The 10 km selected are homogeneous in terms of structure (same structural design and rolling course). In 2017, the rolling course was 15 years old, and the pavement structure itself was 37 years old; the section was rehabilitated, and we aimed to find out how to schedule resurfacing over the lifespan of the new pavement to optimize its sustainability performance. The *IRI* before resurfacing was provided by the data monitoring company per subsection of 10 m over the 10 km, and the raw data are provided in the Supplementary Materials for reproducibility purpose. It shows a very low average *IRI* of 0.37 mm/m (see details of the figures and lifetime models in the Supplementary Materials).

2.2.2. Traffic Data

We calculated highway traffic data using the toll payment data provided by the operator. The class 1 toll corresponds to passenger vehicles, class 2 to large vehicles with a total height of more than 2 m and less than 3 m (i.e., only large utility vehicles and caravans), class 3 to trucks with 2 axles, and class 4 to trucks with 3 axles or more [19]. These correspond to the 4 categories of vehicles in our method: Passenger Cars (PC), Light Commercial Vehicles (LCV), Small Heavy Vehicles (SHV), and Large Heavy Vehicles (LHV). The section carries an Annual Average Daily Traffic (AADT) of around 11,000 vehicles in each direction, consisting mainly of around 85% PCs, 5% LCVs, and 10% trucks, 9% of which having 3 axles or more (LHV). We consider traffic composition to be time-fixed, but the traffic itself to increase by 0.44% each year based on Cofiroute's data. Vehicles are considered to have the following number of tires: 4 tires for PCs, LCVs, and SHVs, and 10 tires for LHVs, 5-axle trucks being the most frequent in France. Average speeds are considered equal to 118 km per hour for LV and 88 km per hour for HV (respectively, 73 and 55 miles per hour).

2.3. *Selection of Resurfacing Strategies for Comparison*

We consider a new pavement commissioned on 1 January 2017. Based on the statistical study made on the rolling course ages (Figure 1), we then compare five maintenance programs that vary according to the range of the operator's resurfacing possibilities. We choose to assess the consequences of each of these resurfacing programs over an empirical road surface lifespan of 39 years. The scope of our assessment includes neither the construction of the initial rolling course in 2017 nor a 39-year resurfacing, since it is considered that the whole pavement will need to be rehabilitated at that time, including the rolling course. The treatment in these five scenarios consists in milling the road surface and replacing it with a 2.5 cm-thick asphalt concrete (VTACO), which is the main rolling course found on highways in France, as shown in a survey carried out with French road operators (see Supplementary Materials).

The characteristics of the five scenarios assessed are described in Table 1. The "REF" scenario corresponds to the reference, classically called the "Business-as-usual" (BAU) scenario. "Max_Field" corresponds to an ambitious practice scenario, i.e., the shortest resurfacing periods observed among French concessionaires. On the contrary, "Invest_Min" represents the longest resurfacing period statistically analyzed in Figure 1. This scenario could cause some problems in terms of waterproofness that could potentially put the structure at risk. Waterproofness could be ensured by crack filling. More problematic issues related to a drop in skid resistance could occur, and could also be corrected by performing sand blasting. Ideally, it would have been interesting to consider these minor maintenance operations, but the road experts highlighted their very low cost and probable negligible impact on the environment and traffic disruption.

Table 1. Specification of different resurfacing strategies tested over the lifespan of the section.

Scenario Code	Details	Number of Resurfacing Nk	Surface Lifespan
Invest_Min	Minimum investment	1	19 years 6 months
REF	Reference	2	13 years
Invest+	Increased investment	3	9 years 9 months
Max_Field	Ambitious approach	4	7 years 10 months
Invest_Max	Maximum investment	6	5 years 7 months

3. Tailored Method Parameterization and Data Development

In this section, we present the different parameters needed or tailored to apply the holistic assessment method presented in part I of this double article [4] to this specific French highway case study. We also present new data, especially the Life Cycle Inventories (LCIs) developed for the French market, to calculate the LCA indicators with better robustness.

The different sections follow the same order as the equations developed in part I for ease of reading.

3.1. International Roughness Index (IRI) Progression Models

All the impact calculations in the assessment method depend on the surface condition of the road section in question. The company that collects this highway's condition data provided the *IRI* data for the section. These data are "two-track", in other words, we were provided with two datasets per direction: one dataset for the left-hand strip and one for the right-hand strip, for the slow lane only, corresponding to the highest traffic densities (cross-sweep). Then, we calculated the surface profile of the pavement over time. The average *IRI* for the section in year 1 is calculated by averaging the scores for the left- and right-hand strips. Then, evolutions of the *IRI* are calculated over the whole lifespan of the pavement: changes excluding roadworks, and changes before/after roadworks. Between two resurfacing operations (=excluding roadworks), the *IRI* is considered to change at a rate of +0.05 m/km.yr, based on the historic *IRI* data for the entire highway. The *IRI* evolution between two resurfacing operations thus follows the following Equation (1), with t in year:

$$IRI(t) = IRI(\mathrm{t} = 0) + 0.05\mathrm{t} \tag{1}$$

The effect of resurfacing work on *IRI* is estimated by calibrating the American progression model established by Wang et al. [20] using the *IRI* data for the highway to which our studied section belongs. We used Equation (2), where e = 0.3 and g = 0.15 m/km.

$$IRI(t_{R+}) = e.IRI(t_{R-}) + g \tag{2}$$

The graphic of the *IRI* evolution model developed for the highway section is presented in the Supplementary Materials. It shows an average reference *IRI* equal to 0.77 m/km on this highway, a value that is considered for the calculation of excess fuel and tire consumption.

3.2. Consumption and Emission Parameters

3.2.1. Resurfacing Works Demand

The width ω of the pavement is equal to 11 m, composed of 3.5 m for each lane, as well as 3 m for the emergency lane and 1 m for the left flattened band. The length λ of the section studied is equal to 10,000 m. N_k, the number of resurfacing operations of type k over the assessment period, is indicated for each scenario presented in Table 1.

3.2.2. Excess Vehicle Consumption Parameters

The models to account for excess consumption and wear are presented in part I of this combined article and its Supplementary Materials [4]. Wear factors affecting tires and suspension systems over a vehicle's lifespan are considered for each vehicle category (see combined article's Supplementary Materials). Nevertheless, *IRI* remains below 3 m/km in our different scenarios; no excess suspension wear is observed. The standard lifetime mileage of the tires on highways is indicated in Table S9 of the Supplementary Materials. However, some parameters are still needed for excess fuel consumption calculations.

Excess fuel consumption is calculated based on average consumptions by type of vehicle on highways in France, using the French model CopCETE [21]. CopCETE is a tool developed for the French Ministry of Transportation. It uses COPERT V4 equations [22] and considers French average speeds on specific networks. It simulates the consumptions of different kinds of energies (and 26 types of pollutants emitted). The average consumption per type of vehicle category varies over time with technological evolutions. Technological evolutions are taken into account by considering predictions developed for the French fleet by the governmental research center IFSTTAR [23] (type and quantity of energy consumed).

3.3. Noise Emission Parameters

The average sound powers during the day (Wj) and night (Wn) are calculated using daily traffic data presented in Section 2.2.2.; the hourly repartition in the day and nighttime from the French governmental technical center Sétra are presented in Table 2 [24], with the additivity method of the noise line sources and the general methodology detailed in part I of this double-article publication [4].

Table 2. Equation to calculate q(véh,i), the hourly traffic of light and heavy vehicles on an interurban highway.

Time of the Day	Day: 6 A.M.–10 P.M.	Night: 10 P.M.–6 A.M.
LV	AADT(LV)/18	AADT(LV)/82
HV	AADT(HV)/20	AADT(HV)/39

LV = Light Vehicle; HV = Heavy Vehicle.

To add the line source's noise levels, Sétra estimated [24] that their noise levels can simply be added by considering one single line source in the middle of the pavement, under the condition that the noise receptors are in direct reception and farther than 2.4 times the width of the road platform, i.e., 40 m away from the line source for 2 × 2-lane roads, a condition that is respected for the studied highway.

3.4. Environmental Metrics: Life Cycle Inventories

To calculate the environmental impact of each maintenance strategy, we use equation 13 from the article's part I. The flows of consumption $flow_c$ are listed in the corresponding consumption life cycle inventory (LCI). We specifically develop LCIs for the French context [25], summarized in the Supplementary Materials.

For roadworks, it consists of road resurfacing processes for 1 m^2 of resurfaced pavement, for each of the eight most frequent resurfacing techniques in France, from cradle-to-laid, i.e., from the extraction of raw materials—e.g., bitumen, aggregates—to construction on site. Models have been developed based on statistics calculated from the annual data of a major French road constructor, providing around one-third of the national road construction. Each resurfacing technique consists of a first tack coat with 500 g per square meter of 65% bitumen emulsion (i.e., 325 g/sm of residual bitumen, including SBS polymers and phosphoric acid), covered with a layer of asphalt mixture. Average materials production, transportation, and use of the building machines on-site are considered based on the road company's statistics. The building machines used for the hot asphalt mixtures (HMA) are a sprayer, a roller, and a finisher. HMAs are made of 4.8% of non-modified bitumen, mixed with aggregates at around 165 °C, e.g., the average hot mixing temperature in France. The average road aggregate in France comes at 76.4% from hard rocks and 23.6% from lose rocks, transported mainly by trucks, but also partly by rail and barges. The average French asphalt mixing burner consumes heavy fuel and natural gas in the amount detailed in the Supplementary Materials. Cold techniques (double-layer micro-surfacing, double-layer dressing, double prechipped surface dressing) are made of a cold asphalt mixture (CAM), or, directly, aggregates and emulsion, and are either built with a CAM spreader/vacuum sweeper, a small loader/roller/finisher, or a binder sprayer/gravel spreader/roller on tires. Polycyclic aromatic hydrocarbon (PAH) emissions are also calculated based on simplified convection models (see Supplementary Materials). For dressing and CAM techniques, the 65% bitumen emulsion is also used. One ton of CAM contains 103 kg of this emulsion, e.g., 6.7% of residual bitumen in the mixture. The asphalt mixture density after compaction is considered equal to 2.4 t/m^3.

The environmental impact of the techniques used in this case study and calculated with the characterization IMPACT World+ and ReCiPe are presented in Table 3, including SCACO, very thin or thin asphalt concrete overlays (resp. VTACO and TACO, that are

2.5 and 4-cm thick in France), and double prechipped surface treatment (ST). The impact of other operations can be found in a Supplementary File S1.

Table 3. Damage to the environment from resurfacing operations over 1 m^2.

Damage	Unit	Milling	SCACO	TACO	TACO	VTACO	Double Prechipped ST
Health	DALY	2.49×10^{-5}	3.51×10^{-3}	2.37×10^{-3}	1.52×10^{-3}	9.45×10^{-5}	7.01×10^{-5}
Biodiversity	PDF.m^2.yr	2.77×10	1.18×10^2	8.01×10	5.19×10	3.29×10	2.75×10
Resources	USD	1.53×10^{-1}	1.71×10	1.17×10	7.59×10	4.85×10^{-1}	4.98×10^{-1}

DALY = Disability Adjusted Life Years; PDF = Potentially Disappeared Fraction.

For vehicles, LCIs relate to fuel consumption, as well as tire and suspension wear. LCIs per kilometer traveled for the use of internal combustion engine vehicles take account of forecast technological developments based on IFSTTAR's prediction model for France's vehicle fleet [23]. Trucks remain 100% diesel, while light vehicles show evolving shares of technologies overtime (see Supplementary Materials). These processes consider emissions of 26 types of substances—heavy metals, gaseous pollutants—specified in the Supplementary Materials, using the French Department of Environment's CopCETE software [21]. Linearity between the energy consumed and the quantity of each pollutant emitted are considered per type of energy, due to a lack of a more accurate model. Diesel and petroleum densities are assumed equal to respectively 0.85 and 0.75 kg/L. The environmental damage from fuel supply and combustion is presented in the Supplementary File S1.

Finally, an in-house environmentally extended input–output analysis is developed to assess the damage from using garages to maintain the vehicles, based on statistical garage expenses, activities, and revenue [26], synthesized in the Supplementary Materials. Shock absorber and tire LCIs are also presented in the Supplementary Materials. The environmental impacts of maintenance operations are presented in Table 4.

Table 4. Damage to the environment from vehicle maintenance.

Functional Unit		One Kit of Shock Absorbers for One Vehicle				One Kit of Tires for One Vehicle				Maintenance Service
Damage	Unit	PC	LCV	Small HV	Large HV	PC	LCV	Small HV	Large HV	EUR 1
Health	DALY	1.27×10^{-3}	2.71×10^{-3}	3.19×10^{-3}	9.47×10^{-3}	4.57×10^{-3}	1.42×10^{-2}	2.16×10^{-2}	9.38×10^{-2}	3.88×10^{-8}
Biodiversity	PDF.m^2.yr	1.69×10	3.59×10	4.23×10	1.29×10^2	9.94×10	2.95×10	4.19×10^2	1.80×10^3	1.79×10^{-4}
Resources	USD	2.45×10	5.21×10	6.13×10	1.85×10	7.67×10	2.38×10	2.84×10	1.22×10^2	1.02×10^{-5}

3.5. *Social Metric Parameters*

3.5.1. Road Noise Health Impact Indicator

The residents' health damage due to traffic noise is calculated over the 39 years of the assessment period according to Equation (15) of the article's part I, using noise emission parameters calculated previously.

3.5.2. Time Loss Parameters

- Roadwork time losses

Roadworks are carried out by means of daytime lane closures, with all the traffic of the highway sent the same way, each direction having one lane instead of two. Average resurfacing construction speeds are taken to be 800 m per day, full width.

Light vehicle speeds are reduced to 90 km/h maximum, instead of 130 km/h normally. We consider this maximum speed to equal the average speed during the resurfacing works. The HV speed is not affected by this new speed limit. No congestion, only reduced speeds, is noted on the section during the roadworks, as a result of being a low-traffic highway. In this simple case, the time loss in work zones is considered relational to a reduction in the authorized speed on the section maintained, in free flow. We use the time-saving formula shown in Equation (3) to calculate the time loss by a vehicle of type i passing through a work zone with its length, $Length_{road\ section}$, the reduced speed due to roadworks for the type of vehicle i, $v_{reduced,i}$, and the standard speed on the section, $v_{standard,i}$.

$$Time_{saved,works,\ i} = Length_{road\ section} \cdot \left[\frac{1}{v_{reduced,i}} - \frac{1}{v_{standard,i}} \right] \tag{3}$$

To calculate the time losses due to roadworks for each alternative road maintenance, we use Equation (17) from the double article's part I [4], by calculating the hourly traffic using Table 2.

- Consumptions time losses

The operation duration of maintenance activities OD_j, which is considered to calculate vehicle consumption time losses, is indicated in Table 5, and explained in the Supplementary Materials.

Table 5. Estimates of the duration of the various operations and maintenance activities.

OPERATION Duration	PC	LCV	SHV	LHV
Tire replacement (min/veh)	120	150	150	480
Suspension replacement (min/veh)	330	330	30	30
Fueling (min/100 L)	40	25	1.9	1.1

3.6. Economic Metric Parameters

3.6.1. Discount and Inflation Rates

The default discount rates for the different stakeholders considered are as follows: 2.5% for society, 1% for government and households, and 8% for Cofiroute, the rate of return for this highway operator being 8.28% in 2018. To account for dynamic economic evolutions, inflation rates are estimated based on linear regressions applied to INSEE's time series for monthly inflation rates (see Supplementary Materials), INSEE being the French national institute for statistics and economic studies. The inflation rate for roadworks is selected based on trends in the Consumer Price Index in France since 1966 [27], i.e., 1.4%, in the absence of consistent trends for the costs of roadworks or even construction works. The inflation rate for service station vehicle maintenance costs is taken to be 6% between 1998 and 2015 [28].

3.6.2. User Cost Parameters

The first step in our calculation of user costs over a given period is to calculate for each year, each type of vehicle, and each type of expenditure, the annual expenditure in current euros for all users. To do this, we multiply the volumes of goods consumed by the unit prices of these goods in current euros. We thus arrive at a table containing, for each year, the sum of the costs in current euros for all traffic.

For the case of the maintenance of suspensions and tires, we developed models of kilometric costs, including tax in 2017 euros, which are presented in the Supplementary Materials of the article part I [4]. To transform these costs into current euros, we use the INSEE series statistics, which follow the inflation of goods (see Supplementary Materials).

Fuel price is made up of the price excluding crude oil taxes set by the market, the cost of refining, transportation and distribution, and taxes, following Equation (4). The taxes levied by the government in France are the Value Added Tax (VAT) and the Domestic Consumption Tax on Energy Products (TICPE). These components vary depending on the fuel considered. The VAT rate on the consumption of natural gas and petroleum products is the normal rate as of 24 January 2018, i.e., 20%. VAT applies to the product itself and the

TICPE. Prior to 1 January 2014, VAT was set at 19.6%. The fuel price formula is therefore that of Equation (16).

$$FuelPrice,\ including\ tax_{\ year\ i} = (FuelPrice,\ excluding\ tax + TICPE)_{year\ i} \times (1 + VAT)_{year\ i} \quad (4)$$

The price of diesel and petroleum in January 2017 was, respectively, 1.27 and 1.48 EUR/L, including tax (current euros). Based on statistical analyses and linear regressions carried out on the INSEE statistical series (see Supplementary Materials), fuel markets are quite volatile, but we studied trends in the price of a barrel of oil (Brent), margins and processing, then taxes. Between 1990 and 2018, we find the following annual trends for diesel: +0.007 EUR/L for the "barrel" part and +0.004 EUR/L for the "margin and transformation" part, i.e., +0.011 EUR/L on the price, excluding tax.

The TICPE is likely to vary in France over the next few years for two reasons. First, a tax catch-up was planned between diesel and gasoline between 2017 and 2022, diesel having so far been fiscally advantaged, despite its leading role in atmospheric pollution. This results in an increase of 4.33 cents/L.year for diesel and 2.02 cents/L.year for petroleum for the 2017–2022 period. Then, an ambitious carbon tax increase was proposed by the Quinet commission [29], generating an increase of 1.78 EURcent/L.year after 2023.

3.6.3. Road Operator Costs

In the absence of a quality cost database, we propose to consider as resurfacing costs the current costs of the French techniques, calculated based on the responses collected in our survey of French interurban road managers (Table 6).

Table 6. Costs of resurfacing techniques in constant currency, including tax (2017 euros).

Cost	SCACO	TACO	VTACO	CMA	ST
Average (EUR/m^2)	18	14	10	5	3

From the database of prices of resurfacing techniques inclusive of all taxes, we obtain the basic price for roadworks, deducting 20% for VAT, together with an 8% profit margin, based on the national economic data for the "civil engineering" branch obtained from INSEE.

3.6.4. Macroeconomic Datasets

- General calculation

To assess the impact of a variation in the final demand for a product on the production of the entire economic system, we use the Leontief inversion relationship presented in this article's part I to make our production calculation. Finally, depending on the demand vector of the maintenance scenario to be assessed, this vector is evaluated with the costs in basic prices (which, for simplicity, are considered equivalent to the price, excluding VAT, even if this is an approximation) of consumption in vehicle maintenance, fuel, and civil engineering. The total output of the system studied is obtained by summing all the terms of the vector P.

- Production dataset

We obtained from the Department of National Accounts of INSEE three types of tables for the years 2010 to 2013, with a disaggregation into 138 branches and products. These tables are confidential data and not supplied in this study. The supply–use balance table (ERE) quantifies the effects of imports for each product (total imports, customs duties, territorial adjustment, subsidies, transport, and trade margins), as well as taxes and duties. It also informs about the market aspect (stocks and investments). The Intermediate Inputs table (TEI) indicates the intermediate product consumption of each branch. It is often rectangular, because a branch can produce several products, and presents basic prices, i.e., the price invoiced by the producer. plus any subsidy minus taxes on each product. TES outputs are shown at base prices (i.e., the amount the producer receives from the buyer per unit of good or service produced, minus taxes on products, plus subsidies on products. The base price excludes transport costs, invoiced separately in current euros (EUR2010 to EUR2013). In this modeling, we consider the matrix of technical coefficients obtained with the most recent economic data, i.e., those of the year 2013. We consider that this matrix is stable, which is approximate, especially in the long term. Finally, we do our calculations in current euros.

- Employment dataset

INSEE supplied the table of domestic employment content by branch in thousands of Full-Time Equivalent jobs over one year (FTE.year) of the 2013 French economy in 88 branches (confidential data). According to the advice of INSEE experts, the employment content assumedly depends linearly on the added value of the same sub-branch in accounting for 138 branches—data to which we also had access through INSEE. We thus broke down the 88 branches' employment content into the 138 production branches to get vectors of the same dimensions to perform our calculation. Employment content calculated are respectively 16.4, 14.4, and 7.2 FTE.year per million euros of demand, excluding tax for garage maintenance, road maintenance, and fuel demand. These figures are consistent with the French literature [29,30].

3.6.5. Tax Revenues

The taxation of road fuels has been detailed in the fuel price model, and consists of the TICPE and the VAT, the latter applying both to the price excluding tax of the fuel and the TICPE. HVs from the European Union can benefit, on request, from a flat-rate partial reimbursement of the TICPE. In the second half of 2017, this rate was EUR11.42/100 L. We consider this to be fixed-rate, without adjusting it for inflation in our model, in the absence of forward-looking data on the evolution of this fixed reimbursement.

The taxation of vehicle maintenance concerns tires, as well as suspensions. In our model, both for the maintenance of tires and suspensions, we propose to retain a VAT of 20% on passenger cars, and 0% for other vehicles, although the rate of exempt professional vehicles is probably not 100%.

Finally, taxes on maintenance operations are not taken into account in the calculation of the government's fiscal surplus because there is a company–government transfer; these taxes are said to be "neutral" [31].

3.6.6. Integrated National Economy

Based on the generic equation presented in part I, the overall financial indicator for road maintenance programs must take into account maintenance (manager) and operating (manager, users) expenditure, as well as government expenditure and revenue. We only count tax receipts for the government, in the absence of data on corporate tax receipts and savings made by job creation (savings in unemployment allowances). Rates of return can be calculated in constant or current euros, however, financial and economic profitability would only make sense in "real terms", i.e., in nominal terms taking inflation into account [31]. The assessment of the overall societal cost is based on values excluding tax at constant euros [32]. To calculate the cost, the opposite of the Net Present Value (NPV) is calculated, using a single discount rate for society at 2.5%, which is the risk-free discount rate recommended by the French Ministry (we consider a risk premium of 0% instead of the 2% recommended) [29].

4. Results

The calculations for the application of the method described in part I of this double article [4] to the case study described in this part II article were carried out using Scilab. This free multiplatform software offers an appropriate environment for scientific applications and uses a programming language designed for high-level numerical calculation. The Scilab code is made available on Github [33] and can easily be adapted to MATLAB language for further usage.

4.1. Comparison of Alternative Maintenance Programs

4.1.1. Sustainable Performance Trends

Figure 2 represents the gains achieved by altering the maintenance program from the highway resurfacing practices currently employed in France (reference scenario). It shows the gains from each of the four alternative maintenance programs on the eleven impact indicators, gains normalized by the maximum gain in each impact category. The darker the color of the bars in the histogram, the higher the level of investment required for the maintenance program it represents. Two groups of indicators emerge, which behave in opposite ways in response to a change in the level of investment in maintenance: two-thirds of the indicators show benefits that grow as the amount invested in maintenance rises, whereas one-third of the indicators shows losses that grow with rising investment. The batch of performance indicators improving with increased maintenance encompasses savings for users, domestic employment and production, as well as health protection related to noise and other environmental impacts, natural resources, and biodiversity. The batch of performance indicators deteriorating with increased maintenance consists of four indicators: tax revenues, savings for the operator, global NPV, and time saved by users.

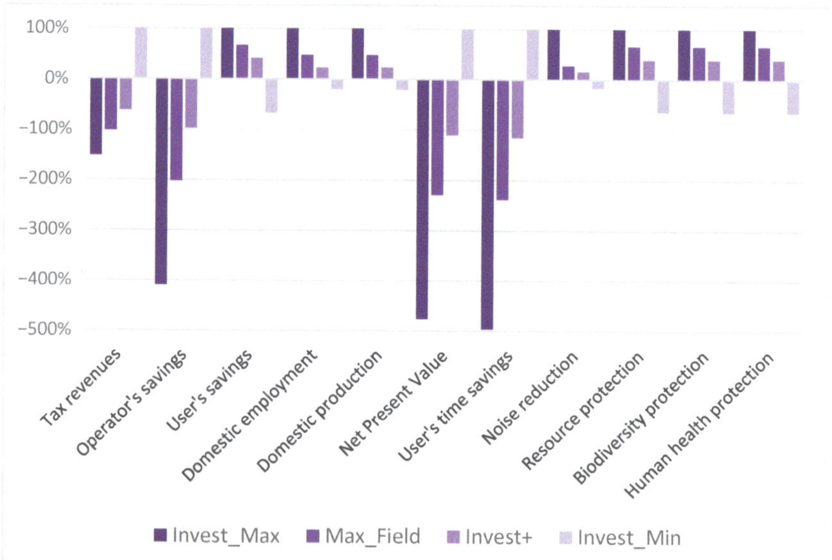

Figure 2. Advantage in choosing the variant relative to the standard program (mathematical normalization by the maximum gain for each indicator).

For the first batch, halving the investment in maintenance compared with current practices generates losses whose absolute value is at least the same as the gains generated by increasing the current maintenance budgets by 50% (noise-related health damage or macroeconomic indicators) and at most equivalent to the level of the gains generated by doubling the maintenance budgets (savings for users, savings in nonrenewable resources, protection of public health). For the second batch, the importance of the loss (in absolute value) caused by tripling the investment in road maintenance even exceeds the gains produced by halving the maintenance budgets.

In this study, the damages to human health due to local pollution from road transportation and traffic noise evolution are compared for the first time. It shows that the evolution of traffic noise due to maintenance strategies has a very low impact on human health damage compared to the evolution of local pollution.

4.1.2. Sustainable Gains' Values

The values of gains and losses over the 39 years of the study period are shown in Table 7 for the financial and economic indicators, and in Table 8 for the social and environmental indicators. These tables show that for the highway and maintenance programs studied, tripling the current level of investment in the maintenance of highway road surfaces would lead to gains of 3.21 million euros for users, 42 FTE years of employment (FTE.yr), almost 9 million euros at 2017 values in French production, the equivalent of almost two DALYs in local environmental noise reduction, and, finally, savings of 120 million US dollars in natural resources, 480 million PDF.m^2.yr in biodiversity, and almost 7700 DALYs as a result of pollution reductions. On the other hand, this tripling in the maintenance budget would lead to losses of 1 million euros in tax revenues, 1.64 million euros for the road operator, 2.13 million euros for society in terms of NPV, and a cumulative time loss of 3.15 years for users. These figures show that the financial gains on a standalone basis exceed by far the losses under a tripled maintenance budget: the gains reach 123 million euros when summing users' gains and natural resources gain, while the losses are limited to 2.64 million euros when summing tax revenue losses and road operator losses. As the positive impacts also include substantial gains in terms of biodiversity and human health protection, a high increase in road resurfacing investment effort is undoubtfully of public interest.

Table 7. Values of the financial and economic gains resulting from a change in the maintenance schedule.

Scenario	Tax Revenues (k EUR)	Operator Savings (k EUR)	User Savings (k EUR)	Employment (FTE.yr)	Production (k EUR 2017)	Total Savings (k EUR 2017)
Invest_Max	-1.06×10^3	-1.64×10^3	3.21×10^3	4.24×10	8.83×10^3	-2.13×10^3
Max_Field	-7.10×10^2	-8.07×10^2	2.16×10^3	2.03×10	4.25×10^3	-1.02×10^3
Invest+	-4.31×10^2	-3.88×10^2	1.32×10^3	9.75×10^2	2.06×10^3	-4.90×10^2
Invest_Min	6.96×10^2	3.99×10^2	-2.16×10^3	-8.35×10^2	-1.82×10^3	4.47×10^2

Table 8. Values of the social and environmental gains resulting from a change in the maintenance schedule.

Scenario	Time Saved (Days)	Noise Reduction (DALY)	"Resource" Gains (USD)	"Biodiversity" Gains (PDF.m^2.yr)	"Health" Gains (DALY)
Invest_Max	-1.15×10^3	1.79×10	1.20×10^8	4.80×10^8	7.69×10^3
Max_Field	-5.52×10^2	5.00×10	7.96×10^7	3.19×10^8	5.12×10^3
Invest+	-2.67×10^2	2.92×10	4.78×10^7	1.92×10^8	3.08×10^3
Invest_Min	2.32×10^2	-2.91×10	-7.67×10^7	-3.08×10^8	-4.95×10^3

Conversely, halving the investment in resurfacing the 10 km highway section studied over 39 years would lead to a loss of 2.16 million euros to users, 8.4 FTE.yr in France, and 1.82 million euros at 2017 values in national production, and cost the equivalent of 0.3 DALYs from increases in local environmental noise, 77 million US dollars in non-renewable resources, 308 million PDF.m^2.yr in biodiversity, and 4950 DALYs from other health damage. In return, this decision would result in an extra 0.70 million euros in tax revenues to the state, savings of 0.40 million euros for the operators, a cumulative time saving of almost 8 months for users, and savings to society of 0.45 million euros. In this decreasing resurfacing frequency scenario, gains reach less than one million euros, while losses account for almost 80 million euros. This comparison consolidates the conclusion that we should increase highway resurfacing to adopt a more sustainable pathway, based on this case study.

4.1.3. Key Factors of the Composite Indicators

The indicators that we call "composite" depend both on the intensity of roadworks and the vehicle consumption in fuel, tires, and suspensions. They are, namely, the gains in time for users, domestic jobs and production, NPV, as well as the three environmental gains. We propose to study the composition of these gains and losses (excluding NPV) in only three alternative scenarios to show the trends in the contribution variability, depending on the maintenance policy direction chosen. The results are presented in Figure 3. The number and name of the maintenance strategy is indicated in the legend after the name of the composite indicator. They show that, for the section of highway studied, the environmental gains and losses depend exclusively on fuel consumption. Road construction operations and tire replacement have a negligible contribution. Conversely, the macroeconomic gains or losses, as well as those in user time, depend mainly on roadwork operations; they bring around 80% of the gain for the scenarios study. Then, the influence of fuel consumption ranges from 5% to 20% of the impact changes. Tire wear has minimal impact on time wasted or gained (<2%), but accounts for up to 7% of the total absolute values of job gains and losses, in the minimum investment scenario. We recall that suspensions do not appear among the contributors, since the *IRI* in this case study remains under 3 m/km over the pavement lifespan, whichever scenario considered, i.e., under the *IRI* threshold to see early deterioration of shock absorbers.

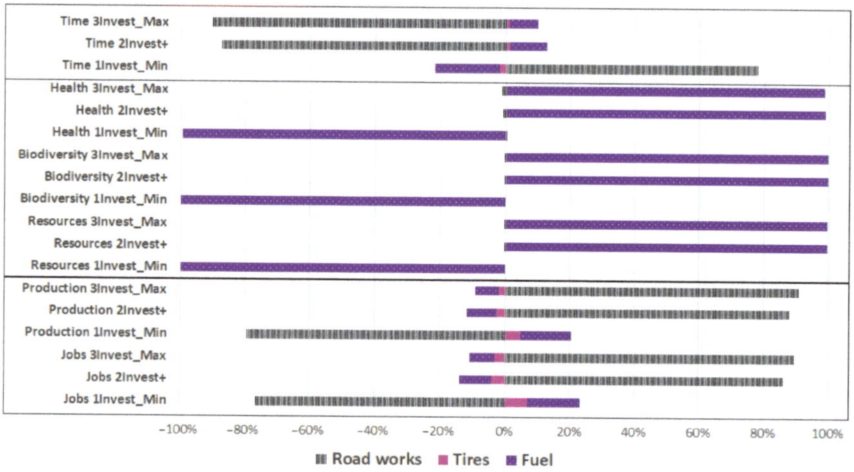

Figure 3. Composition of the composite gains for three of the alternative maintenance scenarios.

Nevertheless, at the scale of the road section, the key factors depend on the level and composition of the traffic, and the infrastructure considered (maintenance schedule, speed of deterioration in *IRI*, and vehicle speeds). The main contributors may be different on different road networks.

4.2. *Additional Economic Analyses*

4.2.1. Multi-Actor Financial Savings Consideration

Figure 4 gives a continuous smoothed representation of the discrete savings generated by changing the road surface lifespan for the set of financial stakeholders {operator + users + government}, to show the trend in the gains depending on maintenance frequency. Here the personal discount rate of each type of the three stakeholders are considered, as discussed before. Investing more in resurfacing generates global financial savings, up to 0.65 million euros for four additional resurfacings over 39 years, while investing less leads to financial losses, up to 1.05 million euros, when one resurfacing operation is canceled over the highway's lifespan. The figure shows a global financial optimum between one and four extra resurfacing operations (e.g., for a road surface lifespan of between 5.6 and 9.75 years). Nevertheless, this optimum applies under the current tax system, which is a politically adjustable variable.

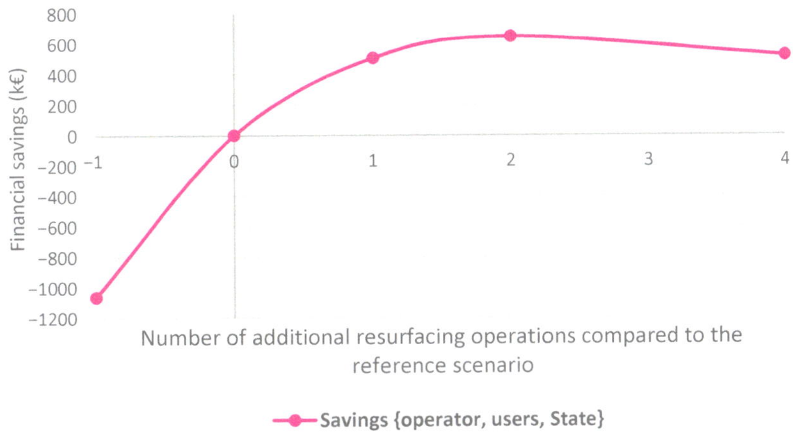

Figure 4. Multi-actor financial gains associated with changes in resurfacing strategies (reference: 13 years).

4.2.2. Cost Effectiveness of Maintenance for Users

We calculated the ratio of the savings made by users from the additional amount invested in resurfacing by the highway operator. This discrete cost effectiveness is represented by continuous smoothing, relative to the additional resurfacing operations to the standard in Figure 5.

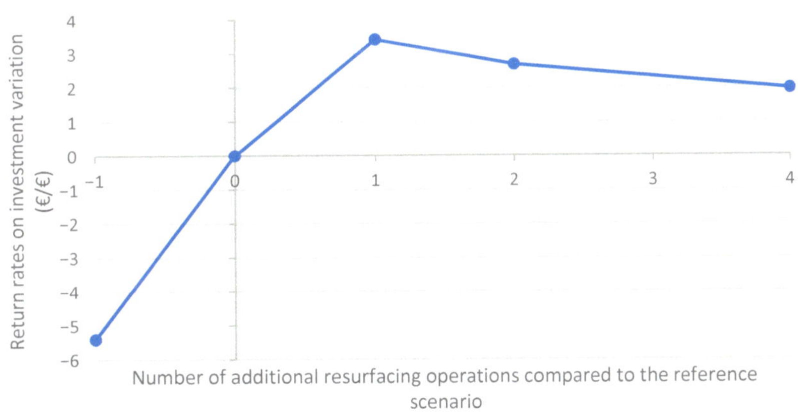

Figure 5. Users' savings for each additional euro invested in maintenance by the road operator (relative to the standard maintenance level).

This figure indicates that each additional euro invested in maintenance generates a saving of between EUR 2 and EUR 3.5 for users. From a practical point of view, maximum cost effectiveness in this case study is achieved by moving from two to three resurfacing treatments over 39 years, i.e., from a road surface lifespan of 13 years to slightly less than

10 years. Mathematically, however, the optimum could be situated between two and four resurfacing operations over 39 years. In the case of a reduction in maintenance, each euro saved by the operator in the minimum investment scenario costs users up to EUR 5.4.

4.2.3. Cost Effectiveness of Maintenance in Health Terms

The notion of cost effectiveness as applied to the financial aspect of maintenance can be extended to public health. Some governments use Values of Statistical Life (VSL) in their socio-economic calculations; in France, the Ministry uses the value of EUR 115,000 (at 2010 values) per year of life [29]. One DALY is equivalent to one year of quality life lost; by equating one DALY of impact on human health with EUR 115,000 at 2010 values, one can calculate the cost effectiveness of investment in resurfacing. By discounting the average VSL for 2017–2056 at a rate of 1%, i.e., leading approximately to EUR 89,000/DALY at the end of the period, one can estimate a "health cost effectiveness", i.e., the cost effectiveness in terms of global public health of marginal investment in maintenance for the road surfaces on our section of highway. Figure 6 shows the ratio of savings made in terms of monetized human lives on the additional amount invested in resurfacing by the highway operator, according to the number of additional resurfacings carried out relative to the standard (i.e., two over 39 years). This "global health return" on investment in highway resurfacing is significantly higher than the financial return; each additional euro invested in maintenance leads to a saving of between EUR 420 and EUR 710 in terms of human life value (respectively, for four to one additional resurfacing operations over 39 years). The maximum return in this example is achieved by moving from two to three resurfacing treatments over 39 years, i.e., from a road surface lifespan of 13 years to slightly less than 10 years. In Figure 6, we can also see that one euro saved in the minimum investment scenario relative to the standard scenario generates a loss of EUR 1100 in human life value.

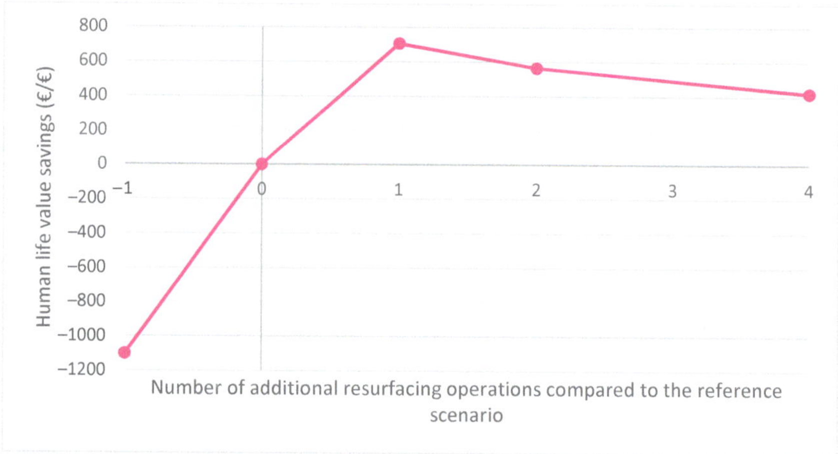

Figure 6. Total health impact per additional euro spent on resurfacing by the operator, depending on the number of additional resurfacing operations carried out over the lifespan of the road.

4.3. Sensitivity of the Marginal Gains

4.3.1. Sensitivity to the Resurfacing Technique

Based on the same resurfacing schedule scenarios, we studied the sensitivity of the results to the road technique used on French highways: 2.5 cm VTACO (baseline) vs. 6 cm SCACO, 4 cm TACO, 1.5 cm VTACO, or double prechipped surface treatment. Since the *IRI* progression model does not differ within these asphalt concrete techniques [20], the objective is indirectly to analyze roughly the variation in gains and losses as a function of the quantities of resurfacing materials used. We present the marginal gain sensitivity to the technique thickness between the baseline scenario and the minimum and maximal investment scenarios in Figure 7, based on the 2.5 cm-thick reference technique. It highlights three groups of indicators with similar behavior. Under the line of equation y = 1, the marginal gains are less affected by this technique than the baseline one (in absolute value), and vice versa. Three indicators are not affected by a modification of the technique in our model: user savings (same *IRI* degradation laws), time savings (single operation pace), and tax revenue (no VAT on road works). However, we also notice that the environmental indicators are not affected either by the resurfacing technique used; they are mainly influenced by fuel consumption, as shown above, as the direct environmental impact of road works is negligible. On the other hand, four types of indicators are highly sensitive to the technique used: the two macroeconomic gains, the global NPV, and the gain for the operator. The model also accounts for the difference in the acoustic category between techniques: surface treatment and thick asphalt concrete in class R3, thin asphalt concrete in class R2. The absolute value of the gains is more sensitive to the thickness when the maintenance rate is slowed down compared to the reference (multiplication factor for the thick asphalt concrete alternative up to 2.20, instead of 1.95, in the maximum investment scenario, on employment, production, road operation savings, and global NPV).

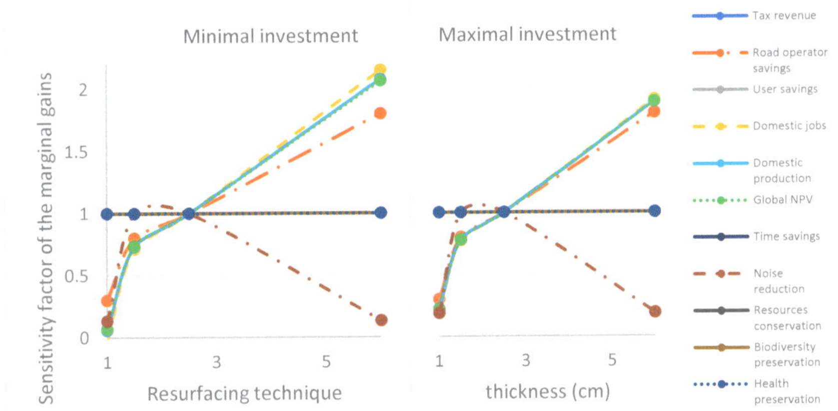

Figure 7. Sensitivity of the marginal gains to the thickness of the resurfacing technique, in the scenarios of minimum and maximum investments.

4.3.2. Sensitivity to the Traffic Level

We test the sensitivity of the results, ceteris paribus, to the AADT, with a factor of ½, 2, and 3 applied to the traffic of the highway studied (i.e., base case with approximately 10000 vehicles per day). The results are presented in Figure 8, under the minimal and the maximal investment scenarios. Under the maximal investment scenario, four groups of indicators behave similarly under changes in the level of traffic. First, there is a group of indicators perfectly proportional to traffic, which includes tax revenue, savings for users, time saved, reduction of the health impact of noise. Then, a second group gathering environmental indicators behaves very similarly. It is roughly linear to traffic, with a slightly increased sensitivity for the health gain relative to the biodiversity gain, and even higher gain in terms of consumption of non-renewable resources. A third group of marginal gains, weakly sensitive to the level of traffic, emerges; it includes the gain in employment, production, and overall economy (NPV). Finally, the manager's investment indicator is obviously not sensitive to traffic, insofar as we do not take into account the impact of works on toll receipts in the case of motorways, nor the effect of the state of the road on the route choice (highways vs. national roads vs. departmental roads). With the minimum investment scenario, we find the same four groups, with the behavior of group n°2 even closer to the behavior of group n°1 (almost identical). Finally, the gains of the third group are more sensitive to traffic in the minimum investment scenario than in the maximum investment scenario.

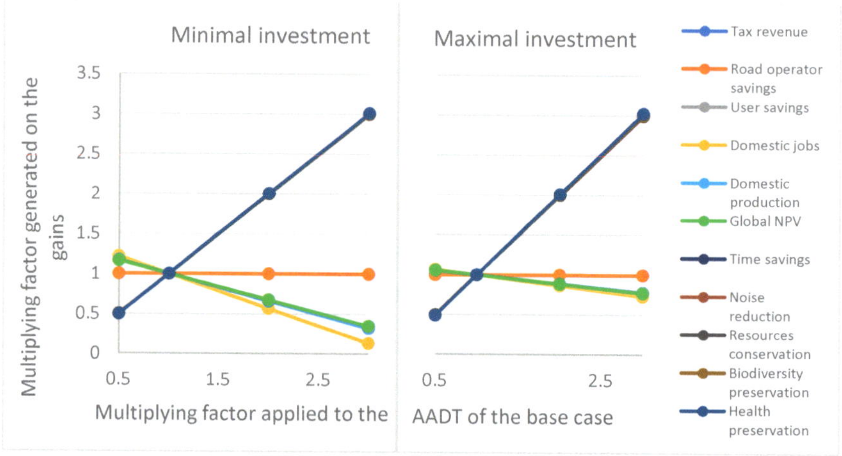

Figure 8. Sensitivity of the marginal gains to the traffic level in the scenarios of minimum and maximum investments.

4.3.3. Sensitivity to the Rolling Course Deterioration Speed

We now study the sensitivity of the indicators to the speed of degradation of the *IRI* as a function of time. In the base scenario, this slope is set at 0.05 m/km per year, and we evaluate the sensitivity of our calculation algorithm to *IRI* slopes affected by a factor of ½, 2 and 3, yielding respective *IRI* time slopes of 0.025 m/km.year, 0.10 m/km.year, and 0.15 m/km.year. Statistical studies of the *IRI* levels simulated according to our coating degradation laws and the effect of resurfacing work were also carried out to test the likelihood of the chosen variants. For each sensitivity analysis, we studied, in particular, the mean, median *IRI* values, and the standard deviation for each series of annual values.

This sensitivity study is presented for the minimum and maximum investment scenarios in Figure 9. Three groups of indicators with similar behavior concerning the evolution of the rate of degradation appear. First, a group of indicators appears almost proportional to the rate of degradation, which includes tax revenue, savings for users, time saved, and environmental gains. A second group of indicators, slightly sensitive to the speed of deterioration, emerges; it includes employment and production indicators, as well as the overall economy. The sensitivity to the degradation rate of this group increases when the resurfacing frequency is reduced. Finally, the manager's investment gains and the health impact of noise are not sensitive to the rate of degradation. In fact, the noise level should depend on the surface condition, but our acoustic emission model uses temporal statistical formulas that are uncertain at old age because of a lack of data on old rolling courses (i.e., >15 years), and without consideration of the IRI.

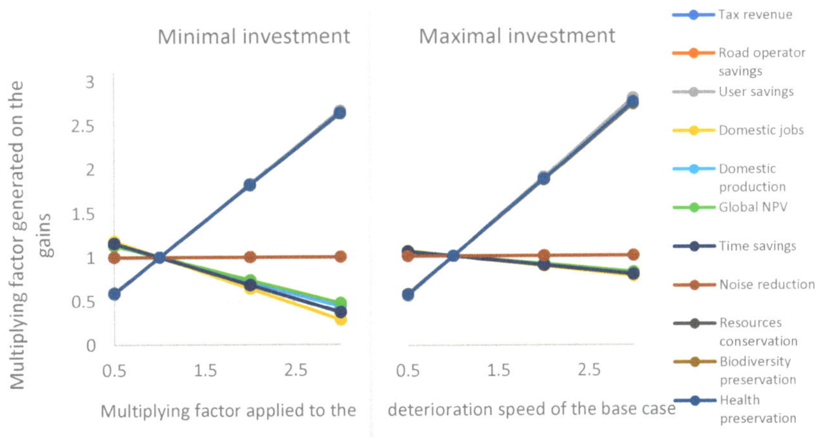

Figure 9. Sensitivity of the marginal gains to the road surface deterioration speed in the scenarios of minimum and maximum investments.

5. Discussion

5.1. Addressing the Assessment Reliability

Uncertainties and variability are essential issues in simulation models developed for decision support purposes. There are uncertainties across the whole modeling chain, from the input data to the performance results. Until these uncertainties can be characterized with appropriate mathematical approaches, the lessons obtained from case studies carried out by means of this method need to be applied with caution. In terms of variability, it would be in the public interest for the method to be applied to roads that are not privately operated, and/or with characteristics differing from highways, e.g., traffic level or operator discount rate, in order to draw trends in terms of optimum resurfacing and maintenance strategies for the country's different road network types.

5.2. Aligning the Interests of the Road Maintenance Stakeholders

Roughly two-thirds of the performance indicators change positively with a higher frequency of highway resurfacing, while the other third changes negatively. We discuss this result and propose several avenues for performance improvement of maintenance programs. Highway strategic management schemes—and notably the relationship between a public authority and a private concessionaire—need to be modified to benefit the environment and macroeconomic outcomes. Nevertheless, the savings generated by job creation—reduced payouts of unemployment compensation and increased payroll taxes—are outside the scope of the current method, but their inclusion could offset the tax revenue loss. The additional cost to the highway operator implied by an investment increase can be justified by the public service provided: savings to users, reduction in the impact on health and noise. With respect to road concessions, a fair system of remuneration needs to be introduced to nudge investment efforts towards the multi-criteria social optimum, negotiated in the contracts between government and operators based on sustainability criteria. The overall benefit indicator recommended for highway maintenance studies in France [29] is a questionable indicator, since it sets a single discount rate and, therefore, in our view, does not represent the reality of the behaviors of the different economic agents. Moreover, two versions of this indicator exist: a 100% financial indicator (classic NPV), and a socioeconomic NPV (SE-NPV) that includes monetized environmental and social impacts. Thus, this indicator aggregates various impacts through a weighting that is politically subjective—relying on monetization, i.e., a capitalistic approach of the society—not allowing for alternative weighting representing other socioeconomic ethics [34]. Moreover, it does not encompass PVI consequences or most of the environmental impacts of road maintenance. Finally, our case study results also show that the time lost under more intensive maintenance scenarios could be reduced by improving roadwork management. It, therefore, seems possible to obtain a more consensual "optimum" maintenance strategy, so that public action becomes consistent with viable, bearable, and equitable development.

6. Conclusions

We have developed a holistic method to evaluate the sustainability of transportation policies and, especially, road resurfacing strategies [4]. Here, it is applied to a French highway section, providing elements to build a three-pillar sustainability strategy for road maintenance. Compared with the reference scenario that represents the standard French highway resurfacing scheme, increasing resurfacing frequency—thus maintenance investment efforts—has a positive effect on the majority of the 11 performance indicators: macroeconomic, environmental, and social benefits. In terms of pollution, each additional euro invested in resurfacing could generate a gain in human health equivalent value representing several hundred euros, depending on the level of traffic. Increased maintenance is also positive for road noise and user savings, potentially sparing several euros for each additional euro invested, depending on traffic levels. On the other hand, increased maintenance generates financial losses for the operator and the government in terms of tax revenues, mainly due to fuel savings generated by a smoother road surface. Thus, when resurfacing policies are modified, some actors gain, others lose. The business model of road maintenance, therefore, needs to be revised to align the currently diverging interests of the stakeholders. In particular, the optimization of high-traffic roads requires increased maintenance to reduce the damage to the environment of transportation, by reducing the energy consumption of vehicles.

Supplementary Materials: The following supporting information can be downloaded at: https://www.mdpi.com/article/10.3390/su14127336/s1, Supplementary Materials File S1.

Author Contributions: Conceptualization: A.d.B. and F.L.; methodology: A.d.B. and F.L.; software: A.d.B.; validation: A.d.B.; formal analysis: A.d.B.; investigation: A.d.B.; data curation: A.d.B.; writing—original draft preparation: A.d.B.; writing—review and editing: A.d.B. and F.L.; visualization: A.d.B.; supervision: F.L. and A.F.; project administration: F.L. and A.F.; funding acquisition: F.L. All authors have read and agreed to the published version of the manuscript.

Funding: This research was funded by industrial chair ParisTech-VINCI in "eco-design of buildings and infrastructure".

Data Availability Statement: The code used to calculate results is available on the GitHub repository accessible at the following link: https://github.com/Anne2B/PhD. Data that are not made available in the Supplementary Materials (e.g., national account tables) were obtained from various third parties and are available from the authors on a case-by-case basis, if the permission is given from the third parties (e.g., INSEE). A word document gives details on figures, modes and datasets used, and Supplementary Materials File S1 are also provided.

Acknowledgments: The authors would like to thank the industrial chair ParisTech-VINCI in "eco-design of buildings and infrastructure" which funded the Ph.D. Thesis from which this work was derived.

Conflicts of Interest: The authors declare no conflict of interest. The VINCI group and its subsidiaries had no role in the design of the study, in the collection, analyses, or interpretation of data, in the writing of the manuscript, or in the decision to publish the results.

Abbreviations

AADT	Annual Average Daily Traffic
BAU	Business-As-Usual
DALY	Disability Adjusted Life Years
FTE	Full-Time Equivalent
HV	Heavy Vehicle

INSEE	National Institute of Statistics and Economic Studies (=Institut national de la statistique et des études économiques)
IRI	International Roughness Index
LCV	Light Commercial Vehicle
LHV	Large Heavy Vehicle
LV	Light Vehicle
NPV	Net Present Value
PC	Passenger Car
PDF	Potentially Disappeared Fraction
PVI	Pavement-Vehicle Interactions
SCACO	Semi-coarse asphalt concrete overlay
SHV	Small Heavy Vehicle
TACO	Thin asphalt concrete overlay
TICPE	Domestic Consumption Tax on Energy Products (=Taxe intérieure de consommation sur les produits énergétiques)
VAT	Value added tax
VTACO	Very thin asphalt concrete overlay

References

1. ITF. Key Transport Statistics—2020 Data. International Transport Forum—OECD. 2021. Available online: https://www.itf-oecd.org/sites/default/files/docs/key-transport-statistics-2021.pdf (accessed on 24 November 2021).
2. IATA. World Air Transport Statistics 2021. International Air Transport Association. 2021. Available online: https://www.iata.org/en/publications/store/world-air-transport-statistics/ (accessed on 24 November 2021).
3. Mattinzioli, T.; Sol-Sánchez, M.; Martínez, G.; Rubio-Gámez, M. A critical review of roadway sustainable rating systems. *Sustain. Cities Soc.* **2020**, *63*, 102447. [CrossRef]
4. De Bortoli, A.; Féraille, A.; Leurent, F. Towards Road Sustainability—Part I: Principles and Holistic Assessment Method for Pavement Maintenance Policies. *Sustainability* **2022**, *14*, 1513. [CrossRef]
5. Statista. Length of the Roads in France from 2000 to 2018 (in Kilometers). 2021. Available online: https://www.statista.com/statistics/1108181/road-network-france-length/ (accessed on 24 November 2021).
6. Statista. Public Road and Street Mileage in the United States from 2000 to 2019. 2021. Available online: https://www.statista.com/statistics/183417/united-states-public-road-and-street-mileage-since-1990/ (accessed on 24 November 2021).
7. Statista. Total Length of Motorways in Europe (EU-28) from 1990 to 2019 (in Kilometers). 2021. Available online: https://www.statista.com/statistics/449781/europe-eu-28-timeline-of-total-motorway-length/ (accessed on 24 November 2021).
8. Saxe, S.; Kasraian, D. Rethinking environmental LCA life stages for transport infrastructure to facilitate holistic assessment. *J. Ind. Ecol.* **2020**, *24*, 1031–1046. [CrossRef]
9. Santos, J.; Bryce, J.; Flintsch, G.; Ferreira, A.; Diefenderfer, B. A life cycle assessment of in-place recycling and conventional pavement construction and maintenance practices. *Struct. Infrastruct. Eng.* **2015**, *11*, 1199–1217. [CrossRef]
10. Chatti, K.; Zaabar, I. *Estimating the Effects of Pavement Condition on Vehicle Operating Costs*; Transportation Research Board: Washington, DC, USA, 2012. Available online: http://onlinepubs.trb.org/onlinepubs/nchrp/nchrp_rpt_720.pdf (accessed on 1 August 2014).
11. Pearson, D. *Deterioration and Maintenance of Pavements*, 1st ed.; ICE Publishing: London, UK, 2011. [CrossRef]
12. LCPC-Sétra. Aide à la Gestion de L'entretien des Réseaux Routiers—Volet Chaussées—Méthode. ISBN 5552720800163. LCPC. Oct. 2000. Available online: https://www.decitre.fr/livres/aide-a-la-gestion-de-l-entretien-des-reseaux-routiers-5552720800163.html (accessed on 19 March 2014).
13. LCPC. Catalogue Des Dégradations de Surface Des Chaussées. 1998. Available online: http://portail.documentation.equipement.gouv.fr/dtrf/notice.html?id=Dtrf-0001954 (accessed on 9 May 2013).
14. Lepert, P. Gestion technico-économique des infrastructures routières = Technical and economic management of road infrastructure. *Bull. Lab. Ponts Chaussées* **2006**, *261–262*, 3–23.
15. Lorino, T.; Lepert, P.; Riouall, A. Application à la campagne IQRN des méthodes statistiques d'analyse de l'évolution des chaussées. *Bull. Lab. Ponts Chaussées* **2006**, *261–262*, 25–41.
16. DGITM. *Politique pour L'entretien des Chaussées du Réseau Routier National non Concédé. Volume 1-Guide pour L'entretien des Chaussées, 2014*; Direction Générale des Infrastructures, du Transport et de la Mer; DGITM: Nanterre, France, 2014.
17. De Bortoli, A. Assessing environmental impacts of road projects: The recent development of specialized eco-comparators in France. In Proceedings of the World Road Congress 2015, Seoul, Korea, 2–6 November 2015.
18. Ministère de la Transition Ecologique. Largeur de Routes Sur le Réseau Routier National. 2021. Available online: https://www.data.gouv.fr/fr/datasets/largeur-de-routes-sur-le-reseau-routier-national/ (accessed on 19 March 2022).
19. ASFA. Les Classes de Véhicules, Association des Sociétés Françaises D'autoroutes. Available online: https://www.autoroutes.fr/FCKeditor/UserFiles/File/ASFA%20CLASSE%20VEHICULE%20BD.pdf (accessed on 25 April 2018).

20. Wang, T.; Harvey, J.; Kendall, A. *Network-Level Life-Cycle Energy Consumption and Greenhouse Gas from CAPM Treatments*; Research Report UCPRC-RR-2014-05; University of California Pavement Research Center UC Davis, UC Berkeley: Berkeley, CA, USA, 2013.
21. CopCete, un Outil Pour le Calcul Des Emissions du Trafic Routier. Cerema. Available online: http://www.cerema.fr/fr/actualites/copcete-outil-calcul-emissions-du-trafic-routier (accessed on 25 February 2019).
22. Gkatzoflias, D.; Kouridis, C.; Ntziachristos, L.; Samaras, Z. COPERT 4—Computer Programme to Calculate Emissions from Road Transport. User Manual Version 9.0. February 2012. Available online: http://emisia.com/sites/default/files/COPERT4v9_manual.pdf (accessed on 27 March 2017).
23. André, M.; Roche, A.-L.; Bourcier, L. Statistiques de Parcs et Trafic Pour le Calcul Des Emissions de Polluants Des Transports Routiers en France. Rapport Ifsttar-LTE. 2013. Available online: https://www.ademe.fr/sites/default/files/assets/documents/ADM00013842_ADM_ATTACHE1.pdf (accessed on 20 September 2018).
24. Sétra. Road Noise Prediction—1—Calculating Sound Emissions from Road Traffic. SKU1916164304. 2009. Available online: https://www.cerema.fr/fr/centre-ressources/boutique/road-noise-prediction (accessed on 25 February 2017).
25. De Bortoli, A. CHAPITRE 5—Construction de données environnementales spécifiques [CHAPTER 5—Regionalization and construction of specific Life Cycle Inventories]. In *Pour un Entretien Routier Durable: Prise en Compte de L'interaction Chaussée-Véhicule Dans L'aide à la Décision Des Politiques de Resurfaçage—Illustration Par un Cas Autoroutier Français*; Université Paris Est—Ecole des Ponts ParisTech: Paris, France, 2018; pp. 343–412.
26. De Bortoli, A. *Pour Un Entretien Routier Durable: Prise en Compte Des Conséquences de L'interaction Chaussée-Véhicule Dans L'aide à la Décision Des Politiques de Resurfaçage—Illustration Par un Cas Autoroutier Français [Toward Sustainable Road Maintenance: Taking into Account Vehicle-Pavement Interactions into the Decision-Making Process—Illustration by a French highway Case Study]*; Université Paris Est—Ecole des Ponts ParisTech: Paris, France, 2018.
27. INSEE. Series 001769682 Seasonally Adjusted Consumer Price Index—Base 2015—All Households—France—All Items | Insee. May 2018. Available online: https://www.insee.fr/en/statistiques/serie/001769682#Tableau (accessed on 28 May 2018).
28. INSEE. Series 000671268 Monthly HICP—All Households—France—Base 2005—European Classification: Maintenance and Repairs of Personal Vehicles—Stopped Series | Insee. 2016. Available online: https://www.insee.fr/en/statistiques/serie/000671268 (accessed on 28 May 2018).
29. Quinet, E. L'évaluation Socioéconomique Des Investissements Publics—Rapport Final. Commissariat Général à La Stratégie et à la Prospective. Tome 1. 2013. Available online: https://www.strategie.gouv.fr/sites/strategie.gouv.fr/files/atoms/files/cgsp_evaluation_socioeconomique_29072014.pdf (accessed on 12 April 2014).
30. Quirion, P. L'effet Net Sur L'emploi de la Transition Energétique en France: Une Analyse Input-Output du Scénario Négawatt. CIRED, Document de Travail No 46-2013. 2013. Available online: http://www2.centre-cired.fr/IMG/pdf/CIREDWP-201346.pdf (accessed on 22 January 2018).
31. Blanquier, A. *Sélection Des Investissements Aux Niveaux National et Régional*; Dunod: Paris, France, 1984.
32. MEDDE. Calcul D'analyse Financière. Fiche Outil Sur l'Evaluation Socio-Economique Des Projets de Transport. 2014. Available online: https://www.ecologique-solidaire.gouv.fr/sites/default/files/VI.1.pdf (accessed on 6 June 2018).
33. De Bortoli, A. *Anne2B/PhD: V1.0.0.PhD_de_Bortoli_2018_Holistic_Road_Resurfacing*; Zenodo: Meyrin, Switzerland, 2022. [CrossRef]
34. Arnsperger, C.; Van Parijs, P. *Éthique Economique et Sociale*; La Découverte: Paris, France, 2007.

sustainability

MDPI

Article

Towards Zero CO_2 Emissions from Public Transport: The Pathway to the Decarbonization of the Portuguese Urban Bus Fleet

Paulo J. G. Ribeiro [1,*] and José F. G. Mendes [2]

1 Centre for Territory Environment and Construction, School of Engineering, University of Minho, 4800-058 Guimaraes, Portugal
2 Department of Civil Engineering, School of Engineering, University of Minho, 4800-058 Guimaraes, Portugal; jfgmendes@civil.uminho.pt
* Correspondence: pauloribeiro@civil.uminho.pt

Abstract: The emission of GHG has been steadily increasing in the last few decades, largely facilitated by the transport sector, which has been responsible for more than two-thirds of the manmade emissions in Europe. In cities, one of the possible solutions to decrease the emissions from fossil fuel engines is to replace vehicles with electric ones. This solution can be applied to the urban public fleet, namely by replacing urban buses with electric vehicles. Thus, this research work focuses on the Portuguese case study, which serves as an example of achieving zero CO_2 emissions from buses by 2034. This timeframe of replacing the current bus fleet, mostly powered by fossil fuels, with a fully electric fleet is proven to bring financial, environmental, and health benefits to the population. The pathway to the decarbonization of urban public transport will unequivocally contribute directly to the accomplishment of several UN Sustainable Development Goals (SDGs), such as the promotion of affordable and clean energy and sustainable cities and communities, as well as to the increasing climate action (SDGs 7, 11, and 13, respectively). In addition, it will provide an opportunity for the replacement of existing buses that are generally less efficient than electric buses, from both an energy and an environmental point of view. As a result of the methodology, the Portuguese urban bus fleet would be totally replaced by electric buses by 2034 (83% battery-electric and 17% hydrogen-electric), which results in zero CO_2 emission from this type of public transport.

Keywords: CO_2 emission; urban bus fleet; decarbonization of public transport; electric buses

check for updates

Citation: Ribeiro, P.J.G.; Mendes, J.F.G. Towards Zero CO_2 Emissions from Public Transport: The Pathway to the Decarbonization of the Portuguese Urban Bus Fleet. *Sustainability* **2022**, *14*, 9111. https://doi.org/10.3390/su14159111

Academic Editor: Antonio Comi

Received: 7 June 2022
Accepted: 19 July 2022
Published: 25 July 2022

1. Introduction

The decarbonization of transport aims to contribute to the resolution of two very demanding social problems: the degradation of air quality due to pollution and the harmful effects of climate change [1].

The transport sector is the one that most depends on the use of fossil fuels, whose combustion results in the emission of carbon dioxide (CO_2) and other greenhouse gases (GHG). The road transport sector is a major air pollutant in Europe, it is responsible for 72% of the total manmade emissions to the atmosphere [2]. Besides, over 50% of the manmade CO_2 emissions are from the transport sector, specifically passenger vehicles [3].

Furthermore, the combustion process of motor vehicles causes the release or formation of other polluting species such as particulate matter (PM), nitrogen oxides (NOx), or ozone (O_3). In urban areas, both private and public transports are a significant source of air pollutants, that are emitted as exhaust gases, and are combined with noise to have a degrading effect on environmental quality and human health [4–6]. All the pollution emitted from the use of fossil-fueled vehicles in cities results in the death of more than 7 million people every year worldwide [7]. In Europe, air pollution is considered the biggest environmental risk, which induces degradation in people's lives and the premature death of about 400,000 people [5].

On the other hand, with the Paris Agreement at the Conference of the Parties (COP) of the United Nations Framework Convention on Climate Change (UNFCC), in 2015, the countries involved assumed the commitment to limit global warming by the end of the century to below 2 °C, trying to make an effort to reach the target of 1.5 °C. For this, countries should achieve carbon neutrality by 2050. Thus, to decrease the emission of air pollution from road vehicles, the European Union set an ambitious climate goal to reduce CO_2 emissions by 40% in 2030 and 60% in 2040 [8,9]. To achieve these goals "clean technology" has to be developed and deployed, such as electric vehicles, which have the potential to attain sustainable results in combination with renewable and nonpolluting electric energy sources [10–15].

Some measures throughout the years have been taken in order to reduce the amount of air pollution emitted by vehicles in Europe, such as the establishment of emission standards for vehicle engines, which are the Euro vehicular standards. Although, the commitment to shift public transport from fossil-fueled engines to electric-powered batteries is a step forward toward the total decarbonization of the transport sector. The reduction and elimination of the use of fossil fuels as a source of energy for the movement of vehicles, through the adoption of renewable and sustainable sources, is the key to the definitive solution to the problems of GHG emissions and air pollution from transport, especially in what concerns urban public transport [16].

Focusing on the public transport sector, electrification has emerged as a leading option for decarbonizing ground transportation [17–22]. Nowadays, the leading commercial options are battery-electric vehicles (BEVs) and hydrogen fuel-cell vehicles (HFCVs) [23]. For BEVs, the propulsion is ensured exclusively by an electric motor, using electricity stored in an onboard battery that is charged through its own dedicated charging equipment. On the other hand, in HFCVs, propulsion is provided exclusively by an electric motor, using electricity generated onboard by a fuel cell powered by compressed hydrogen (H_2) and using oxygen from the atmosphere. As with BEVs, this type of vehicle has no pollutant emissions associated with their trips, they only generate steam.

In Portugal, the current bus fleet, composed mostly of fossil-fueled vehicles, is seeing some development in the sense of becoming more sustainable in the future. Efforts have been made to shift the bus fleet to electric vehicles as a pathway to becoming a zero-emission public transport across the country. As of 2021, 55 urban buses that run in Portugal are electric, and the investments for this area will allow more electric vehicles to come, even full electrification of the fleet by 2034.

The present work is an extension of the urban bus fleet replacement and scraping methodology developed by Ribeiro and Mendes [24]. In the mentioned work, the authors presented a detailed study on the age distribution and fuel type of three categories of buses (i.e., minibuses, standard and articulated buses) to develop a methodology enabling the replacement and scrapping of all fossil-fuelled buses in Portugal in a timeframe of four-teen years. Such a study only focused on the timeline of the replacement of the old urban bus fleet with zero-emission engines (i.e., BEVs and HFCVs) and the estimated costs asso-ciated with its renewal. In the initial paper, the investment in the new bus fleet was esti-mated, based on the following indicators: Total Investment for Zero Emissions, Reference Value (maintenance of fossil fuels and CNG), and the difference between these two—the Cost of Decarbonization. However, the previous work [24] did not quantify the possible reduction in emissions achieved by the replacement and scraping of the entire Portuguese urban bus fleet with electric engines throughout the fourteen-year timeframe.

Thus, the present work aims to present a distinct methodology to estimate CO_{2e} emis-sions for different buses, based on the fuel type and the CO_{2e} emission factor for each EU-RO emission class. Using the same Portuguese case study presented in Ribeiro and Mendes [24], this paper takes a completely different approach by analysing and assessing the impact of the negative externalities of the fleet operation. For this purpose, CO_{2e} emissions are estimated yearly for fourteen years to provide all pathways to the decarbonization of the Portuguese urban bus fleet, with a special focus on the avoided CO_2 emissions, and their respective costs. This research introduces a novel approach for studies on investment in fleet scraping and replacement considering the price of carbon emissions in the European market.

Following this perspective, the present research work aims to:

(i) Develop a novel methodology to quantify the possible CO_2 reduction, i.e., carbon neutrality, from the complete replacement of the urban bus fleet in Portugal from diesel and gas/CNG to fully electric buses, based on criteria for scraping and replacing the urban bus fleet with electric buses defined by Ribeiro and Mendes [24];

(ii) (ii) Present the results of a full decarbonization of the urban bus fleet in Portugal as a way to demonstrate the feasibility of the methodology presented, namely in the re-duction in CO_2 emissions and corresponding costs.

The paper is organized as follows. Section 2 presents a methodology for an urban bus fleet decarbonization process that includes the estimation of CO_2 emissions and the re-spective decarbonization costs. In Section 3, a brief characterization of the urban bus fleet in Portugal is made. In Section 4, the criteria for scraping and replacing the urban bus fleet according to vehicle age are presented, based on the results presented in Section 3. Section 5 addresses the timeframe for the replacement of the bus fleet with electric engines throughout the years. Section 6 presents the main results of the application of the pro-posed methodology for the replacement of the bus fleet, namely CO_2 emission reduction for the entire bus fleet (minibuses, standard and articulated buses), the total avoided CO_2 emissions and the respective economic benefit of this reduction. Finally, Section 7 presents the main conclusions of this work.

2. Urban Bus Fleet Decarbonization Methodology

After addressing the issue of decarbonization of the transport sector, and in particular urban public transport, within the scope of the theme of combating climate change and reducing atmospheric pollution, the following methodological points are addressed in the next sections:

(i) Characterization of the urban road public transport fleet in terms of vehicle typology, age, engine, and emission classes;

(ii) Presentation of a scenario for the scraping and replacement of buses, based on maximum age and zero-emission engine criteria;

(iii) Calculation of the trajectory over time from the reduction in CO_2 emissions to the total decarbonization of the urban bus fleet for Portugal.

The methodology for the gradual decrease in GHG emission is divided into two different phases. First, the calendar for the replacement of the bus fleet is made to identify the number of urban buses that still generate CO_2 emissions and the number of electric ones in the fleet. Next, the trajectory of the emission reduction is obtained according to the emission factor and the urban bus fleet. Figure 1 shows a flowchart of the methodology presented for this paper.

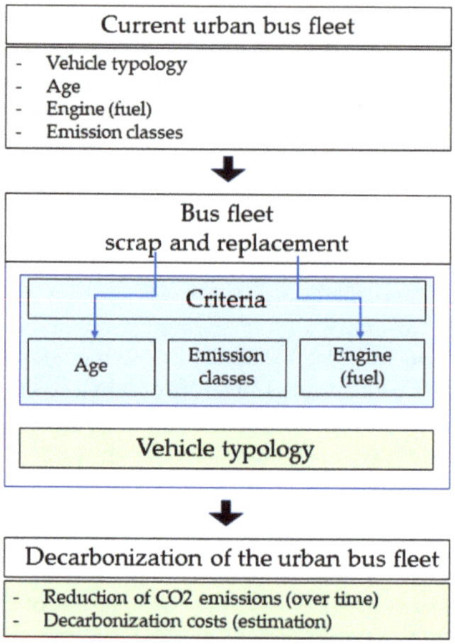

Figure 1. Flowchart of the methodology for the bus fleet decarbonization.

CO_2 Emission Factors for Urban Buses

A crucial part of the bus fleet decarbonization relies on the knowledge of the emissions made each year by the fossil-fuel engines. In order to achieve the current and future CO_2 emission of the urban bus fleet, the CO_2 emission factor for each type of bus must be correlated to the number of kilometres traveled each year. Table 1 shows the current factor that is used according to the classification of the engine for standard buses, which can range from Euro 0 to Euro VI [25]. The CO_2 emission factor for minibuses and articulated buses is derived from the weighting of the values from standard buses according to their fuel consumption.

In addition to the emission factor for buses according to their Euro classification, the annual kilometres traveled must be known in order to calculate the emissions from each bus type and Euro category. In Portugal, minibuses travel on average 30,000 km per year, while standard buses, which are the most common type of bus in the country, travel 50,000 km per year, which is the same amount that articulated buses travel. The annual kilometres traveled by each type of bus can be seen in Table 2. Although, as the kilometres traveled changes every year, a more accurate length of annual travel is used to calculate the CO_2 emission in the results section.

Table 1. CO_2 emission factor for standard buses.

Vehicle Typology	Type of Fuel	CO_2 Emission Factor (g/km)						
		Euro 0	Euro I	Euro II	Euro III	Euro IV	Euro V	Euro VI
Standard Buses	Diesel	1687	1503	1441	1446	1280	1299	1374
	Gas/CNG	-	-	1525	1250	1100	1100	1100
	Electric	-	0	0	0	0	0	0
Minibuses	Diesel	808	720	690	693	613	622	658
	Gas/CNG	-	-	731	599	527	527	527
	Electric	-	0	0	0	0	0	0
Articulated Buses	Diesel	1933	1722	1651	1657	1466	1488	1574
	Gas/CNG	-	-	1747	1732	1260	1260	1260
	Electric	-	0	0	0	0	0	0

Table 2. Annual kilometres traveled by buses in Portugal.

Type of Bus	Annual Trips (km/year)
Minibus	30,000
Standard	50,000
Articulated	50,000

3. Urban Bus Fleet in Portugal

The urban buses, then, are categorized into three different categories, which are: (i) minibuses—capacity of 20 to 50 passengers; (ii) standard buses—capacity of 50 to 105 passengers; and (iii) articulated buses—capacity higher than 105 passengers [26]. Data from 2020 shows that in Portugal there are currently 14,390 buses of all types, although only 5633 buses meet the urban bus criteria. The urban bus fleet comprises 509 mini (10%), 4808 standards (85%), and 316 (5%) articulated buses, with an average age of 11.6, 16.3, and 15.6 years, respectively, being the average age of the entire urban bus fleet which is 15.9 years old (Figure 2). A detailed description of the characteristics of the Portuguese urban bus fleet is provided elsewhere [24].

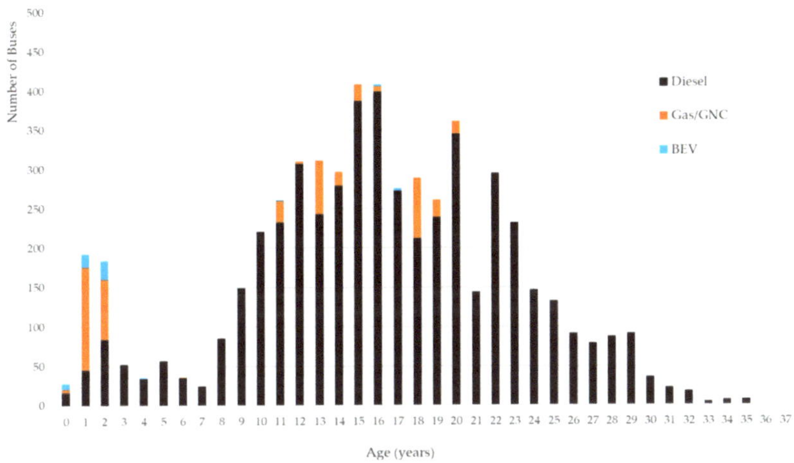

Figure 2. Age distribution for the urban bus fleet.

Bus Emission Classification

The European Union has strict legislation that regulates the emission of pollutants from vehicles. Thus, manufacturers and companies must improve their engines to meet the specification set in Europe [27]. The Euro regulation that addresses pollutant emissions from vehicle engines is a clear commitment to the decrease in air pollution across Europe, whose benefits expand beyond the borders of the European Union, since most G20 members and several emerging economies in Asia and Latin America set standards based on the European-devised system [28].

The buses considered for purchase in the European Union follow the emission protocols set by the Euro standards since 1988, when Euro 0 was created, with the last update in 2014 with the advent of Euro VI [29].

In Portugal, most of the buses fall into Euro III and Euro II standards, which means that the engines need to be replaced with less pollutant ones. On the other hand, there is a growing number of buses in the Euro V and Euro VI categories. Table 3 and Figure 3 show the number of buses by the type of fuel and by each category of Euro standard in Portugal.

Table 3. Bus classification under Euro standards in Portugal.

Emission Level	Minibus		Standard		Articulated	
	Diesel	CNG	Diesel	CNG	Diesel	CNG
Euro VI	73	3	230	208	20	1
Euro V	149	0	486	27	77	0
Euro IV	112	0	688	61	29	28
Euro III	148	0	1309	127	53	0
Euro II	11	0	926	16	79	0
Euro I	4	0	508	0	23	0
Euro 0	0	0	176	0	6	0
Total	497	3	4323	439	287	29
Total fossil-fueled	500		4762		316	
Total electric	9		46		0	
Total vehicles	509		4808		316	

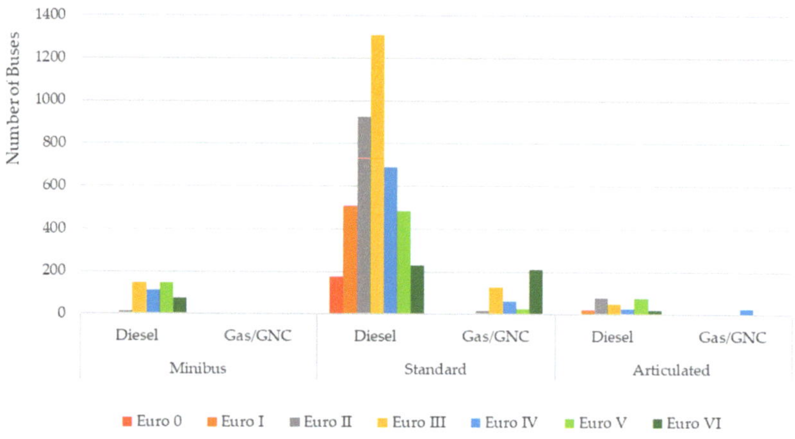

Figure 3. Number of buses by category and type of fuel.

The distribution of the Portuguese bus fleet over Euro emission standards is not very different from the general panel in Europe, as 28% of all buses rely upon Euro III standards, 25% under Euro V, and 18% under Euro II [30]. However, the bus fleet must be renewed to reduce air pollution, which culminates in the replacement of fossil-fueled buses with electric ones that do not emit CO_2 into the atmosphere. Thus, the next section of this paper presents a methodology and pathway to the full decarbonization of the Portuguese bus fleet.

4. Criteria for Scrap and Replacement of the Urban Bus Fleet According to Vehicle Age

The criteria for the bus fleet scrap and replacement in Portugal are presented to stimulate the sustainability of the urban public transport sector. Other criteria could have been selected, but the authors have already adopted and fully justified their choice, as dis-cussed in a previous paper [24]. Thus, a trajectory of scraping and replacement is defined to assure the complete decarbonization of the urban bus fleet, which will have a maximum age of fourteen years old by the end of the total replacement of fossil-fueled buses with electric buses. The age of scrap of buses is fixed at 14 years because it is two times the life cycle of an electric battery, whose replacement would cost as much as a new electric bus. After the first seven years of usage, only the battery is replaced in order to decrease the residue caused by the eventual disposal of the bus structure. However, after the second period of seven years, the bus in its totality is replaced.

Thus, for minibuses, all fossil-fueled and electric vehicles will be replaced by BEVs. Articulated vehicles fueled by fossil fuels will be replaced by HFCVs since BEVs are not suitable for this type of vehicle, which needs a bigger battery in order to move a large vehicle, and standard urban buses will be replaced by both BEVs and HFCVs, following the percentages presented in Table 4.

Table 4. Criteria for scraping and replacement of the bus fleet.

	Minibus		Standard			Articulated	
Year	Scrap	Replace	Scrap	Replace		Scrap	Replace
		BEV		BEV	HFCV		HFCV
2021	≥21 years	100%	≥21 years	100%	0%	-	-
2022	≥20 years	100%	≥20 years	95%	5%	-	-
2023	≥19 years	100%	≥19 years	90%	10%	-	-
2024	≥18 years	100%	≥18 years	85%	15%	≥21 years	100%
2025	≥17 years	100%	≥17 years	80%	20%	≥19 years	100%
2026	≥16 years	100%	≥16 years	75%	25%	≥17 years	100%
2027	≥15 years	100%	≥15 years	70%	30%	≥15 years	100%
2028	≥14 years	100%	≥14 years	65%	35%	≥14 years	100%
2029	≥14 years	100%	≥14 years	60%	40%	≥14 years	100%
2030	≥14 years	100%	≥14 years	55%	45%	≥14 years	100%
2031+	≥14 years	100%	≥14 years	50%	50%	≥14 years	100%

Starting from 2021, all urban buses are replaced when they reach a predetermined age, starting with the oldest buses at a replacement rate that is economically feasible for the companies, until all buses older than fourteen years can be replaced. In the following years, the rule stays the same, which will allow all buses fueled by fossil fuels to be replaced by BEVs and HFCVs by 2028. However, for articulated vehicles, due to their size and weight, it is considered that an electric battery is not the most suitable option due to increasing operating costs from the weight of the batteries required for this type of bus. On the other hand, the market is now starting to develop articulated buses fueled by hydrogen. Thus, it is considered that the replacement only should start in 2024 to provide more time for the industry to prepare and respond to the market needs.

It is important to mention that for articulated buses, due to their dimension and weight, VEBs are not recommended, instead, for this type of bus, the recommended engine is the HFCVs. Thus, the replacement of articulated buses can be scheduled to start only in 2024, when hydrogen fuel-cell vehicles will be available in the market. On the other hand, it is expected that all fossil-fueled buses in Portugal will be replaced by BEVs and HFCVs by 2034.

In addition, the replacement of fossil-fueled engines for electric engines in buses will determine the improvement of charging stations and electric power stations as city and transport infrastructures, since the charging of bus batteries needs special power treatment.

5. Timeframe for the Replacement of the Bus Fleet

According to the projections for this research work, it is possible that in Portugal all bus fleets become zero emissions by 2034. The replacement of the current fossil-fueled bus fleet can occur following the criteria of the fourteen-year-old replacement, which means that when a bus reaches the age of fourteen years, it is going to be replaced by a new one that can contribute to the zero-emission criteria (i.e., a new fleet of electric buses). For this, all types of urban buses are going to be replaced, such as minibuses, standard buses, and articulated buses. More detailed information regarding the three categories of the urban bus fleet can be ob-tained elsewhere [24].

Considering the bus fleet in Portugal in 2020 (5633 vehicles), it is expected to take up to fourteen years to completely replace the entire urban bus fleet with electric engines.

Figures 4 and 5 show how the replacement would take place considering the entire fleet from 2020 to 2034.

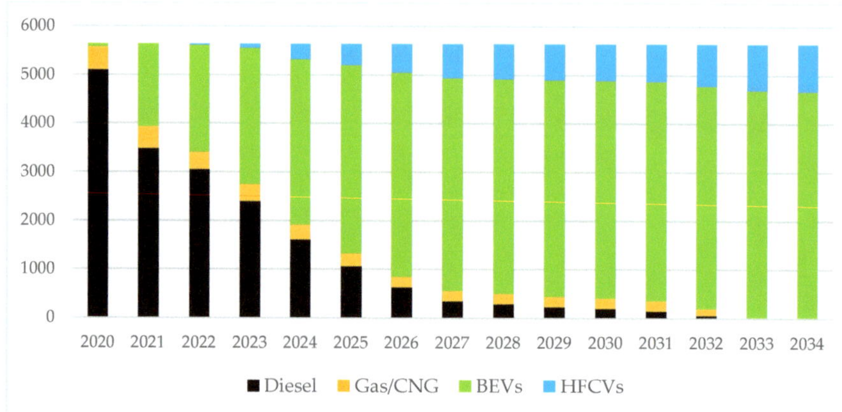

Figure 4. Replacement of the bus fleet with electric engines throughout the years.

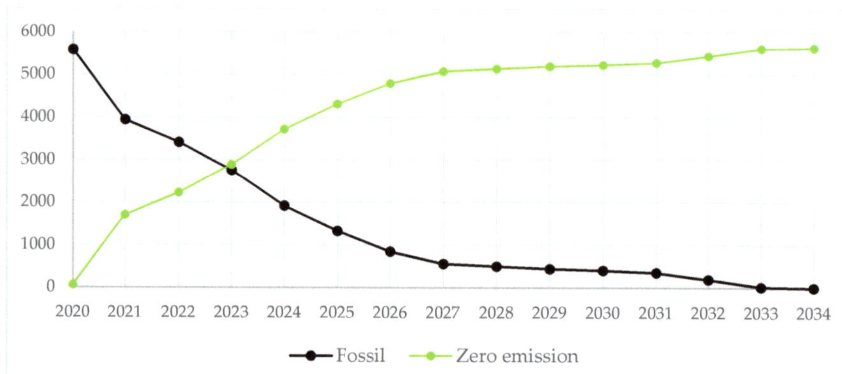

Figure 5. Accumulated vehicles (fossil-fueled vs. zero-emission).

As it is possible to infer from Figures 4 and 5, only in 2034 will diesel vehicles not be used in the urban bus fleet, which will represent the total conversion of the engines to electric-powered ones. This rate of replacement will define the decarbonization of the public transport of the entire country, which depends on the CO_{2e} emission factor of each engine.

6. Results

This section shows the total annual CO_{2e} (amount of gases equivalent to the quantity of carbon dioxide in the atmosphere) emissions for each type of bus and the differences in GHG emissions according to the number of buses that run with fossil fuels (emissions factor) and the kilometres traveled. Year by year, it is possible to see a decrease in pollutant emissions because of the replacement of fossil-fueled buses with electric vehicles.

For minibuses, it is possible to notice that the CO_{2e} emission from diesel-fueled vehicles is greater than vehicles fueled by Gas/CNG. This difference in emission is because there are more diesel buses available that travel greater distances every year. The power of the decarbonization achieved by the replacement of the minibus fleet can be seen as early as the first five years, in which it is possible to observe a 56% reduction in CO_{2e} emission, which represents fewer than 5451 tons of CO_{2e} in the atmosphere. Table 5 and Figure 6 show year by year the difference in CO_{2e} emission by minibuses according to the rate of vehicle replacement with sustainable technology.

Table 5. CO_{2e} emission reduction for minibuses.

Year	Number of Vehicles		Km Traveled		Emission CO_{2eq} (ton)		
	Diesel	Gas/CNG	Diesel	Gas/CNG	Diesel	Gas/CNG	Total
2020	497	3	14,910,000	90,000	9676	47	9723
2021	482	3	14,460,000	90,000	9361	47	9409
2022	473	3	14,190,000	90,000	9174	47	9222
2023	367	3	11,010,000	90,000	6971	47	7018
2024	314	3	9,420,000	90,000	5917	47	5964
2025	222	3	6,660,000	90,000	4224	47	4272
2026	152	3	4,560,000	90,000	2917	47	2964
2027	77	3	2,310,000	90,000	1517	47	1564
2028	68	3	2,040,000	90,000	1343	47	1391
2029	56	3	1,680,000	90,000	1106	47	1153
2030	32	3	960,000	90,000	632	47	679
2031	21	3	630,000	90,000	415	47	462
2032	16	3	480,000	90,000	316	47	363
2033	0	0	0	0	0	0	0
2034	0	0	0	0	0	0	0
Total	-	-	-	-	53,568	617	54,185

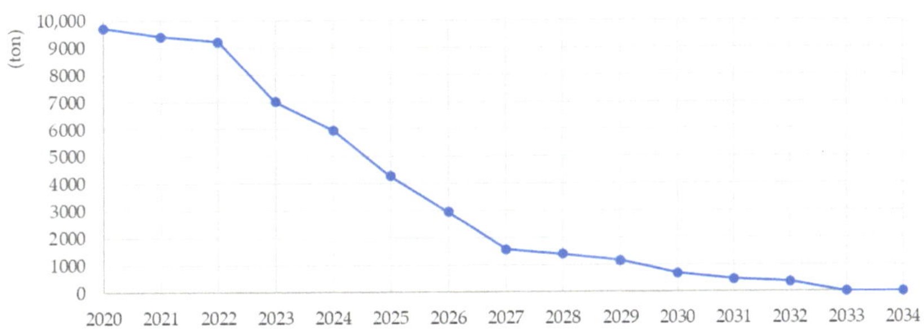

Figure 6. Decrease in CO_{2e} emission by minibuses.

For standard buses, the difference in CO_2 emission is even greater than for minibuses. This type of bus is the most used in the country, which contributes to more pollution from the burning of both diesel and gas/CNG fuels. As it can be seen in Table 6, the drops in CO_2 emission from standard buses represent a great decrease in air pollution from public urban transport. Only in the first five years of bus replacement there is a drop of more than 80% in the amount of CO_2 emitted into the atmosphere. From Figure 7, it is possible to see a considerable drop in emissions from 2020 until 2026, when the number of fossil-fueled standard buses reaches a number below 500. The following years face a less accentuated decrease, however, the replacement of the remaining buses with electric engines allows reaching the milestone of zero CO_2 emissions from standard buses in 2034, when all standard bus fleet will be composed of BEVs and HFCVs.

Table 6. CO_{2e} emission reduction for standard buses.

Year	Number of Buses		Km Traveled		Emission CO_{2e} (ton)		
	Diesel	Gas/CNG	Diesel	Gas/CNG	Diesel	Gas/CNG	Total
2020	4323	439	216,150,000	21,950,000	305,780	25,438	331,217
2021	2713	423	135,650,000	21,150,000	186,039	24,218	210,257
2022	2295	324	114,750,000	16,200,000	155,818	18,030	173,848
2023	1750	317	87,500,000	15,850,000	116,415	17,593	134,007
2024	1149	278	57,450,000	13,900,000	75,079	15,290	90,369
2025	716	235	35,500,000	11,750,000	47,367	12,925	60,292
2026	394	208	19,700,000	10,400,000	26,453	11,440	37,893
2027	250	208	12,500,000	10,400,000	17,100	11,440	28,540
2028	200	207	10,000,000	10,350,000	13,740	11,385	25,125
2029	156	207	7,800,000	10,350,000	10,717	11,385	22,102
2030	152	207	7,600,000	10,350,000	10,442	11,385	21,827
2031	114	207	5,700,000	10,350,000	7832	11,385	19,217
2032	35	132	1,750,000	6,600,000	2405	7260	9665
2033	10	4	500,000	200,000	687	220	907
2034	0	0	0	0	0	0	0
Total	-	-	-	-	975,873	189,393	1,165,265

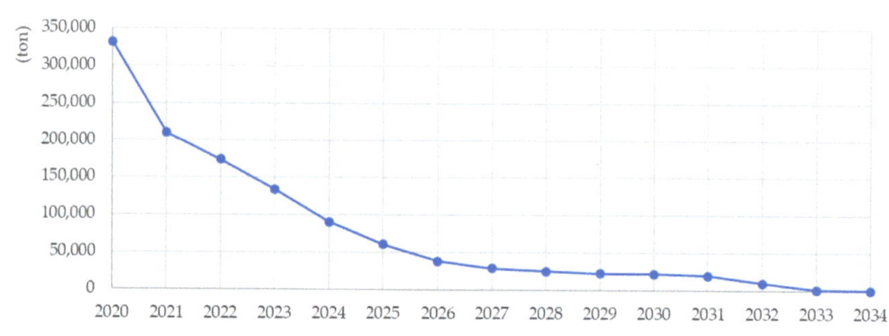

Figure 7. Decrease in CO_{2e} emission by standard buses.

For articulated buses, the decrease in CO_2 emission is no different, however, as the replacement of the vehicles only starts in 2024, the pollution only starts to drop this year, as can be seen in Table 7 and Figure 8. Only in the first years of the bus fleet replacement more than 50% of pollution is reduced, when the decrease is more accentuated. The next years are marked by the reduction in the total emission of CO_2 by articulated buses until the year 2034, when the entire articulated bus fleet will be replaced with electric engines that create zero emissions to the atmosphere.

Table 7. CO_{2e} emission reduction for articulated buses.

Year	Number of Buses		Km Traveled		Emission CO_{2e} (ton)		
	Diesel	Gas/CNG	Diesel	Gas/CNG	Diesel	Gas/CNG	Total
2020	287	29	14,350,000	1,450,000	22,900	1827	24,727
2021	287	29	14,350,000	1,450,000	22,900	1827	24,727
2022	287	29	14,350,000	1,450,000	22,900	1827	24,727
2023	287	29	14,350,000	1,450,000	22,900	1827	24,727
2024	146	29	7,300,000	1,450,000	11,086	1827	12,913
2025	122	29	6,100,000	1,450,000	9136	1827	10,963
2026	87	1	4,350,000	50,000	6559	63	6622
2027	20	1	1,000,000	50,000	1574	63	1637
2028	20	1	1,000,000	50,000	1574	63	1637
2029	19	1	950,000	50,000	1495	63	1558
2030	13	1	650,000	50,000	1023	63	1086
2031	10	1	500,000	50,000	787	63	850
2032	10	0	500,000	0	787	0	787
2033	6	0	300,000	0	472	0	472
2034	0	0	0	0	0	0	0
Total	-	-	-	-	126,093	11,341	137,434

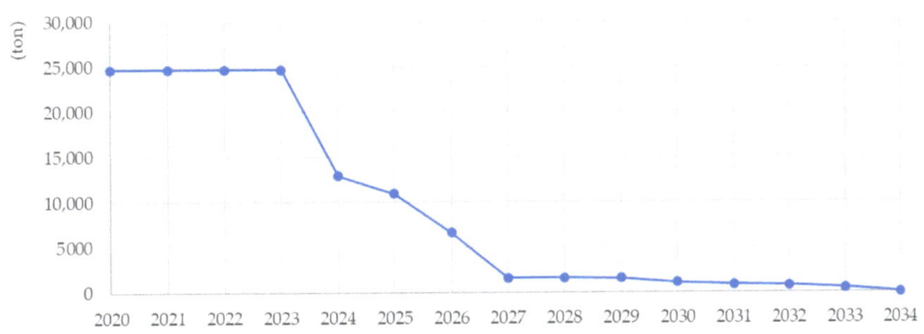

Figure 8. Decrease in CO_{2e} emission by articulated buses.

After evaluating the reduction in CO_2 emission from different types of buses according to their specific rate of replacement for electric engines, the total reduction in all buses in Portugal shows a bigger picture of the benefits of the zero-emission pathways to achieve sustainable urban public mobility. The results in Table 8 and Figure 9 show the amount of CO_2 that is not emitted as a result of the replacement of the bus fleet. At the end of 2034, when all buses will run on electricity, 1,356,884 tons of CO_{2e} will be emitted in a timeframe of fourteen years. This same number also tells us that in the next fourteen years after the total zero-emission bus fleet replacement, the benefits in air quality will be considerable, since the expected pollution from urban public transport will be zero.

In addition to the environmental benefits of the pathway to zero emission by the public urban bus fleet, the avoided emissions also represent savings for the economy of the country. Table 9 shows the value of avoided emissions to reflect what would be the cost of non-decarbonization and if the carbon emitted was valued at the price of the ton of carbon emitted on the European market (ETS). The value is calculated by multiplying the volume of emissions avoided by the estimated market price, starting from 50 euros in 2021 to 120 euros in 2030 [31].

Table 8. CO_2 emission reduction for the entire bus fleet.

Year	Number of Buses		Km Traveled		Emission CO_{2e} (ton)		
	Diesel	Gas/CNG	Diesel	Gas/CNG	Diesel	Gas/CNG	Total
2020	5107	471	245,410,000	23,490,000	338,355	27,312	365,667
2021	3482	455	164,460,000	22,690,000	218,300	26,092	244,392
2022	3055	356	143,290,000	17,740,000	187,892	19,905	207,796
2023	2404	349	112,860,000	17,390,000	146,285	19,467	165,752
2024	1609	310	74,170,000	15,440,000	92,082	17,165	109,246
2025	1060	267	48,560,000	13,290,000	60,727	14,800	75,527
2026	633	212	28,610,000	10,540,000	35,929	11,550	47,479
2027	347	212	15,810,000	10,540,000	20,191	11,550	31,741
2028	288	211	13,040,000	10,490,000	16,657	11,495	28,153
2029	231	211	10,430,000	10,490,000	13,319	11,495	24,814
2030	197	211	9,210,000	10,490,000	12,098	11,495	23,593
2031	145	211	6,830,000	10,490,000	9034	11,495	20,529
2032	61	135	2,730,000	6,690,000	3508	7307	10,815
2033	16	4	800,000	200,000	1159	220	1379
2034	0	0	0	0	0	0	0
Total	-	-	876,210,000	179,970,000	1,155,534	201,350	1,356,884

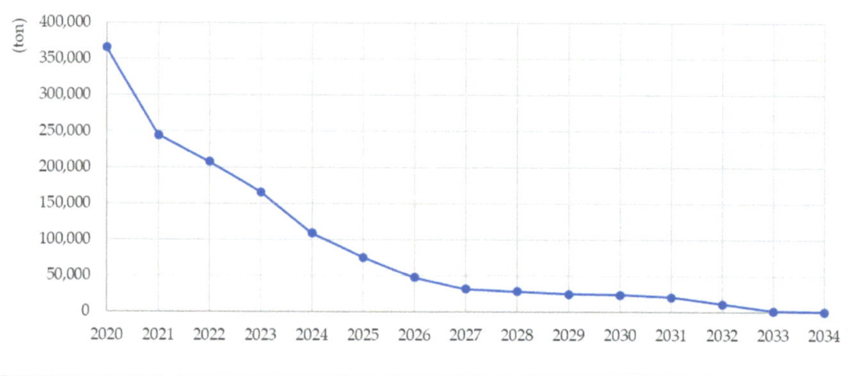

Figure 9. The total decrease in CO_2 emissions from buses.

Both Table 9 and Figure 10 show the benefits of the total electrification of the bus fleet in Portugal, namely in economic terms as well as in environmental terms. At the end of the fourteen years of the bus fleet replacement, almost EUR 420,000,000 could be saved in CO_2 emissions. In addition, the kilometers not traveled by fossil-fueled buses in the same timeframe represent a reduction of more than 4 million tons of CO_2 emitted to the atmosphere.

Table 9. Prevented CO_2 emissions and their respective cost.

Year	Number of Buses	km Not Traveled by Fossil Fuel Engines (km)	Avoided CO_2 Emissions (ton)	Price of a Ton of CO_2 (EUR)	Avoided Emissions (EUR)
2020	5578	-	-	-	-
2021	3937	81,750,000	121,274	50	6,063,719
2022	3411	107,870,000	157,870	58	9,121,398
2023	2753	138,650,000	199,915	66	13,105,518
2024	1919	179,290,000	256,421	73	18,804,179
2025	1327	207,050,000	290,140	81	23,533,593
2026	845	229,750,000	318,187	89	28,283,323
2027	559	242,550,000	333,926	97	32,279,497
2028	499	245,370,000	337,514	104	35,251,491
2029	442	247,980,000	340,853	112	38,251,258
2030	408	249,200,000	342,074	120	41,048,859
2031	356	251,580,000	345,138	120	41,416,535
2032	196	259,480,000	354,852	120	42,582,223
2033	20	267,900,000	364,288	120	43,714,514
2034	0	268,900,000	365,667	120	43,880,019
Total	-	2,977,320,000	4,128,118	-	417,336,125

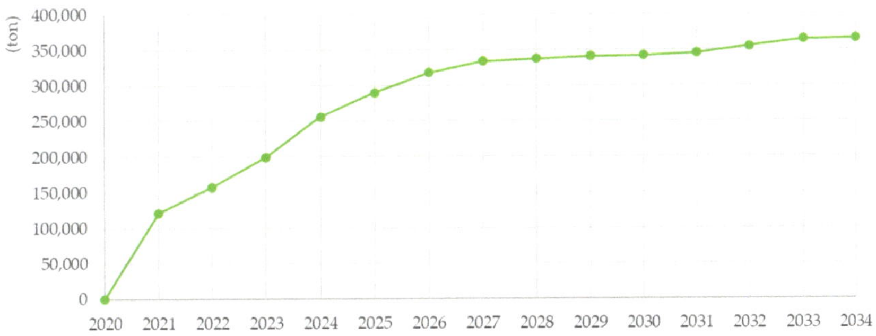

Figure 10. Total avoided emissions by the replacement of the bus fleet.

7. Conclusions

The replacement of fossil fuels with electricity in urban public transport can make a very significant contribution to reducing GHG emissions, within the framework of the Portuguese National Energy and Climate Plan objectives [32], as well as to reducing atmospheric pollution in urban areas.

A previously published paper [24] addressed the methodological procedure for the replacement and scraping of urban buses and the associated costs, whereas the present paper mainly focuses on the environmental benefits (i.e., reduction in CO_2 emissions) of a sustainable decarbonization process for the Portuguese urban bus fleet.

This research work innovatively shows and justifies the need to act in order to reduce the impact of public transport in view of the climatic challenges that humankind now faces.

In Portugal, the urban bus fleet was composed of 5633 vehicles on 31 December 2020, of which 509 were minibuses, 4808 were standard buses, and 316 were articulated buses. The average age of the bus fleet was 15.9 years, with 23% of the fleet over 21 years old. In the urban bus fleet (2020), 91% of vehicles were powered by diesel, 8% by natural gas, and 1% by electricity. Among fossil-powered vehicles, 39% met Euro IV or later standards, while 61% met Euro III or earlier standards.

From the methodology proposed in this research work, the replacement of the bus fleet follows the criteria of the complete decarbonization of the bus fleet by 2034, considering the maximum age of 14 years for vehicles. This would result in a decrease in the average age of vehicles from 15.9 years (2020) to a minimum of 4.3 years in 2026, reaching 9.9 years in 2034.

In the replacement of the fleet of 5,633 vehicles, 4,675 (83%) would be battery-electric, while 958 (17%) would be hydrogen-electric. After the first five years of the replacement of the bus fleet (in 2026), annual CO_2 emissions would be reduced by 87%. By 2034, after which GHG emissions would be zero, the volume of avoided CO_{2eq} emissions would be 4.1 million tons. Considering the price per ton of carbon to vary from EUR 50 to EUR 120, avoided emissions would have a reference value of EUR 417 M.

Following what was described in this research work, it is also important to measure the impact and how the decrease in other pollutants will occur with the replacement of the urban bus fleet, which will be the theme for future works.

In short, the replacement of the bus fleet in Portugal represents both environmental and economic benefits for all. The total zero-emissions bus fleet, despite the investments needed for it to occur, will bring, in the long term, savings and quality of life for its users and the population in general.

Author Contributions: Conceptualization, J.F.G.M. and P.J.G.R.; methodology, J.F.G.M. and P.J.G.R.; formal analysis, J.F.G.M. and P.J.G.R.; investigation, J.F.G.M. and P.J.G.R.; resources, J.F.G.M.; data curation, J.F.G.M. and P.J.G.R.; writing—original draft preparation, J.F.G.M. and P.J.G.R.; supervision, J.F.G.M. All authors have read and agreed to the published version of the manuscript.

Funding: This research was funded by Fundação Mestre Casais.

Institutional Review Board Statement: Not applicable.

Informed Consent Statement: Not applicable.

Data Availability Statement: Not applicable.

Acknowledgments: We would like to thank the data contribution from the Instituto da Mobilidade e dos Transportes, I.P.

Conflicts of Interest: The authors declare no conflict of interest.

References

1. Varga, B.O.; Mariasiu, F.; Miclea, C.D.; Szabo, I.; Sirca, A.A.; Nicolae, V. Direct and Indirect Environmental Aspects of an Electric Bus Fleet under Service. *Energies* **2020**, *13*, 336. [CrossRef]
2. European Environmental Agency Greenhouse Gas Emissions from Transport in Europe. Available online: https://www.eea.europa.eu/ims/greenhouse-gas-emissions-from-transport (accessed on 28 March 2022).
3. OECD/IEA. *Saving Oil in a Hurry*; IEA: Paris, France, 2005; ISBN 9264109412.
4. Jakub, S.; Adrian, L.; Mieczysław, B.; Ewelina, B.; Katarzyna, Z. Life Cycle Assessment Study on the Public Transport Bus Fleet Electrification in the Context of Sustainable Urban Development Strategy. *Sci. Total Environ.* **2022**, *824*, 153872. [CrossRef] [PubMed]
5. Kumbaroğlu, G.; Canaz, C.; Deason, J.; Shittu, E. Profitable Decarbonization through E-Mobility. *Energies* **2020**, *13*, 4042. [CrossRef]
6. Ribeiro, P.; Fonseca, F.; Santos, P. Sustainability Assessment of a Bus System in a Mid-Sized Municipality. *J. Environ. Plan. Manag.* **2020**, *63*, 236–256. [CrossRef]
7. WHO. New WHO Global Air Quality Guidelines Aim to Save Millions of Lives from Air Pollution. Available online: https://www.who.int/news/item/22-09-2021-new-who-global-air-quality-guidelines-aim-to-save-millions-of-lives-from-air-pollution (accessed on 28 March 2022).
8. European Commission. *EU Action on Climate Change*; European Commission: Brussels, Belgium, 2019.
9. Perissi, I.; Jones, A. Investigating European Union Decarbonization Strategies: Evaluating the Pathway to Carbon Neutrality by 2050. *Sustainability* **2022**, *14*, 4728. [CrossRef]
10. Göhlich, D.; Kunith, A.; Ly, T. Technology Assessment of an Electric Urban Bus System for Berlin. *WIT Trans. Built Environ.* **2014**, *138*, 137–149. [CrossRef]
11. Nanaki, E.A.; Koroneos, C.J.; Roset, J.; Susca, T.; Christensen, T.H.; De Gregorio Hurtado, S.; Rybka, A.; Kopitovic, J.; Heidrich, O.; López-Jiménez, P.A. Environmental Assessment of 9 European Public Bus Transportation Systems. *Sustain. Cities Soc.* **2017**, *28*, 42–52. [CrossRef]

12. Kenworthy, J.R.; Svensson, H. Exploring the Energy Saving Potential in Private, Public and Non-Motorized Transport for Ten Swedish Cities. *Sustainability* **2022**, *14*, 954. [CrossRef]
13. Sovacool, B.K.; Noel, L.; Kester, J.; Zarazua de Rubens, G. Reviewing Nordic Transport Challenges and Climate Policy Priorities: Expert Perceptions of Decarbonisation in Denmark, Finland, Iceland, Norway, Sweden. *Energy* **2018**, *165*, 532–542. [CrossRef]
14. Ribeiro, P.; Dias, G.; Pereira, P. Transport Systems and Mobility for Smart Cities. *Appl. Syst. Innov.* **2021**, *4*, 61. [CrossRef]
15. Dzikuć, M.; Miśko, R.; Szufa, S. Modernization of the Public Transport Bus Fleet in the Context of Low-Carbon Development in Poland. *Energies* **2021**, *14*, 3295. [CrossRef]
16. Liu, J.; Cui, J.; Li, Y.; Luo, Y.; Zhu, Q.; Luo, Y. Synergistic Air Pollutants and GHG Reduction Effect of Commercial Vehicle Electrification in Guangdong's Public Service Sector. *Sustainability* **2021**, *13*, 11098. [CrossRef]
17. Ku, A.Y.; Souza, A.; McRobie, J.; Li, J.X.; Levin, J. Zero-Emission Public Transit Could Be a Catalyst for Decarbonization of the Transportation and Power Sectors. *Clean Energy* **2021**, *5*, 492–504. [CrossRef]
18. Lajunen, A.; Lipman, T. Lifecycle Cost Assessment and Carbon Dioxide Emissions of Diesel, Natural Gas, Hybrid Electric, Fuel Cell Hybrid and Electric Transit Buses. *Energy* **2016**, *106*, 329–342. [CrossRef]
19. Li, X.; Castellanos, S.; Maassen, A. Emerging Trends and Innovations for Electric Bus Adoption—A Comparative Case Study of Contracting and Financing of 22 Cities in the Americas, Asia-Pacific, and Europe. *Res. Transp. Econ.* **2018**, *69*, 470–481. [CrossRef]
20. Kwon, Y.; Kim, S.; Kim, H.; Byun, J. What Attributes Do Passengers Value in Electrified Buses? *Energies* **2020**, *13*, 2646. [CrossRef]
21. Lefèvre, J.; Briand, Y.; Pye, S.; Tovilla, J.; Li, F.; Oshiro, K.; Waisman, H.; Cayla, J.M.; Zhang, R. A Pathway Design Framework for Sectoral Deep Decarbonization: The Case of Passenger Transportation. *Clim. Policy* **2021**, *21*, 93–106. [CrossRef]
22. Verbrugge, B.; Hasan, M.M.; Rasool, H.; Geury, T.; el Baghdadi, M.; Hegazy, O. Smart Integration of Electric Buses in Cities: A Technological Review. *Sustainability* **2021**, *13*, 12189. [CrossRef]
23. Islam, A.; Lownes, N. When to Go Electric? A Parallel Bus Fleet Replacement Study. *Transp. Res. Part D Transp. Environ.* **2019**, *72*, 299–311. [CrossRef]
24. Ribeiro, P.J.G.; Mendes, J.F.G. Public Transport Decarbonization via Urban Bus Fleet Replacement in Portugal. *Energies* **2022**, *15*, 4286. [CrossRef]
25. Agência Portuguesa do Ambiente. *National Inventory Report 2021 Portugal*; Agência Portuguesa do Ambiente: Lisbon, Portugal, 2021.
26. Ferraz, A.; Torres, I. *Transporte Público Urbano*, 2nd ed.; Rima: São Carlos, SP, Brazil, 2004.
27. EEA. Road Transport 2019. In *EMEP/EEA Air Pollutant Emission Inventory Guidebook 2019—Update October 2020*; European Environmental Agency: Copenhagen, Denmark, 2019; Volume 53, pp. 1689–1699.
28. Mulholland, E.; Miller, J.; Braun, C.; Jin, L.; Rodríguez, F. *Quantifying the Long-Term Air Quality and Health Benefits from Euro 7/VII Standards in Europe*; International Cuoncil on Clean Transportation: Berlin, Germany, 2021.
29. UNECE. UNECE to Adopt New Ceilings of Emissions for Trucks and Buses. Available online: https://unece.org/transport/press/unece-adopt-new-ceilings-emissions-trucks-and-buses (accessed on 31 March 2022).
30. UITP. *Bus Systems in Europe: Towards a Higher Quality Of Urban Life and a Reduction of Pollutants and CO_2 Emissions*; UITP Europe: Brussels, Belgium, 2015.
31. OECD. *Effective Carbon Rates 2021: Pricing Carbon Emissions through Taxes and Emissions Trading*; OECD Publishing: Paris, France, 2021.
32. República Portuguesa. *Plano Nacional Energia-Clima*; República Portuguesa: Lisbon, Portugal, 2019.

sustainability

MDPI

Article

Different Scenarios of Electric Mobility: Current Situation and Possible Future Developments of Fuel Cell Vehicles in Italy

Guido Ala, Gabriella Di Filippo, Fabio Viola *, Graziella Giglia, Antonino Imburgia, Pietro Romano, Vincenzo Castiglia, Filippo Pellitteri, Giuseppe Schettino and Rosario Miceli

Department of Engineering, University of Palermo, 90133 Palermo, Italy; guido.ala@unipa.it (G.A.); gabriella.difilippo@unipa.it (G.D.F.); graziella.giglia@unipa.it (G.G.); antonino.imburgia01@unipa.it (A.I.); pietro.romano@unipa.it (P.R.); vincenzo.castiglia@unipa.it (V.C.); filippo.pellitteri@unipa.it (F.P.); giuseppe.schettino@unipa.it (G.S.); rosario.miceli@unipa.it (R.M.)

* Correspondence: Fabio.viola@unipa.it

Received: 6 December 2019; Accepted: 9 January 2020; Published: 11 January 2020

check for updates

Abstract: The diffusion of electric vehicles in Italy has started but some complications weight its spread. At present, hybrid technology is the most followed by users, due particularly to socioeconomic factors such as cost of investment and range anxiety. After a deep discussion of the Italian scenario, the aim of the paper is to recognize whether fuel cell technology may be an enabling solution to overcome pollution problems and respect for the environment. The opportunity to use fuel cells to store electric energy is quite fascinating—the charging times will be shortened and heavy passenger transport should be effortless challenged. On the basis of the present history and by investigating the available information, this work reports the current e-mobility state in Italy and forecasts the cities in which a fuel cell charging infrastructure should be more profitable, with the intention of granting a measured outlook on the plausible development of this actual niche market.

Keywords: e-mobility; electric vehicles; battery electric vehicles; socio-technical transition; future of e-mobility

1. Introduction

In the last twenty years, great attention has been paid to the problem of climate change and to the decrease of greenhouse gas emissions in the atmosphere. Most of the carbon dioxide, CO_2, emissions in the environment are due to road transportation consuming fossil fuels [1]. The carbon dioxide emissions from road transportation have increased considerably and in 2013 they were even 50% higher than in 1990 [2].

In 2016, global CO_2 emissions due to fuel combustion were about 32 Gt CO_2, substantially similar to 2015 levels. These data show that emissions have more than doubled since the beginning of the 1970s and have increased by around 40% since 2000. Most of these increases are relative to the growth in economic production. Although emissions are relatively stable between 2013 and 2016, the initial International Energy Agency (IEA) investigation exposed that in 2017 emissions increased by about 1.5%, directed by the growing demand of China and India and the European Union [3].

Total energy consumption in 2018 increased by almost double compared to 2010 and CO_2 emissions grew by 1.7% within a year, reaching a new negative record (33 Gt) [4]. According to the study, the demand for all types of fuels has grown, driven by natural gas, which has been the fuel of "first choice"; its demand has in fact increased by almost 45% compared to the total energy market. Fossil fuels follow, with a growth of about 70% for the second consecutive year. The fact is that the final balance sheet is worrying, because due to the higher energy consumption, the IEA warns CO_2 emissions are increasing.

The energy production of coal-fired power plants continues to be the main cause of the deterioration, given that it represents 30% of all CO_2, emissions related to energy (10 Gt). In Asian region China and India are the main producers of CO_2 emissions and together with the United States they account for 85% of the net intensification in emissions, while pollution levels in Germany, Japan, Mexico, France and the United Kingdom are decreasing. Overall, the global average annual CO_2 concentration in the atmosphere was more than 400 ppm in 2018, up 2.4 ppm from 2017. This is a significant increase over pre-industrial levels, which ranged from 180 to 280 ppm. Despite the growth in coal consumption, the IEA notes, however, that the transition to gas was accelerated in 2018, avoiding the use of nearly 60 million tons of coal and the dispersion in the air of 95 million tons of CO_2. Without this result, the increase in emissions would have been over 15% higher [4].

A change of course is needed but different sociopolitical aspects prevent the development and spread in the communities of sustainable vehicles [5–7]. In technical literature different papers deal with the acceptation of new technology cars also using complex correlative algorithms to face "the egg and hen problem" [8], considering incentives for the purchase, the total cost of ownership of electric vehicles and highlighting the advantages of road use offered by the various municipalities [9] and also, since a car is a prestige good, of neighbors' opinions [10]. The authors believe that a possible way to solve problems related to the environment could be the use of hydrogen-powered cars, whose charging methods are similar to those of traditional cars but which require huge investments, especially for charging stations, which should be accurately planned in position and time [11–14]. So, this paper, after having discussed the situation of the electrification of transport and the perception that the population has in different countries and especially in Italy [9], investigates which communities are more ready to accept a complex technology such that of fuel cells.

The paper is divided in the following paragraph. Paragraph 2 refers to the situation of electrification of transport. Paragraph 3 focuses its attention on the sociopolitical aspects in the acceptance of electric vehicles and on the pollution situation in Italy. Paragraph 4 explains the used algorithm to identify cities more ready for new technology. Paragraph 5 refers to the negative aspects that must be considered and overcome. Then, conclusions arrive.

2. Current Scenarios of Electric Mobility

The mobility agenda in Europe [15] supports the transition to low-emission and zero emission vehicles with targets to be achieved by 2025. In 2025, in fact, the average CO_2 emissions of new heavy vehicles will have to be 15% lower than the 2019 level and, for 2030 an indicative reduction target of at least 30% is proposed in comparison with 2019.

In the communication "Europe on the move—an agenda for a socially just transition towards clean, competitive and interconnected mobility for all", the Commission presented several legislative initiatives concerning transport on the road. Mobility is the main economic sector in the world and in Europe there is a continuous growth of transportation activities, so that between 2010 and 2050 passenger transport should increase by about 42%, while freight transport by 60%. It is widely recognized that electric and automated vehicles will be an important part of achieving the goals set in these documents.

Previous studies [16] show that the use of electric vehicles (EVs) compared to petrol-driven vehicles can save (around 60%) greenhouse gas emissions throughout or in most EU Member States, related to the assessed consumption of EVs. Compared to diesel instead, electric vehicles show an average greenhouse gas savings of about 50% in some EU member states [17].

Even though EVs are not entirely free of environmental influence, due to greenhouse gas (GHG) emissions both during the production process and its end of life, studies on the impact and life cycle of EVs have suggested that these may have greenhouse gas emissions overall lower than conventional internal combustion (ICE) vehicles [18]. However, the process towards the adoption of EVs is still long. Even in the countries where the embracing of EVs by the users is remarkable, these cannot presently be considered a true opponent of traditional automobiles and buses with internal combustion engines because their diffusion is still marginal. One of the biggest obstacles to the expansion of electric cars is the reloading anxiety of users, as the infrastructure of charging stations is not yet sufficiently widespread and the autonomy of cars is limited. The development of a recharging network is therefore one of the keys to allow the spread of EVs [19]. From 2005 to today, the use of an electric motor for the mobility has exceeded its own first detachment friction, surpassing a critical line, by benefiting from various developments (technical and not) whose influence is growing increasingly importance—high oil prices, carbon constraints and an increase in organized car sharing and integration with other mobility forms, as micro-mobility. The progress of full electric vehicle and hybrid technology is subjected to changes in refueling infrastructure, variations in mobility, vicissitudes in the global automotive market, variations in energy prices, climate policy and improvements in the electricity sector. Particular care is devoted to the collaboration of technical alternatives, such as full electric vehicles, hybrid ones and hydrogen fuel cells vehicles [20]. The EVs sector is constantly growing in almost all parts of the world and the current situation is characterized by several disconnected projects in many countries, Figure 1. An attempt is made to assess the tendency of direct current charging systems to verify the feasibility of a transnational corridor infrastructure that guarantees accessibility to all automobiles. A predictive algorithm for the study of the trend for the electric vehicle was implemented to comprehend the market and infrastructure growth in the coming [21,22].

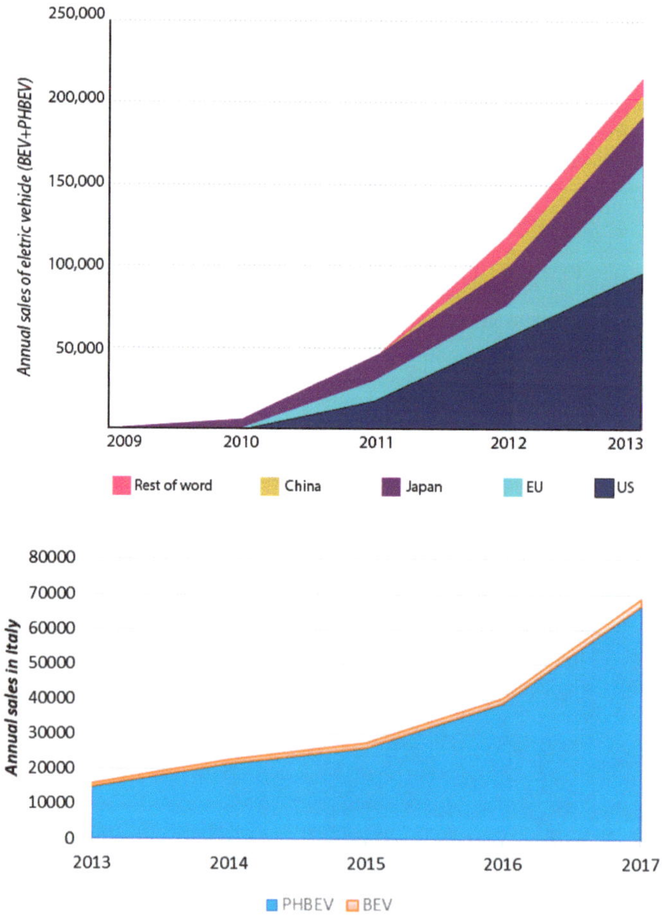

Figure 1. Electric vehicle (EV) sales growth from 2009 to 2013—figure reproduced from Reference [21] and focus on Italian sales [22].

Furthermore, in the near future other obstacles to the implementation of EVs will have to be overcome, such as the standardization of charging and refueling stations. This, with attention to connector types and adopted charging methods, must be addressed immediately. Between the several and diverse standards, guidelines and rules, there are two dual categories of charging method for electric vehicles (Figure 2)—high-power direct current DC and alternating current AC charging mode [11].

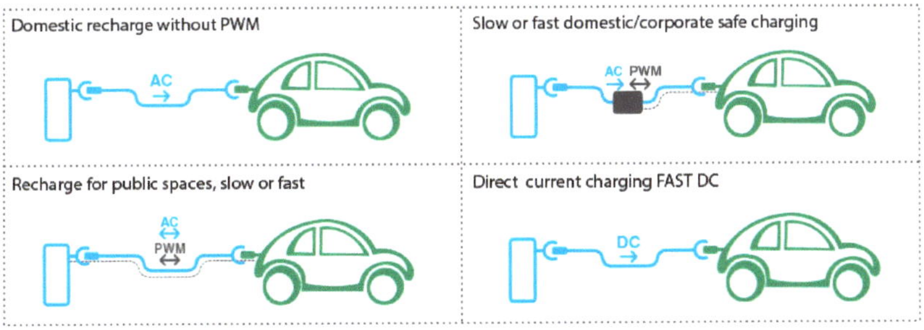

Figure 2. Different ways of electric vehicle charging methods.

Currently the European standard IEC 62,196 [23] considers four charging modes based on the speed of charge (slow or fast), the protection systems and the types of connectors. As for the fast-charging DC stations, the Tokyo Electric Power Company (TEPCO) in Japan has developed the quick-charge DC connector, also called CHAdeMO. In 2014, Japan and Europe had the most widespread CHAdeMO charging installations in the world, with 2129 and 1372 rapid loaders each, respectively.

Vehicle-to-Grid Paradigm

In addition to studies on the progress and convenience of plug-in hybrid vehicles (PHEV) and batteries (BEV) ones, attention must also be paid to the growing relation and interaction between the developing smart grid and the electric vehicles, seen as a resource of energy (Figure 3). Currently, many car manufacturers have invested significant resources in the development and production of new electric vehicle models; however, the development of an efficient recharging network, as the presence in city of different charging columns but also in the link connections between different cities, is one of the keys to allow the spread of this new way of vehicles. In different countries a question is arising, could the actual electric grid support the EVs revolution? The vehicle-to-grid paradigm (V2G) should be a solution.

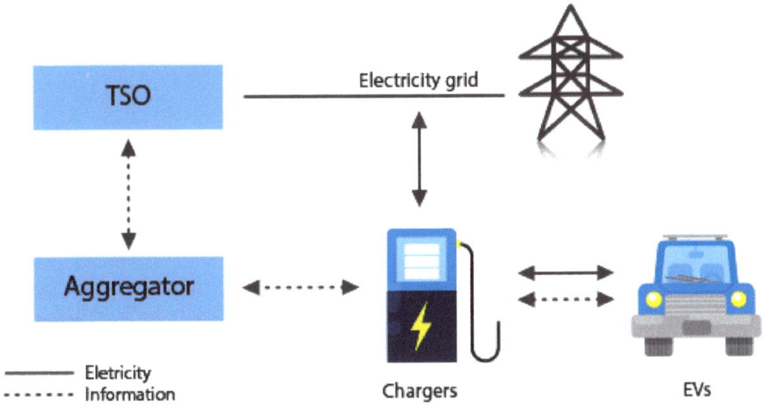

Figure 3. Basic scheme of vehicle-to-grid (V2G).

In the energy area, two significant progresses will concern electric mobility, the development of renewable energy technologies (but especially their acceptance by the consumer) and the emergence of smart grid systems, for which the energy flow is bilateral. The irregular availability of most renewable energies requires the storage of electricity, usually made with dedicated batteries. With the assistance of advanced smartgrid-based electricity management systems, batteries can be employed in the storage of electric energy and aid as "rotating energy reserve", when the peaks of demand occur if the BEVs are inactive. For electric energy providers, electrification of vehicles offers a way to solve peak demand, by offering a support to the stability of the local network, without having to intervene distant production plants, so reducing the weight on the network structure during peak hours [24]. In addition to generating demand and sales, batteries in EV should support utilities in decrease inadequacies and system variations built into today's network. For electric utilities, a synergistic connection among smartgrids and battery vehicles and renewable energy sources is clear. Besides, when BEV are progressively joined into smartgrids, great amounts of data and information will be available to those involved in infrastructure and communication systems, the whole sector of energy will be included. New chances and opportunities for companies operating in the automotive sector will rise, also influencing the competitive places, market shares, business models and strategies of current car manufacturers, the next electric and connected cars can be built by large corporations of the information technology.

The protocol vehicle-to-grid, V2G, realized together with the other protocols to have a connected car (vehicle to vehicle V2V, vehicle to everything V2X) enables not the spreading of information to improve travel times but the bidirectional flux of energy. This is one of the smarter technologies that allows the feeding of energy to the grid, due to the capability to simplify two-way communication between EV and the electricity grid to spread and obtain energy when the EVs are connected to the network (Figure 4).

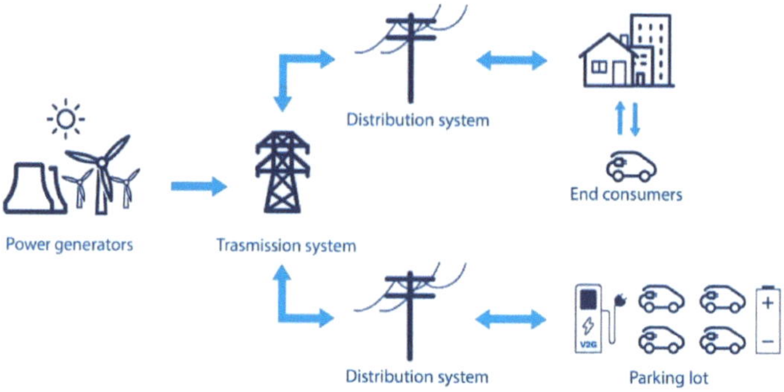

Figure 4. Schematic of the vehicle-to-grid (V2G) concept, it is noted the bidirectionality introduced in the smartgrid.

The enormous number of electric vehicles should be used in different possibilities of auxiliary services, for example frequency balance, voltage regulation, load leveling (downstream filing), peak load leveling, lines congestion mitigation and energy storing [25].

Since the charger is bidirectionally connected to the network, the charger receives (EV charging mode) and supplies (EV discharge mode) electric energy to and from the net, respectively. In fact, not only can the two-way charger participate in V2G, the unidirectional loader is also capable of running the V2G service, which is absorbing electric energy from the network [26]. As a matter of fact, the growing market share of EVs displays a global fear for climate changes. Numerous policies were introduced by the administration to promote the growth and implementation of electric vehicles, such as conventional ICE car purchase limitations in big cities and subsidies for domestic electric vehicles and plug-in hybrid electric vehicles. However, the stress on the electricity net becomes a problematic issue when there is a high diffusion of the EVs recharging request.

In a study conducted in Indonesia [27], the feasibility of V2G in the national electricity grid is evaluated (Figure 5). In the Indonesian case, as the network's current capacity to regulate supply and demand is very restricted, the huge EV charge further aggravates the condition due to the lack of energy storage. The auxiliary services of electric vehicles have led to the idea of using electric vehicles to support the network, in particular with the increase in the number of electric vehicles. Load leveling and frequency re-balance have been carefully observed. This study also analyzes the impacts of the adoption of electric vehicles and V2G contribution rates, which are driven by a certain incentive from the transmission service operator (TSO). The outcomes show that the energy received and released by electric vehicles after their recharge and discharge is feasible to provide the Indonesian network if electric vehicles are correctly controlled.

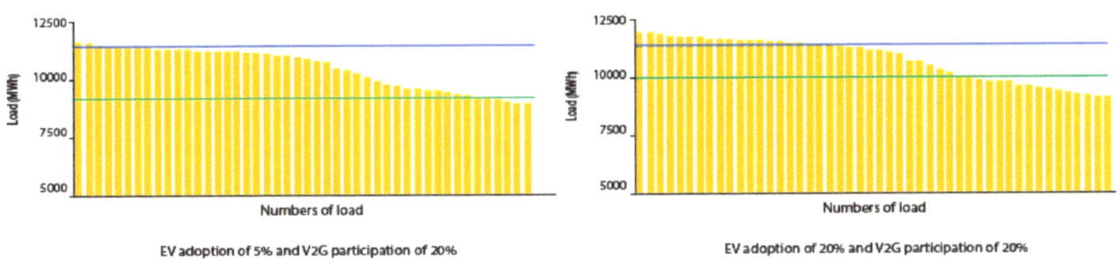

Figure 5. The load leveling amount under different EV acceptance and V2G participation, a quality windows in feeding the grid is given by the two lines—data from Huda et al., 2019 [27].

It is also known that renewable energy resources (RES) are promising solutions for energy issues and EVs and V2G are the appropriate support systems for RES, to overcome the problem of their intermittent nature [28]. Along with advances in V2G technology, the implementation of electric vehicle charging stations on electricity distribution nets is facing new challenges (Figure 6). As already underlined, the recharging infrastructures with V2G capacity should be columns in the energy management of the network and offer positive effects on the stability of the same net.

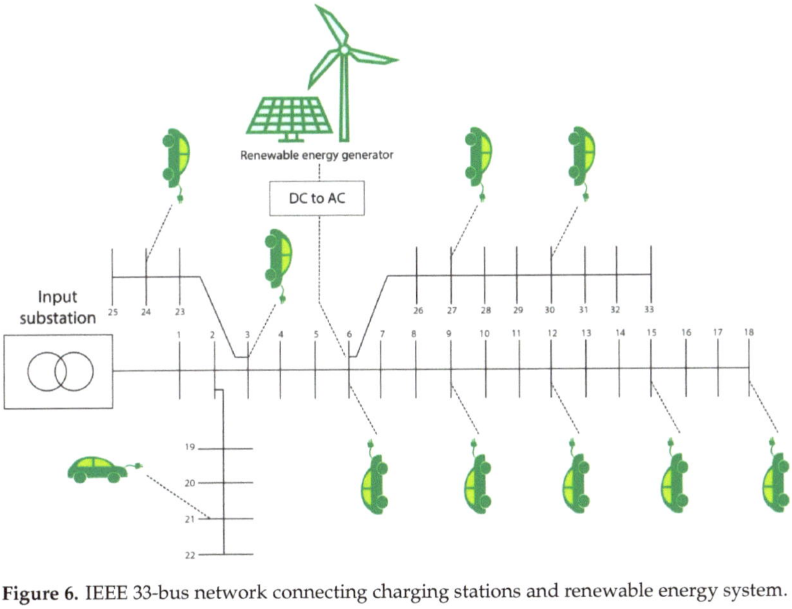

Figure 6. IEEE 33-bus network connecting charging stations and renewable energy system.

Vehicle-to-grid should be used correctly in a domestic energy managing system as a storage element to decrease energy costs [29]. It is also a promising answer to support the diffusion of renewable energy generation; in particular, it could regulate the intermittent service of RES [30]. The key contribution of V2G is to reduce peak demand and provide peak demand through the discharge of electric vehicles (Figure 7). This procedure provides positive technical, environmental and economic impacts on EVs. Electric vehicles can be considered as local energy reserves, thus not using very distant "rotating" sources, therefore subject to transport losses. Both for shaving the peak and for filling the valley at the same time V2G would be applied [31]. Therefore, it can support the network from different points of view, such as load shifting and congestion managing. Following common strategies, electric vehicles charge the surplus of RES during off-peak periods and send this energy to the grid when renewable energy decreases. It could be effectively treated with V2G technology. As a negative problem, battery degradation and its perception by users, is one of the problems related to the participation in V2G program and the cost of battery degradation limits undoubtedly the economic advantages, if providers do not support and incentive the participation in the program. Battery degradation is associated with energy emission and is highly sensitive to discharge depth. To make the operation of a battery storage system for V2G systems economically feasible, it is necessary to design an efficient power electronics converter that must be supported by an adequate control strategy [32]. The appropriate solution to reduce the cost of battery degradation can be intelligent charging [33].

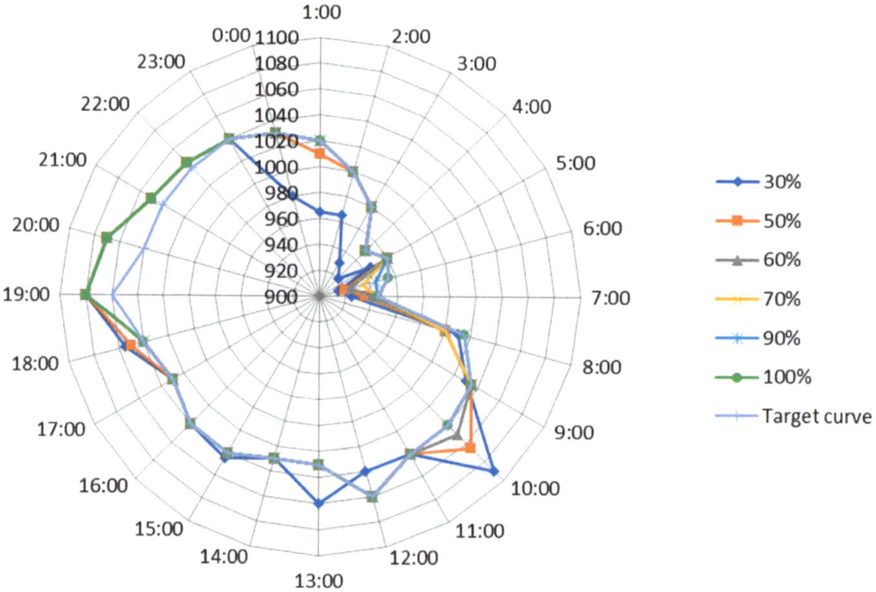

Figure 7. Amount of connected electric vehicles on peak shaving and valley filling, data from Wang and Wang, 2013 [33].

3. Sociopolitical Aspects and Care to the Environment

Other aspects to be taken into consideration in the process of transition to sustainable mobility are those related to the sociopolitical context, since the national contextual factors can favor certain emerging technologies. The commitment and public support for the definition of new development paths play a crucial role in the transition to sustainable transport technologies. Socio-technological schemes consist of formations of social and technological elements. Geels [34] has shown that technical and social transitions are the result of co-evolution among various actors and clusters of actors, for example organizations, legislative bodies, economic corporation, natural resources and physical artifacts such transportation structures. Previous research demonstrates how local policy instruments such as subsidies for the purchase, regulation of the use of public car parks and the encouragement of public procurement contribute to the development of this system [35].

A study carried out in several European cities shows how the rate of adoption of electric vehicles is influenced by numerous factors, both locally and nationally. This study presents a qualitative comparative analysis (QCA) of incentive e-mobility policies and supports in 15 European cities in order to identify political configurations at urban and local level that lead to favorable results in the promotion of the consequent adoption of electric vehicles [36]. It emerges how particular configurations are sufficient for favorable results to occur, such as the total cost of ownership of electric vehicles in combination with encouraging the installation of domestic battery chargers or recharging points on private parking lots in addition to the creation of a public recharging network in combination with other factors that discourage the use of conventional cars. The results of this study, even if mostly empirical, confirm that the adoption of successful electric vehicles is connected to a systemic political approach that encourages electric vehicles by simultaneously discouraging conventional cars, using both fiscal and specific local policy measures. This means that isolated measures are unlikely to work. This analysis has similarities with other recent results. For example, Wang et al. [37] employed correlation analysis approach and multiple linear regression analysis in order to discover the relationship between incentive policies and other socioeconomic factors with the adoption of electric vehicles in 30 countries considering the year 2015. Such study states that the positive and statistically significant factors are the practicability of the road (access to bus lanes), the density of tariffs (the number of battery chargers correct for the population), the price of fuel, while direct incentive and subsidies are not the only reason for the enormous difference in the absorption of electric vehicles among different countries. Comparable results have been found by Yong and Park. [38]. The current results highlight that the social-technological transitions have complex nature and particularly they depend on the location. A starting point for the current study can be learned, within the same country, where the incentive policy is the same, different behaviors can be diversified in cities with the same number of inhabitants but dissimilar infrastructure conditions.

Authors tend to prefer empirical studies, discrete choice models using stated preference data are frequently adopted in previous studies, conducted on consumer preferences, by considering a small group of items such the type of fuel vehicle, that is, ICE, BEV, PHBEV. The study of forecasting models can lead to provide the right political implications for particular countries like Korea [39].

More detailed studies consider various attributes—price, range, acceleration, top speed, pollution, size, luggage space, operating cost and charging station availability [40]. These parameters can differ enormously from country to country, due to objective aspects such as geography, climate and wealth but also subjective as the goodness of the incentives applied, awareness campaign and driving styles. Again empirically-based choice models have been used in a study based on the impressions of different Canadian citizens [41], to learn the choice models in a community. The behavior is influenced by incentives and disincentives as carbon taxes, gasoline vehicle deterrents and single occupancy vehicle discouragements. The analysis has shown different levels of technological change. The study of persuasive techniques on a socio-cultural environment is interesting. Other items, to predict the sale of EVs, were studied with the vector regression; the prediction of sales in automotive markets employs economic parameters—gross domestic product, consumer price index (CPI), interest rate, unemployment rate and gas prices with automobile sales [42]. Authors in Reference [43] after providing an interesting review on prediction markets in different countries, offer a discussion on short-term and long-term forecasting. Incentive policies are investigated, the subsidy-based and the tax-based policies. The former embraces purchase, charging as well as maintenance subsidies. The latter is demonstrated in tax on vehicle purchase, circulation and electricity cost. Authors show the effects of policy in the vehicle fleet in China.

Substantially the major impediments encountered by new vehicles users are restricted to two fields—one economic and other technical [9]. With an economy of scale, the economic bottleneck can be reduced. The costs of EVs are constantly decreasing and also those of the charging stations. From a technical point of view, the most important parameters in catching EV customer preferences are the number of kilometers the vehicles can travel between recharging and the number of stations that have the capability to recharge the vehicles. Fuel cell vehicles should be the right solution of the problem.

3.1. Fuel Cell: Forecasts for the Development of Hydrogen Technology for the Reduction of Air Pollution

In this scenario of transition towards sustainable mobility linked to the use of electric vehicles, Fuel Cell EV (FCEV) technology is introduced, where the battery in the vehicle is recharged by hydrogen stored in a special tank. The advantage of this system is that it has zero emissions, since the only waste product is water vapor. However, pollution is not completely eliminated because of the way the hydrogen is produced. The latter does not exist naturally in its natural state and to produce it is necessary to consume more energy to produce it. Therefore, the overall environmental impact of hydrogen mobility depends on the energy source used to produce it. There are in fact several methods to produce of hydrogen—the methane reforming of natural gas vapor, the biomass gasification, electrolysis and hydrogen derivation from existing industrial plants. Production can be located on-site or in central production units [44].

The use of different technologies, such as fuel cells, aims to overcome the problems related to charging anxiety, since the FCEV have a greater autonomy and a shorter recharge time. Fuel cell vehicles are more appropriate for long-term units, since their autonomy is much longer than PHEV or BEV but require special charging infrastructure. Their diffusion will depend critically on the costs of fossil fuel (oil), the progress of ICE fuel vehicles' effluents and the CO_2 regulations. The first hydrogen car with Full Cell System zero emission technology is the Toyota Mirai. This vehicle can travel a distance of 500 km (in 10 s it accelerates from 0 to 100 km/h) and the recharge time of the hydrogen tank is assessed in an interval of 3–5 min. In commercial catalogues, beyond the Mirai, Honda Clarity and Hyundai Tucson, are present with comparable presentations.

In Europe, Germany is the country that has so far invested the most in hydrogen cars and filling stations. The most served areas (disclosed by Fleet Europe) are those of Frankfurt, Stuttgart and Munich. The situation in Scandinavia is different, even though Denmark is a positive exception, with refueling stations for hydrogen cars located throughout the country. Sweden, on the other hand, is "hydrogen free", while France has no widely distributed stations (they are mainly in the north), as well as the UK and the situation in Spain is similar to that of our country. In Italy there are only, for cars and buses, three functioning filling stations, among which the Centro Alto Adige of Bolzano stands out, which is the connection point between our country and the rest of Europe.

Forecasts indicate 2025 as the year in which Fuel Cell and traditional electric will reach a substantial break-even [45]. In a study conducted by Harrison et al. an analysis of the European electro-mobility market was presented, with the aim of obtaining information on what could inhibit the success of market penetration of electric vehicles. The results of this study provided a forecast of the market diffusion of PHEV, BEV and FCEV. The authors hypothesize the evolution of BEV sales quotas between 2015 and 2050 in different scenarios. Based on the observations made, the BEV and the PHEV show a similar market penetration even if more successful for the PHEV until 2030, a period in which the technologies should become mature and the objectives have been achieved, leaving room for the less mature FCEV. The latter shows a slightly quick sales growth between 2025 and 2045, which growths further from 2045 when the new targets will be visible to the manufacturer.

3.2. Focus on Italian Air Pollution and Hydrogen Stations

It is well known that e-mobility therefore is a significant technology that has a core purpose—decrease the direct emissions of exhaust pipes and improve the air quality in the metropolises. Different are the means employed in several countries—some Asian States are fronting the fast diffusion of vehicles and are developing the construction own EVs, contemplating all the segments from micro-mobility such bicycle, electric scooters to heavy bus but old European countries face the problematic issue of the conversion from a traditional use of long range internal combustion engine, for which they hold the largest number of patents, to electric motor fresh technology and this necessitates very high standards for EVs to be compared with actual long range mobility.

As specified in Reference [46], in 2015, sellers in California introduced in their showrooms the first commercially available FCEV. Recharging stations, devoted to sustenance private clients in refilling, correspondingly arrived at the identical time, significantly circumventing the problematic of the "*chicken and the egg*", a philosophical dilemma establishing in this case a paradox without charging station no one buys an EV and without EVs no one constructs a recharge station. In Italy, the scenario is dissimilar from Californian one; until now few refilling infrastructures are dedicated to experimental FC bus and the participation of private users or customers is not active.

The use of reduced exhaust pipe emission vehicles and better battery vehicles can give high benefits and help for Italy since pollution of air surpasses the limit levels recommended by the World Health Organization (WHO) (particulate matter PM2.5 10 μg/m^3, PM10 20 μg/m^3) [47], in Figure 8.

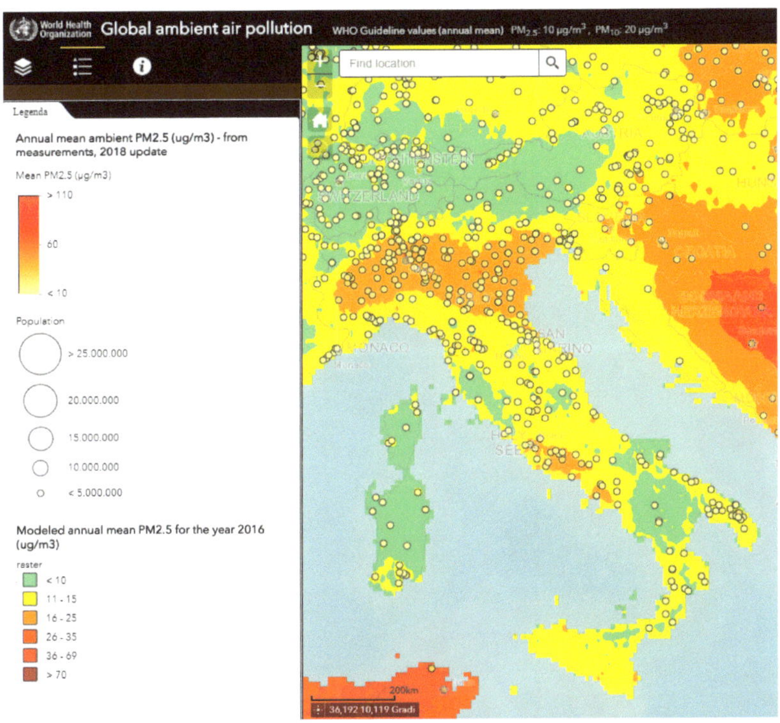

Figure 8. Chart of the PM2.5 distribution among Italy, green areas are under the limitations of 10 μg/m^3, yellow areas in the range of 11–15 μg/m^3, orange areas are in the windows of 16–25 μg/m^3. Air pollution maps are present at http://maps.who.int/airpollution.

By taking into account an additional exhaustive database of WHO for the year of 2016 [48], Figure 9 refers the limits of PM2.5 and PM10 for different Italian cities. Different cities overcome the limits suggested by WHO, a possible clustering in the exceeding the levels, takes four set levels, so recognizing cities on which an urgent action is required to improve air quality.

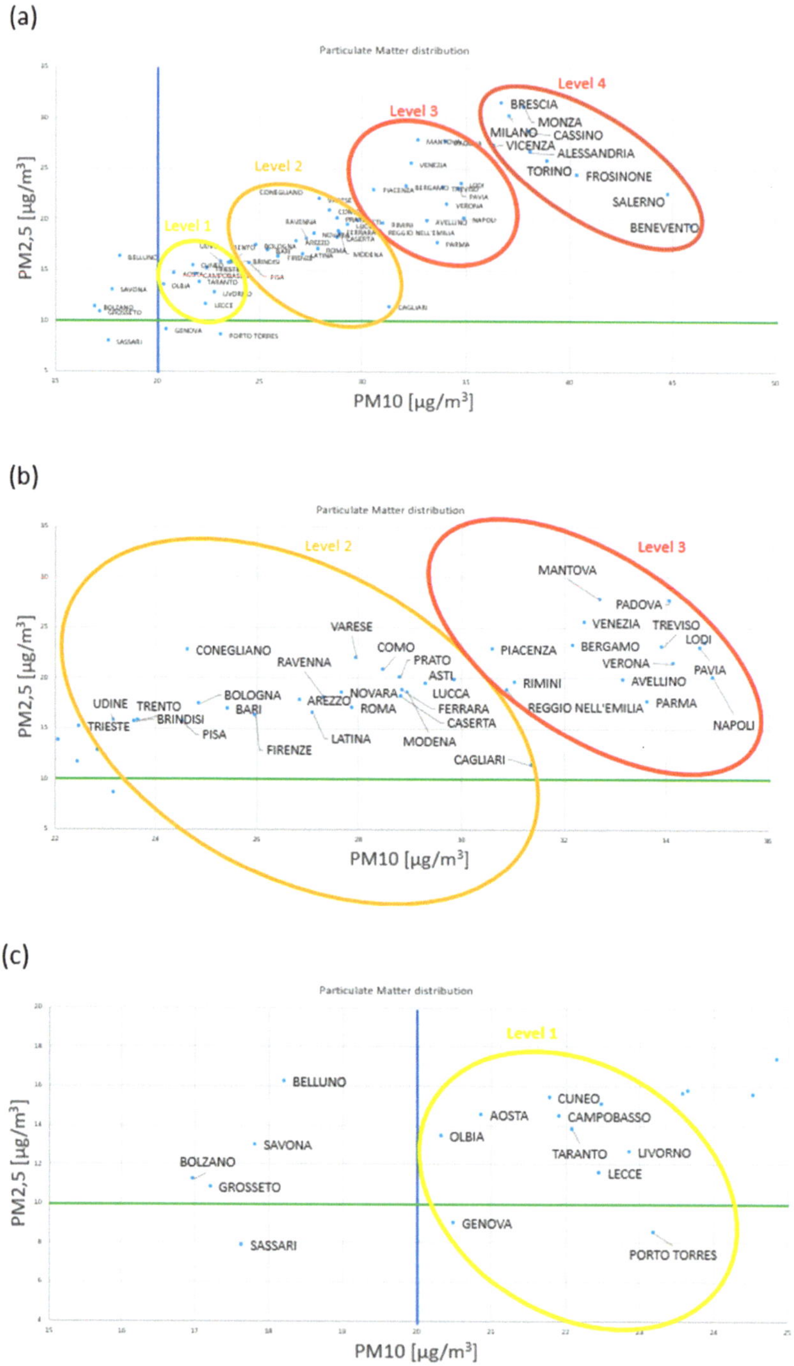

Figure 9. Cluster of particulate matter distribution in Italy. Cities are grouped in four levels (**a**), an enlargement is shown in (**b**,**c**). Green and blue lines dash the limits of WHO. http://www.who.int/phe/health_topics/outdoorair/databases/cities/en/.

A method to contrast the rising occurrence of particulate matter, is the use of hydrogen to fill of energy the vehicles but the derivation of hydrogen gains principal importance—if it is obtained from deposit of natural gas or if as the fuel are employed hydrocarbons, emissions of particulate matter should be reduced by 50% compared to conventional use of ICEs; once the hydrogen is gained by using renewables or nuclear sources, productions should be reduced by 90% [49,50].

In the old Continent, at the present time, the diffusion of such vehicles is very limited. To investigate the growth of FCEVs attention should be dedicated to the buses. In Europe the few hydrogen charging stations were realized for experimental plans, then for support private customers. Buses employing Ballard FCs technology run in London and about 61 buses are operating in all Europe. Due to the experimental achievement of FC buses in Europe, China, where the car problem grows due to dependence on a Western model, it has developed the largest hydrogen cell bus (HFCB) project in the world, with 300 buses operating or being expected to operate in Foshan [51]. A contract signed with Ballard [52] realized a novel assembly line, in Yunfu. Units with 90 kW of FC power components, started from July 2017 to achieve to the premeditated 300 buses. The cost of single bus is around $600.000, employment of 7.05 kg of hydrogen is considered for a range of 100 km.

The Italian region at present claims over eight planned hydrogen charging stations, Figure 10 but few of the planned infrastructures really operate [53]. Bolzano, Carpi, Milan, Pontedera, Trento, Verona, Rome and Capo d'Orlando are the cities involved in different projects that realized a recharge station. This number is very limited to really favor the growth of FCEVs. In order to discuss a similar delay in supporting of this enabling technology, the tardive measures adopted by the Government may be indicated. A technical issue forbade the diffusion of FCEV until March 2017; before such date FCEVs could only use as refueling systems only ones that not exceed the threshold of 350 bar, while the modern and much evolved generation of hydrogen vehicles, employs 700 bar tanks, like the Toyota Mirai.

Figure 10. Chart of the Italian hydrogen charging stations (in green the distributors in operation, in red those inactive) https://www.mobilitah2.it/distributori.

With Legislative Decree 257 of 16 December 2016, entered into force on March 2017, FC supply infrastructures embraced the pressure of 700 bar. In such a way, Italy incorporated the European directive 94 of 22 October 2014, on the deployment alternative fuel infrastructures (DAFI), so launching a clean fuel strategy.

At present, among the FC infrastructures set in Italy, only the plant in the H2 Alto Adige technological center of Bolzano reaches 700 bars. So, it can be highlighted that the network of infrastructures is lacking in Italy but it is required in order to guarantee a satisfactory coverage and to expand the sales of the FCEV themselves. Also, the population should be pushed to understand and adopt this new sustainable technology which release only water from the exhaust pipe.

The policy of the competent ministry is to undertake a path to build a suitable supply network through the whole national region by 2025.

The objective of next sections is to offer an outline of the Italian inclination to the adoption of consolidated PHEV and BEV and to extrapolate the feasibility of FC refilling infrastructures in Italy, by employing models, that try to point-out the necessities for the construction of hydrogen refueling stations. By considering the following key features such the knowledge of the current scenario of EV charging infrastructures, EVs adoptions during years, cities dimensions and density, mean personal income in the cities, a map of cities ready to adopt the FC technology can be drafted and also the trend to increase the amount of charging infrastructure year by year can be obtained.

4. Chart of the Present Scenario

As stated in Section 3, different forecasting method can be chosen. The models available to deal with the prediction of the development of EV adoption are substantially based on empirical reproductions. Singular spectrum analysis (SSA), which was developed to model univariate time series, in its application with financial and economic data has shown acceptable results [54]. While SSA is non-parametric and data-driven technique, vector autoregressive model (VAR), can face multiple time series [55] and was successfully used in the dynamic couplings between EV sales and economic indicators [43]. A step forward was presented in Reference [29], in which, in addition to the forecasting model based on the algorithm of random-coefficient logic model, the impact on the power grid is projected. Key features are in limiting access to charging stations, limiting maximum driving distance and in introducing a high vehicle price negatively to influence the consumer choice of electric vehicles on the automobile market. The adoption of FCEVs is still in its infancy, so in order to create a robust prediction in this study it has been preferred to use an algorithm that does not have multiple parameters. On the other hand, as specified in the previous paragraphs, the main bottleneck that blocks the adoption of EVs is represented by the technical aspects as the distance to travel and the availability of charging stations.

By following a metric traced in Reference [46], in this paper the SERA (Scenario Evaluation, Regionalization and Analysis) model is used. By referring to the same terminology, an early adopter's metric (EAM) is followed. This is an analysis in which EV adoption is used to predict FCEV one. The metric is able to define and outline areas where there can be the development of *early adopters* (literally *"first users,"* like seeds) of FCEVs, by taking into account socioeconomic factors, also dependent on the wealth of the city, such as population density, having consequences in the historical sales of electric and hybrid cars and varying on geographical region, town by town.

In Italy, different population densities characterize metropolitan areas. As a consequence, the amount of refueling infrastructures essential to guarantee the access of *"early adopters"* to a reliable support network, differs also in the same city, more in the high-density part of the city, less in the other one. But not only, if an amount of stations is required, it is based on the quantity of EVs existing in a territorial part of the city and on the geographical extension itself, just to ensure consistent covering. Again, by using the same terminology adopted in the SERA model, first charging infrastructures are called *enabling stations*.

In order to explain the SERA model, two Italian cities, Brescia and Parma, can be taken as an example. The two cities share the same number of inhabitants, Parma, due to the dimension of the city, has a density of inhabitants nearly four times lower than Brescia, so a major quantity of infrastructures is needed.

By using the *"Urban Market Sequencing"* model, an analysis of the propensity for innovation of the inhabitants is made, in order to find the importance of urban markets, for the FC infrastructures.

By using a Cartesian coordinates system, the number of early adopters for enabling stations was indicated on the vertical axis, that is, the probable adopters per refilling infrastructure, while in the abscissas the quantity of early adopters per square meter, that is, the density of early users, was specified. The number of users is therefore related to the difficulties in using the charging stations, whether they are few in number or distant from each other. These limitations have to be understood as how early adopters tackled problems such low urban density and low number of charging infrastructures. Thus, the efficiency of placing a limited quantity of infrastructures coping the concentration of early adopters of a specific urban market, can be evaluated.

The databases present in References [22,53] provided suitable data for the analysis.

The *"priority"* in assessing positive urban markets is shown in Figure 11. Cities were clustered in three grouping levels, that is, Level 1, on the top right, Level 2 in the middle, Level 3 on the bottom left. Level 1 cities were grouped since these cities provide a higher presence of early adopters, against ballasting factors such as the limited number of charging infrastructures or city extension. An excellent option is to finance and construct charging infrastructures in these cities, due to the great quantity of sales of electric and hybrid cars.

Among the "greenest" cities, for which the development of environmental sustainability is straightforwardly realizable, there are Como, Pordenone, Trento, Turin, Varese, Milan, Bergamo and Bologna. These cities also share geographical position—they are in the North of Italy and, above all, in the Lombardy region.

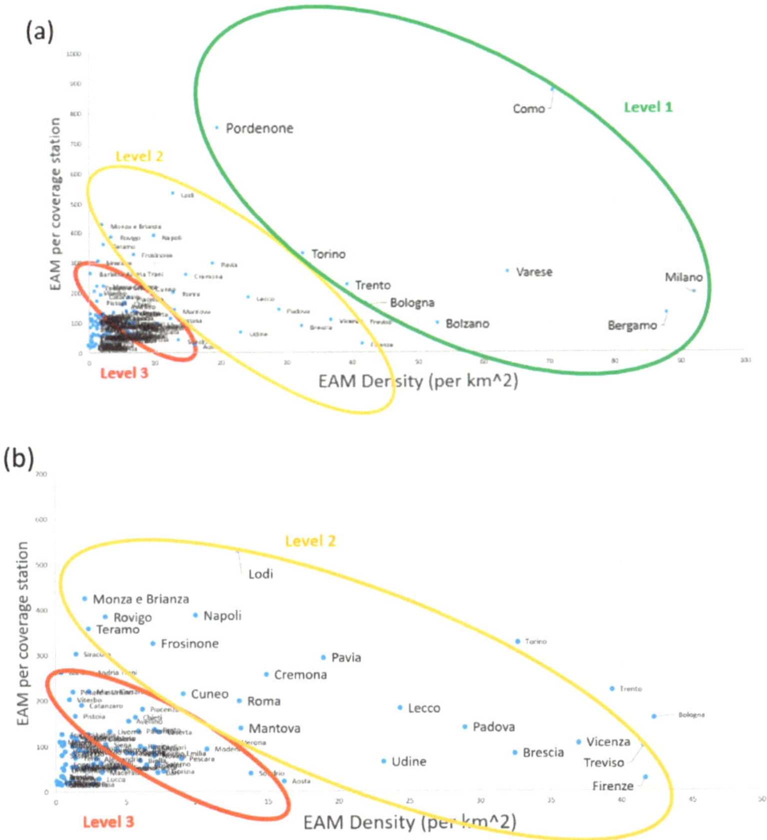

Figure 11. *Cont.*

(c)

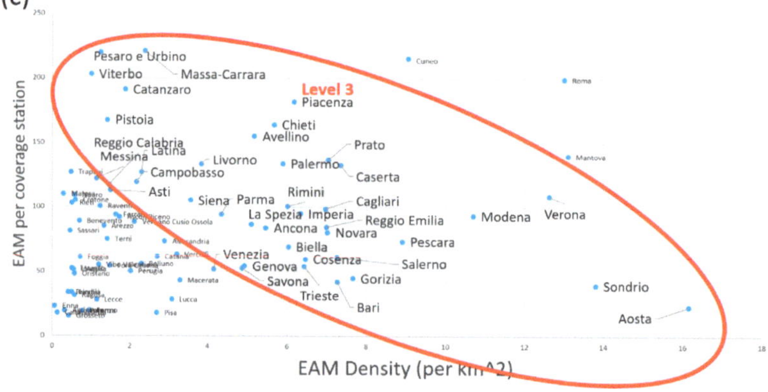

Figure 11. "Early adopters metrics" (**a**), in abscissas the number of early adopters per coverage station, that is, divided by the dimension of cities, in ordinates the amount of early adopter's density weighted by amount of charging infrastructures. (**b**,**c**) an enlargement of (**a**); level 1 represents the cities that are more ready to adopt the Fuel Cell technology; level 2 represents the cities that are behind in the adoption of ecological vehicles; level 3 shows the cities in which there is the presence of obstacles (economic or infrastructure) that limits so much the adoption of ecological vehicles.

In Figure 11 Level 2 shows the cities that soon but not immediately, can embrace the electric revolution. These cities suffer from structural or economic deficiencies, which have ballasted the acceptance of the new model of vehicle. Level 3 represents that set of cities in which some impediments inhibit the development and spread of electric vehicles.

The incentive policies must therefore be extended from a national level, common throughout the Italy, keeping an eye on the local characteristics that will make it difficult to adopt and embrace a new vehicle technology, if they are not resolved first. For example, the reduction in the cost of purchasing an electric vehicle, a global incentive, can attract a larger proportion of potential users in the richer regions; on the contrary a policy of access to limited traffic areas, preferential lanes and free parking spaces can be more attractive locally, even in the less wealthy areas.

Upcoming Scenarios

The chosen SERA model is not limited to a static analysis but it also predicts the upcoming scenarios. SERA uses a method of space-time placing of the infrastructures; a deterministic algorithm, called "*Station Counts*," allows the estimation of the quantity of infrastructures to be realized in next years, in a precise urban zone, based on the dynamic recordings of new EVs bought in the previous years. The method applies a forward finite difference scheme, refining with a time domain study, the unknown relationship between the presence of charging stations and the use of electric vehicles.

Basing on this scheme and by taking into account the data associated to the EV sales and the amount of charging infrastructures already present, the algorithm enables the prediction of the increase in the number of infrastructures from the initial ones, city by city.

The initial point of the analysis is based on a hypothetical value, so as defined in Section 3.2; such hypothesis is used to resolve the *chicken and the egg* problem.

Thus, by considering a spending capacity to realize the hydrogen stations equal to that already spent to realize the charging infrastructures, the initial number N_0 of starting hypothetical FC infrastructures is 1/25 of the number of electric recharging plants existing in the Italian regions in 2016.

By taking $D(t)$ the recordings of EV sales in year t, $D(t+1)$ the recordings in year $t+1$ and $N(t)$ the number of recharging plants realized in the year t. An empirical parameter α adopts the rate of 2.5, while the $Qave\_max(t)$ takes the rate of 8000 [46].

By following a forward difference scheme, the amount of plants after one year is:

$$N(t+1) = N(t) + \beta\, W(t+1), \tag{1}$$

where

$$W(t+1) = \frac{D(t+1) - D(t)}{Q_{ave}(t)}, \tag{2}$$

$$N(t=0) = N_0, \tag{3}$$

$$Q_{ave} = \frac{D(t)}{N(t)}, \tag{4}$$

$$\beta = \alpha^{\left(\frac{Q_{ave}(t)}{Q_{ave\_max}(t)} - 1\right)}. \tag{5}$$

In Equation (1) the number of EVs depends on the number of EVs the previous year, increased by a factor β multiplied for the so called "*willingness*" of the citizen to assume a new ecological lifestyle and buy an EV.

The willingness $W(t+1)$ in (2), obtained with a dynamic function, is a time dependent function, realized with the forward finite difference scheme employing the number of bought EVs taking into account two different years.

A weighing factor is used in (2) to reduce the willingness, it is expressed in (4) which faces the relation "chicken and egg" problem.

Databases present in References [22,53] are taken into account and particularly to enforce the forward finite difference the recording of years 2014–2017 were employed.

The attitude of various cities to increase the number of FC plants, seen as enabling stations, are plotted in Figure 12. In this analysis only the cities clustered on Level 1 and 2 of Figure 11 have been considered.

Abscissas represents the amount of starting FC stations (also not integer number), given by the number of electric recharging infrastructures divided by a factor of 25, about equal to the cost ratio between the FC station and the electric charging station.

A different behavior is immediately evidenced—some cities start with a considerable number of plants, others with a fractional number.

The ordinate axis hosts the progression of the number of enabling stations, year after year. Such number is amplified as a consequence of novel recordings of vehicles.

Milan and Rome, very big cities, start with a large number of hydrogen infrastructures but fail in double up the amount of enabling stations. Florence keeps quite the hypothetical initial number; Bolzano almost quadruples the starting number [56].

Figure 12b clusters a group of cities that increase the limited initial number of enabling stations but currently do not have structures whose cost can be compared to that of the first FC enabling station.

(a)

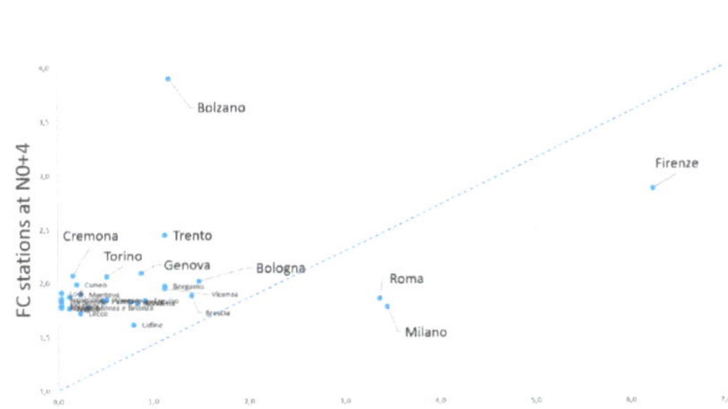

Figure 12. *Cont.*

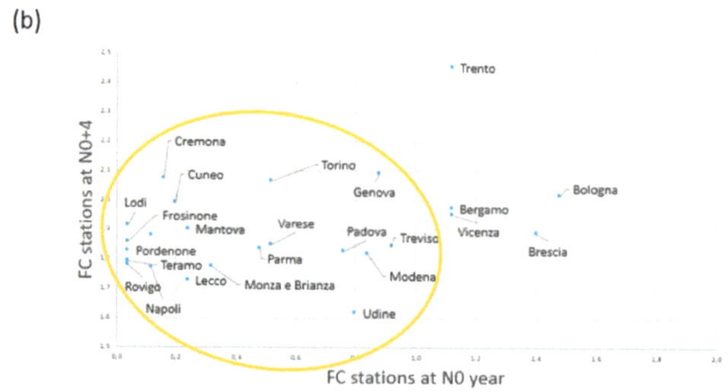

Figure 12. This picture reports the willingness to adopt the FC technology for different cities, Figure (**a**). Among the cities, Bolzano showed the tendency to quadruple the presence in the territory of the FC enabling stations. The considered years are in the windows 2013–2017. Figure (**b**) regroups cities that without having an initial FC enabling station, rapidly increase the number of FC plants.

Substantially Figure 12 can be broken down into two halves—the left upper part shows a tendency to increase the number of enabling stations, while the lower right part tends to preserve the initial number.

Although Figure 12 has been obtained by considering data from past years, the goodness of the choice of a dynamic approach has highlighted the trend of adoption of the new technology. The adoption trend reflects the smart mentality of the population but also the easiness with which modern charging stations can be used; cities with the same number of inhabitants but with different population density or with equal extension but different average income, showed very different behaviors.

5. Drawbacks in Development

The development of a hydrogen economy is not new. This theme pervades scientific and less scientific circles and in the collective imagination it represents the future [57,58].

There are certainly pitfalls that slow down its development.

Let's start with the comparison between a battery and fuel cell vehicles. We consider the same distance to travel on vehicles similar in mass and performance. The differential will be due to how the powertrain gets energy. As reported in Reference [59] a fuel cell vehicle requires more energy to drive the same distance, Figure 13.

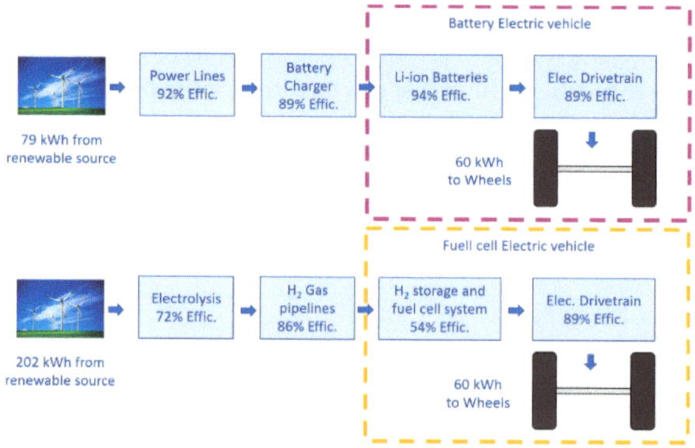

Figure 13. Well-to-wheel energy pathway for battery electric vehicle and fuel cell vehicle. The BEV regeneration capability reduces 60 kWh requirement by 6 kWh, while achieving the same range. For FCEV the pipeline includes losses from compression, expansion, storage and distribution [59].

Such comparison shows the major requirement for a FCEV than BEV in primary energy [59] but also advises that the production and transport chain has a low yield and if no renewable energies are used, a great amount greenhouse gas emission is created, which may be close to those generated by the employment of ICE. Again in Reference [59] a comparison can be made for the cost of the vehicle. The cost is divided by items, such lion battery, fuel cell, storage tank, drivetrain. In 2004 the hypnotized total investment costs are $19.951 for BEV and $29.157 for FCEV.

A recent study actualizes the costs by considering the same parameters (fuel cell stack, batteries, drivetrain) and reports the following costs—ICE $13.784, BEV $37.838, FCEV $90.090 [60]. At the present moment the Mirai has a cost of $58,500, included among those previously hypothesized.

The cost of a charging station is the biggest obstacle for the diffusion of FCEV. Mayer at al. in Reference [61] reports the cost of an investment in refueling station based on data in Reference [62]; a Liquid H_2 pump requires an investment costs of $650,000 for 2015, $650,000 for 2020 and $250,000 for 2050. A recent review suggests a window of costs for the investment, currently it ranges between €0.8 and €2.1 million and it is expected to drop to €0.6 to €1.6 million by 2023 [63]. By considering the more advanced Country in Europe for the adoption of EVs, Norway, similar investment costs are suggested in Reference [64].

Another obstacle to the development of such vehicles is the risk of accidents due to the strong explosiveness of hydrogen. For example, EuroTunnel does not allow "vehicles powered by any flammable gasses", including hydrogen, to use the link between the UK and France. However this problem is to be considered also in all battery-powered vehicles, since as explained in Reference [65], in case of short circuit, the thermal runaway produces the same explosive hydrogen.

In 2016 the Ministry of Economic Development (MISE) adopted the National Plan for Hydrogen Refueling Infrastructure ("Piano Nazionale di Sviluppo—Mobilità Idrogeno Italia") [66] with the collaboration of the association H2IT [67], a long time missing in Italy. The plan expects FC passenger cars to grow in number from 1000 in 2020 to 27,000 in 2025, 290,000 in 2030 and FC buses to reach 3660 in 2030 from an initial 100 in 2020 and 1100 in 2025. Such goals should be matched by the deployment of 440 strategically placed FC charging stations by 2030, starting from an initial 20 in 2020 and about 200 in 2025 [66].

These factors certainly influence the development of fuel cell technologies in recharging station investment but as has been discussed in the previous sections, some cities are advanced and smart enough to be able to face and overcome these problems.

6. Conclusions

This article aims to study the propensity to adopt ecological and sustainable vehicles in Italy. Specifically, the scenario of FC vehicles was investigated. The objective is to draw which, among the Italian cities, are motivated and organized to adopt the new FC technology which requiring similar investment for the total cost of the ownership of the vehicle (FCEV and BEV have similar prices) but most expensive costs in the infrastructures of refueling stations.

A reason for urgently adopting vehicles with low emissions can be the city air quality, so the first part of the study shows which cities need a rapid intervention to reduce the particulate matters PM10 and PM2.5 mainly due to the use of internal combustion engines. Several cities in the Lombardy region should carry out severe interventions to lower pollution levels.

Second part of the study highlights, by the *"early adopters metric"* method, the propensity for innovation in the use of electric vehicles among Italian cities. Again, Lombardy region, between the twenty Italian ones, should be the most motivated to follow the FC technology.

The third part of the analysis starts from the observation that there are no appreciable FC refueling stations on the Italian territory. So the study highlights what is the response capacity of the cities in increasing their infrastructure investments, which are the cities able to successfully convert initial investments in this technology in a continuous development over the years. To predict the development in the number of refueling infrastructures over the years, the *"willingness"* to accept a novel electric vehicle technology is evaluated by enforcing a mathematical approach. Once supposed the existence of a FC infrastructure (or a part of it), the growing of it is calculated by means of a dynamic scheme. Year by year the model correlates the effects of a new quantity of recharging plants in the city influencing in following year the EV sales and an increased quantity of EVs that requires new charging stations. This "*willingness*" thus represents the development trend of the recharging stations.

The final part of the analysis shows that the cities requiring an immediate action to reduce particulate matter, are not among those most likely to increase investments in recharging stations. Bolzano is the city with the highest tendency to increase the number of FC enabling stations.

Author Contributions: Authors contributed equally to the presented work. All authors have read and agreed to the published version of the manuscript.

Funding: This work was financially supported by the University of Palermo. Some researchers received external funding from the projects Prosib (PROpulsione e Sistemi IBridi per velivoli ad ala fissa e rotante) and Reaction (first and euRopEAn siC eigTh Inches pilOt liNe).

Acknowledgments: The authors kindly thank Unione Nazionale Rappresentanti Autoveicoli Esteri for the provision of the data about the sales of electric vehicles.

Conflicts of Interest: The authors declare no conflict of interest.

References

1. IEA. CO_2 Emissions from Fuel Combustion, 2018 Highlights. Available online: https://webstore.iea.org/co2-emissions-from-fuel-combustion-2018-highlights (accessed on 11 July 2019).
2. IEA. CO_2 Emissions from Fuel Combustion. Paris: IEA Report 2013: P. 566. Available online: www.iea.com (accessed on 11 July 2019).
3. IEA. CO_2 Emissions Statistics. 2017. Available online: https://www.iea.org/statistics/co2emissions (accessed on 11 July 2019).
4. IEA. Global Energy & CO_2 Status of the International Energy Agency. 2018. Available online: https://www.iea.org/geco/ (accessed on 11 July 2019).
5. Huétink, F.J.; van der Vooren, A.A.; Alkemade, F. Initial infrastructure development strategies for the transition to sustainable mobility. *Technol. Forecast. Soc. Chang.* **2010**, *77*, 1270–1281. [CrossRef]
6. Al-Alawi, B.M.; Bradley, T.H. Review of hybrid, plug-in hybrid, and electric vehicle market modeling Studies. *Renew. Sustain. Energy Rev.* **2013**, *21*, 190–203. [CrossRef]
7. Zagorskas, J.; Burinskienė, M. Challenges Caused by Increased Use of E-Powered Personal Mobility Vehicles in European Cities. *Sustainability* **2020**, *12*, 273. [CrossRef]
8. Gnann, T.; Plötz, P.; Wietschel, M. How to address the chicken-egg-problem of electric vehicles? Introducing an interaction market diffusion model for EVs and charging infrastructure. In Proceedings of the Conference: ECEEE Summer Study 2015, Hyères, France, 1–6 June 2015.
9. Viola, F.; Longo, M. On the strategies for the diffusion of EVs: Comparison between Norway and Italy. *Int. J. Renew. Energy Res.* **2017**, *7*, 1376–1382.
10. Schwoon, M. Simulating the adoption of fuel cell vehicles. *J. Evol. Econ.* **2006**, *16*, 435–472. [CrossRef]
11. Miceli, R.; Viola, F. Designing a Sustainable University Recharge Area for Electric Vehicles: Technical and economic analysis. *Energies* **2017**, *10*, 1604. [CrossRef]
12. Chen, T.D.; Kockelman, K.M.; Khan, M. The electric vehicle charging station location problem: A parkingbased assignment method for Seattle. In Proceedings of the In Transportation, Research Board 92nd Annual Meeting, Washington, DC, USA, 13–17 January 2013; Volume 340, p. 131254.
13. Ge, S.; Feng, L.; Liu, H. The Planning of Electric Vehicle Charging Station Based on Grid Partition Method. In Proceedings of the IEEE Electrical and Control Engineering Conference, Yichang, China, 16–18 September 2011.
14. Melaina, M.W. Initiating hydrogen infrastructures: Preliminary analysis of a sufficient number of initial hydrogen stations in the US. *Int. J. Hydrog. Energy* **2003**, *28*, 743–755. [CrossRef]
15. EUROPEAN COMMISSION. Press Release Database. 2018. Available online: https://europa.eu/rapid/press-release_IP-18-3708_it.htm (accessed on 11 July 2019).

16. Harrison, G.; Vilchez, J.J.G.; Thiel, C. Industry strategies for the promotion of E-mobility under alternative policy and economic scenarios. *Eur. Transp. Res. Rev.* **2018**, *10*, 19. [CrossRef]
17. Moro, A.; Lonza, L. Electricity carbon intensity in European Member States: Impacts on GHG emissions of electric vehicles. *Transp. Res. Part D Transp. Environ.* **2018**, *64*, 5–14. [CrossRef]
18. Union of Concerned Scientists. 2018. Available online: https://www.ucsusa.org/clean-vehicles/electric-vehicles/life-cycle-ev-emissions (accessed on 11 July 2019).
19. Fox, G.H. Electric vehicle charging stations: Are we prepared? *IEEE Ind. Appl. Mag.* **2013**, *19*, 32–38. [CrossRef]
20. Dijk, M.; Orsato, R.J.; Kemp, R. The emergence of an electric mobility trajectory. *Energy Policy* **2013**, *52*, 135–145. [CrossRef]
21. Brenna, M.; Foiadelli, F.; Longo, M.; Zaninelli, D. E-Mobility forecast for the transnational e-corridor planning. *IEEE Trans. Intell.Transp. Syst.* **2015**, *17*, 680–689. [CrossRef]
22. Unione Nazionale Rappresentanti Autoveicoli Esteri. Available online: http://www.unrae.it/UNRAE, https://www.colonnineelettriche.it/ (accessed on 9 November 2019).
23. General Requirements, Eur. Std. IEC 62196-1. 2014. Available online: https://webstore.iec.ch/publication/6582 (accessed on 11 July 2019).
24. Aziz, M.; Oda, T.; Ito, M. Battery-assisted charging system for simultaneous charging of electric vehicles. *Energy* **2016**, *100*, 82–90. [CrossRef]
25. Luo, X.; Xia, S.; Chan, K.W. A decentralized charging control strategy for plug-in electric vehicles to mitigate wind farm intermittency and enhance frequency regulation. *J. Power Sources* **2014**, *248*, 604–614. [CrossRef]
26. Mehrjerdi, H.; Rakhshani, E. Vehicle-to-grid technology for cost reduction and uncertainty management integrated with solar power. *J. Clean. Prod.* **2019**, *229*, 463–469. [CrossRef]
27. Huda, M.; Aziz, M.; Tokimatsu, K. The future of electric vehicles to grid integration in Indonesia. *Energy Procedia* **2019**, *158*, 4592–4597. [CrossRef]
28. Mehrjerdi, H. Optimal correlation of non-renewable and renewable generating systems for producing hydrogen and methane by power to gas process. *Int. J. Hydrog. Energy* **2019**, *44*, 9210–9219. [CrossRef]
29. Wu, X.; Hu, X.; Moura, S.; Yin, X.; Pickert, V. Stochastic control of smart home energy management with plug-in electric vehicle battery energy storage and photovoltaic array. *J. Power Sources* **2016**, *333*, 203–212. [CrossRef]
30. Dallinger, D.; Wietschel, M. Grid integration of intermittent renewable energy sources using price-responsive plug-in electric vehicles. *Renew. Sustain. Energy Rev.* **2012**, *16*, 3370–3382. [CrossRef]
31. Wang, Z.; Wang, S. Grid power peak shaving and valley filling using vehicle-to-grid systems. *IEEE Trans. Power Deliv.* **2013**, *28*, 1822–1829. [CrossRef]
32. Garcés Quílez, M.; Abdel-Monem, M.; El Baghdadi, M.; Yang, Y.; Van Mierlo, J.; Hegazy, O. Modelling, analysis and performance evaluation of power conversion unit in g2v/v2g application—A review. *Energies* **2018**, *11*, 1082. [CrossRef]
33. Ahmadian, A.; Sedghi, M.; Mohammadi-ivatloo, B.; Elkamel, A.; Golkar, M.A.; Fowler, M. Cost-benefit analysis of V2G implementation in distribution networks considering PEVs battery degradation. *IEEE Trans. Sustain. Energy* **2017**, *9*, 961–970. [CrossRef]
34. Geels, F.W. The dynamics of transitions in socio-technical systems: A multi-level analysis of the transition pathway from horse-drawn carriages to automobiles (1860–1930). *Technol. Anal. Strateg. Manag.* **2005**, *17*, 445–476. [CrossRef]
35. Egnér, F.; Trosvik, L. Electric vehicle adoption in Sweden and the impact of local policy instruments. *Energy Policy* **2018**, *121*, 584–596. [CrossRef]
36. Held, T.; Gerrits, L. On the road to electrification–A qualitative comparative analysis of urban e-mobility policies in 15 European cities. *Transp. Policy* **2019**, *81*, 12–23. [CrossRef]
37. Wang, N.; Tang, L.; Pan, H. A global comparison and assessment of incentive policy on electric vehicle promotion. *Sustain. Cities Soc.* **2019**, *44*, 597–603. [CrossRef]
38. Yong, T.; Park, C. A qualitative comparative analysis on factors affecting the deployment of electric vehicles. *Energy Procedia* **2017**, *128*, 497–503. [CrossRef]
39. Shim, D.; Kim, S.W.; Altmann, J.; Yoon, Y.T.; Kim, J.G. Key Features of Electric Vehicle Diffusion and Its Impact on the Korean Power Market. *Sustainability* **2018**, *10*, 1941. [CrossRef]

40. Brownstone, D.; Train, K. Forecasting new product penetration with flexible substitution patterns. *J. Econom.* **1998**, *89*, 109–129. [CrossRef]
41. Bruhl, B.; Hulsmann, M.; Borscheid, D.; Friedrich, C.M.; Reith, D. (Eds.) A Sales Forecast Model for the German Automobile Market Based on Time Series Analysis and Data Mining Methods. In Proceedings of the Industrial Conference on Advances in Data Mining Applications and Theoretical Aspects, Miami, FL, USA, 20 July 2009.
42. Horne, M.; Jaccard, M.; Tiedemann, K. Improving behavioral realism in hybrid energy-economy models using discrete choice studies of personal transportation decisions. *Energy Econ.* **2005**, *27*, 59–77. [CrossRef]
43. Zhang, Y.; Zhong, M.; Geng, N.; Jiang, Y. Forecasting electric vehicles sales with univariate and multivariate time series models: The case of China. *PLoS ONE* **2017**, *12*, e0176729. [CrossRef] [PubMed]
44. Stiller, C.; Bünger, U.; Møller-Holst, S.; Svensson, A.M.; Espegren, K.A.; Nowak, M. Pathways to a hydrogen fuel infrastructure in Norway. *Int. J. Hydrog. Energy* **2010**, *35*, 2597–2601. [CrossRef]
45. Tlili, O.; Mansilla, C.; Frimat, D.; Perez, Y. Hydrogen market penetration feasibility assessment: Mobility and natural gas markets in the US, Europe, China and Japan. *Int. J. Hydrog. Energy* **2019**, *44*, 16048–16068. [CrossRef]
46. Muratori, M.; Bush, B.; Hunter, C.; Melaina, M.W. Modeling Hydrogen Refueling Infrastructure to Support Passenger Vehicles. *Energies* **2018**, *11*, 1171. [CrossRef]
47. WHO. *Interactive Air Pollution Maps*; WHO: Geneva, Switzerland, 2016; Available online: http://maps.who.int/airpollution/ (accessed on 9 November 2019).
48. WHO. Global Urban Ambient Air Pollution Database. 2016. Available online: http://www.who.int/phe/health_topics/outdoorair/databases/cities/en/ (accessed on 9 November 2019).
49. U.S. Department of Energy, Office of Energy Efficiency and Renewable Energy. Hydrogen and Fuel Cell Program Record 13005: Well-to-Wheels Greenhouse Gas Emissions and Petroleum Use for Mid-Size Light-Duty Vehicles. Available online: http://go.usa.gov/xW8CH (accessed on 11 May 2013).
50. Kendall, M. Fuel cell development for New Energy Vehicles (NEVs) and clean air in China. *Prog. Nat. Sci. Mater. Int.* **2018**, *28*, 113–120. [CrossRef]
51. Kendall, K.; Kendall, M.; Liang, B.; Liu, Z. Hydrogen vehicles in China: Replacing the Western Model. *Int. J. Hydrog. Energy* **2017**, *42*, 30179–30185. [CrossRef]
52. Ballard Power System, Majsmarken 1, DK-9500 Hobro. 2019. Available online: https://www.ballard.com/ (accessed on 9 November 2019).
53. Hydrogen Analysis Resource Center. Available online: https://h2tools.org/hyarc/ (accessed on 11 September 2019).
54. Golyandina, N.; Nekrutkin, V.; Zhigljavsky, A. (Eds.) *Analysis of Time Series Structure. SSA and Related Techniques*; Chapman & Hall/CRC: Boca Raton, FL, USA, 2010.
55. Dekimpe, M.G.; Franses, P.H.; Hanssens, D.M.; Naik, P.A. Time-Series Models in Marketing. Erim Report. *Int. J. Res. Mark.* **2000**, *17*, 183–193. [CrossRef]
56. Viola, F.; Zaninelli, D.; Ala, G.; Schettino, G.; Castiglia, V.; Miceli, R. Forecasting the diffusion of hydrogen EV refuelling infrastructures in Italy. In Proceedings of the 14th International Conference on Ecological Vehicles and Renewable Energies, EVER, Monte-Carlo, Monaco, 8–10 May 2019.
57. Rifkin, J. *The Hydrogen Economy: The Creation of the Worldwide Energy Web and the Redistribution of Power on Earth*; TarcherPerigee: New York, NY, USA, 2003.
58. Turtledove, H. *Worldwar*; Del Rey: New York, NY, USA, 1994; ISBN 0-345-38241-2.
59. Eaves, S.; Eaves, J. A cost comparison of fuel-cell and battery electric vehicles. *J. Power Sources* **2004**, *130*, 208–212. [CrossRef]
60. Veziroglu, A.; Macario, R. Fuel cell vehicles: State of the art with economic and environmental concerns. *Int. J. Hydrog. Energy* **2011**, *36*, 25–43. [CrossRef]
61. Mayer, T.; Semmel, M.; Morales, M.A.G.; Schmidt, K.M.; Bauer, A.; Wind, J. Techno-economic evaluation of hydrogen refueling stations with liquid or gaseous stored hydrogen. *Int. J. Hydrog. Energy* **2019**, *44*, 25809–25833. [CrossRef]
62. U.S. Department of Energy. *Fuel Cell Technologies Office Multi-Year Research, Development, and Demonstration Plan. 3.2 Hydrogen Delivery*; Office Efficiency Renewable Energy: Washington, DC, USA, 2015.
63. Apostolou, D.; Xydis, G. A literature review on hydrogen refuelling stations and infrastructure. Current status and future prospects. *Renew. Sustain. Energy Rev.* **2019**, *113*, 109292. [CrossRef]

64. Ulleberg, Ø.; Hancke, R. Techno-economic calculations of small-scale hydrogen supply systems for zero emission transport in Norway. *Int. J. Hydrog. Energy* **2020**, *45*, 1201–1211. [CrossRef]
65. Barnett, G. *Vehicle Battery Fires: Why They Happen and How They Happen*; SAE International: Warrendale, PA, USA, 2017; ISBN 978-0-7680-8359-0. Available online: https://www.sae.org/publications/books/content/r-443/ (accessed on 11 January 2020).
66. IEA. Hybrid and Electric Vehicle, Italy-Policies and Legislation. Available online: http://www.ieahev.org/by-country/italy-policy-and-legislation/ (accessed on 11 September 2019).
67. Available online: https://www.h2it.it/wp-content/uploads/2019/12/Piano-Nazionale_Mobilita-Idrogeno_integrale_2019_FINALE.pdf (accessed on 27 December 2019).

sustainability

Article

Impact of the Light-Duty Vehicles' Storage and Travel Demand on the Sustainable Exploitation of Available Resources and Air Pollution Abatement

Mihai Machedon-Pisu * and Paul Nicolae Borza

Department of Electronics and Computers, Transilvania University of Brașov, B-dul Eroilor Nr. 29, 500036 Brașov, Romania; borzapn@unitbv.ro
* Correspondence: mihai_machedon@unitbv.ro

Abstract: Light-duty vehicles are the predominant means of road transport. As the world population is expected to increase significantly in the following decades, so too will the car fleet. Due to the rising population, and the implicitly higher travel demand, the energy demand of cars will increase too, and this will put a strain on current resources, with negative effects on the supply chain, possibly leading to more pollution. Many of the current sustainable transport models and frameworks attempt to predict the vehicle market share for different powertrains and the resulting impact based on scenarios that cater to the automotive market and industry demands. At the same time, most neglect aspects regarding resources' depletion and storage demand. In this sense, this study proposes a coherent testing methodology based on the ratio between demand and supply in order to address the limitations of these studies, mainly related to the sustainable exploitation of available resources, which are analyzed herein in correlation with the current predictions. A sensitivity analysis is provided in order to evaluate the uncertainty of utilized predictions. As a result of this analysis, two novel scenarios for assessing the evolution of the vehicle market share are proposed by the authors. When compared to similar scenarios, it was shown that the proposed scenarios lead to noticeable benefits in reducing dependency on the resources associated with a demand of energy and raw materials and in mitigating air pollution, including related costs.

Keywords: light-duty vehicles; sustainable transport; energy; resources depletion; storage demand; supply chain; air pollution

check for updates

Citation: Machedon-Pisu, M.; Borza, P.N. Impact of the Light-Duty Vehicles' Storage and Travel Demand on the Sustainable Exploitation of Available Resources and Air Pollution Abatement. *Sustainability* **2022**, *14*, 8571. https://doi.org/10.3390/su14148571

Academic Editors: Joel R.M. Oliveira, Hugo Silva, R. Christopher Williams and Zejiao Dong

Received: 12 June 2022
Accepted: 9 July 2022
Published: 13 July 2022

1. Introduction

Transport activities play an important part in the evolution of our society, assuring the mobility of people and goods [1]. In terms of market, environmental, and human health impacts [2], as well as society's travel needs [1,2], light-duty vehicles (LDV) are the leading road transport means when compared to other transport means (more than 80% of the total fleet), such as buses (that exceed 3.5 tons) and heavy-duty vehicles (HDV), e.g., trucks [3]. The sustainability of transport activities is mainly modulated by two factors: (a) the technological and scientific evolution in the automotive domain [4,5]; (b) the improvement of the legislative frame followed by the implementation of appropriate policies [6]. In the case of LDVs, the evolution of the powertrain is relevant, starting from internal combustion engine vehicles (ICEVs) to hybrid vehicles such as hybrid electric vehicles (HEVs) and plug-in hybrid electric vehicles (PHEVs) to battery (all) electric vehicles (BEVs) and to fuel cell vehicles (FCVs) [4,5]. Each type of powertrain requires the improvement of specific components [5]. For example, in the case of electric and hybrid vehicles, energy storage (such as batteries) or generators (such as fuel cells) are essential to achieve high performance [4]. More recently, the development and improvement of autonomous vehicles [7] increased efficiency and safety in operation.

On the one hand, the technological progress is closely related to the evolution of the structure of transport systems and is extensively treated in [4]. On the other, one must take also into account the impact of the LDVs' powertrain evolution on the demand for energy and raw materials in order to address the challenges imposed by the energy transition in [8]. This refers to energy generation including fossil fuels, renewables, and electromechanical sources (in BEVs and FCVs). The on-board energy efficiency, storage capacitance, and type of the LDV's energy converter also play an important role in this sense, as depicted in [9]. Incorporated raw materials can be recovered to a certain degree by recycling and reintroduced in a new industrial cycle, which is a desideratum of circular economy [10]. Starting from the vehicle's production to its end of life cycle, which covers the stages included in a life cycle assessment (LCA)-based analysis, as in [11,12], various studies have assessed the impact of the LDV powertrain related to air pollution [12,13] and the supply chain [14]. As highlighted in [8], the decarbonization of the energy sector can be obtained by opting for renewable energy sources (RESs), as shown in [15].

Many of these considerations are found in the different frameworks and models proposed for sustainable transport in [12,14,16–27]. Dedicated to the decarbonization of the energy sector, the energy–economy simulation models in [16–20] are based on user demand and air pollution abatement, mainly greenhouse gas (GHG) emissions, as detailed in [28,29]. Similar to [16–21], they offer no distinction between the different LDV powertrains when analyzing the various facets of sustainability. However, then, the environmental, economic, and energy impacts of different LDV powertrains are depicted in [12,14,22–27]. Based on various scenarios, the evolution of the vehicle market share for LDV powertrains (ICEVs, HEVs, PHEVs, HEVs, and FCVs) is predicted for all types in [14,23,27] until 2050 and beyond in [24].

Table 1 synthesizes the mentioned sustainable transport studies in terms of data on energy (generation and consumption) including fuels (petroleum, hydrogen), raw materials (supply and demand), storage (battery type and capacity), and on vehicle market share (evolution in time). As seen in Table 1, the main omissions of these studies refer to LDVs' storage characteristics and required raw materials.

Figure 1 offers an image of sustainability which both synthesizes the main characteristics depicted in Table 1 and reflects the methodology to be described in the next chapters.

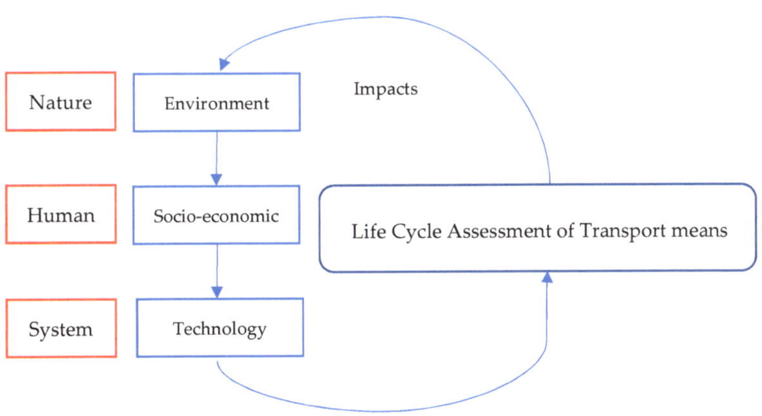

Figure 1. Simplified sustainable transport framework based on the methodology proposed.

Table 1. Comparison between mentioned models and frameworks in terms of data provided based on sustainable transport considerations.

Link	Energy for LDVs		Raw Materials for LDVs		LDV Storage	Vehicle Market Share	
	Gen./Cons.	Fuels/RES	Supply	Demand	Battery Type and Cap.	LDV Powertrain	Period
[12]	Yes/no	Yes/yes	No	No	No	BEV, PHEV [1]	2010–2015
[14]	Yes/yes	Yes/yes	Yes	Yes	Yes	All types	2020–2050
[16]	Yes/yes	Yes/yes	No	No	No	None	-
[17]	Yes/yes	Yes/no	No	No	No	None	-
[18]	No/yes	Yes/no	No	No	No	BEV, PHEV [1]	2017–2030
[19]	Yes/yes	Yes [2]/yes	No	No	No	None	-
[20]	Yes/yes	Yes/no	No	No	No	ICEV, HEV	-
[21]	No/no	Yes/no	No	No	No	All types	-
[22]	No/no	Yes/no	No	No	No	All types	2010–2030
[23]	Yes/yes	Yes/no	No	No	No	All types	2010–2050
[24]	Yes/yes	Yes [2]/yes	No	Yes	No	All types	2020–2100
[25]	No/yes	Yes [2]/no	No	No	No	All except HEV	2015–2050
[26]	Yes/yes	Yes/yes	No	No	No	All except FCV	-
[27]	No/yes	Yes [2]/no	No	No	No	All types	2010–2050

[1] Only regional, [2] Hydrogen included, Gen. = generation, Cons. = consumption.

The paper is structured as follows. The introduction in Section 1 evokes the main LDV powertrain characteristics required in order to analyze the impacts of various sustainable transport models and frameworks mostly related to the exploitation of resources for covering the travel and storage demands and to the decarbonization of the energy sector (GHG emissions).

As a result of this analysis, the paper shortly defines a simplified sustainability framework which represents the basis for the methodology to be presented in the next sections. In Section 2, Materials, an overview of the evolution of LDV sales and production is portraited in correlation with the population and GDP evolutions until 2020, as well as electricity and fuel resources. Future predictions on LDVs' production and travel demand are correlated also with the GDP and population evolution predictions from 2020 to 2050. Based on the LCA analysis, this overview is addressed also in the context of storage demand, which includes both the current battery technologies (mostly based on lithium, nickel, and cobalt) and the new batteries to be implemented (only lithium-based), as well as the decrease in the demand for materials important when recycling, as will be assessed in the next section. Section 3, Methods, includes the proposed methodology used in order to determine the demand over supply ratios (DSRs) for the scenarios proposed in [12,14,16–27], including two novel scenarios proposed herein. The new scenarios proposed in this section are based on a balanced DSR. In Section 4, Results, the most sustainable transport scenarios in terms of DSR that forecast the LDVs' market share evolution are analyzed in terms of the dependency on resources related to energy and raw materials and air pollution abatement. The results are discussed in the final section, Discussion, which highlights the main findings of this study. The uncertainty of the utilized predictions in Sections 2 and 3, which present strong correlations between GDP, population, travel demand, and LDV sales and production evolutions, is also assessed in the final section based on a sensitivity analysis of the projection parameters and their extrapolation.

2. Materials

Based on various reports [30–34], Figure 2 highlights the evolution of population [30], gross domestic product (GDP) [31], and LDV production [32] and sales [33,34] in the last 15–20 years. As seen in Figure 2b, the LDVs' sales and production have a similar slope and show an almost ideal correlation for the study interval. When compared to Figure 2a, correlations can be also found between the evolution in LDVs' production and sales and the evolution of population and GDP.

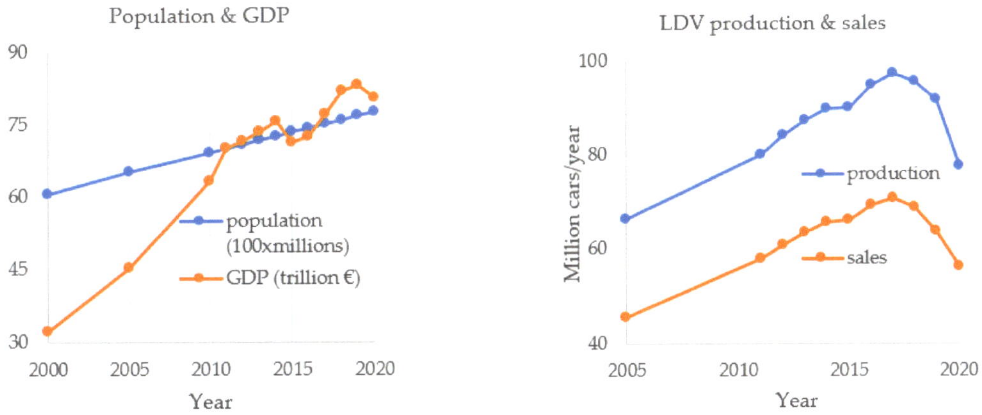

(**a**) (**b**)

Figure 2. Evolution of: (**a**) population and GDP from 2000 to 2020; (**b**) LDVs' production and sales from 2005 to 2020.

Various predictions for population [35–42] and GDP [36,37,43–49] by 2050 can be compared to the estimated evolution of LDVs' production, depicted in Table 2 and Figure 3, based on three evolution scenarios.

Table 2. Values in million cars/year based on the evolution scenarios for LDVs production.

Year	Linear	LN	EXP
2030	109.1791	109.1025	114.0415
2040	123.6438	123.4236	136.5359
2050	138.1086	137.6747	163.4674

LN = natural logarithm, EXP = exponential interpolations.

Figure 3 displays only the prediction formulas for LN (natural logarithm). In Table 3 only LN and EXP (exponential) evolutions are considered since the Linear one is similar in values to LN. Other types of extrapolations, such as spline, e.g., a third or fourth order polynomial, were also tested; however, they present a poor correlation with the population and GDP predictions [35–49] in most cases. The degree of correlation (%) between the LN and EXP predictions in Table 2 and population and GDP predictions is depicted in Table 3. Various studies that predict the evolution of LDVs' production and sales [14,23–25,27] confirm the feasibility of the adapted predictions shown in Table 2 for the low-demand scenario in [14], for the base and alternative cases in [24], and for the scenarios in [27].

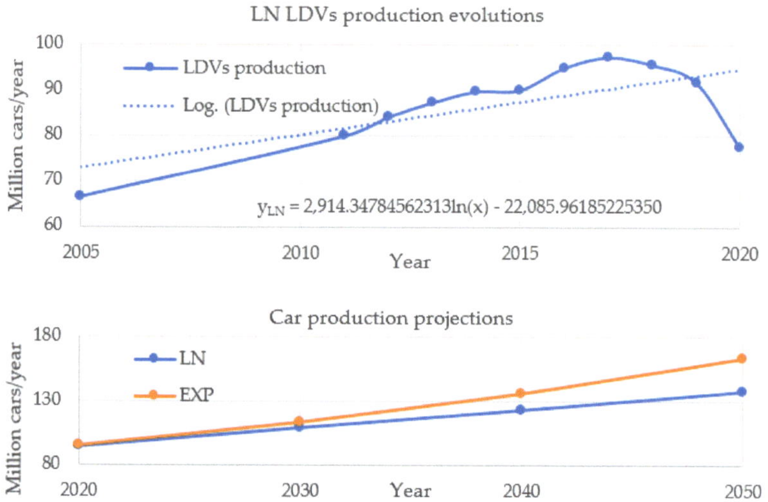

Figure 3. Nonlinear evolution and prediction scenarios (LN and EXP) for LDVs' production including formulas where x = year and y = annual car production.

Table 3. Approximated degree of correlation (%) between Population and GDP predictions and LN and EXP predictions for the LDVs' production from 2020 to 2050 (2020, 2030, 2040, 2050).

Population Predictions Versus LN/EXP LDV Production Predictions										
[35] [1]	[35] [2]	[36]	[35] [3], [37]	[38]	[39]	[40] [1]	[41]	[42] [1]	[42] [1]	[40] [2]
100/100	86/82	−100/−99	100/99	100/100	100/99	100/99	100/99	96/94	99/98	100/99
GDP predictions versus LN/EXP LDV production predictions										
[36]	[43]	[44] [3]	[44] [1]	[37]	[45]	[46]	[47]	[48]	[49]	
−63/−69	100/100	99/100	100/100	99/99	100/100	100/100	99/100	98/99	100/100	

[1] High-, [2] Low-, [3] Medium-valued predictions.

The high demand scenarios in [14] and low stock scenarios in [25] are neglected due to overrated (high) and, respectively, underrated (low) estimated values, which are in contradiction with the population- and GDP-predicted evolutions. The extreme cases that do not match the reasonable assumptions are observed in Table 3 for the low valued predictions in [35] and for [36]. They are also displayed in Figures 4 and 5, which forecast the GDP and population evolutions from 2020 to 2050.

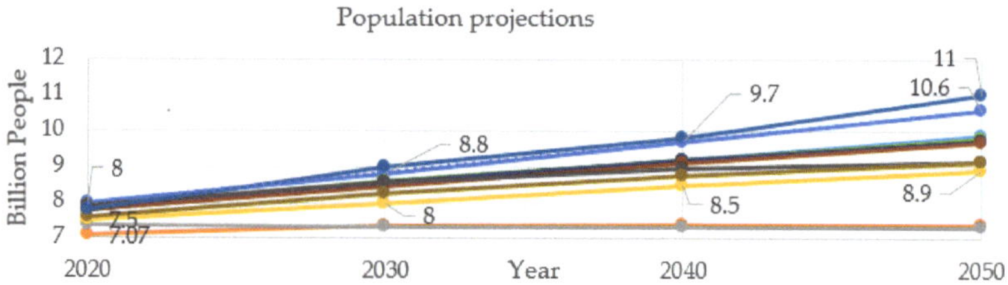

Figure 4. Population predictions from 2020 to 2050 [35–42].

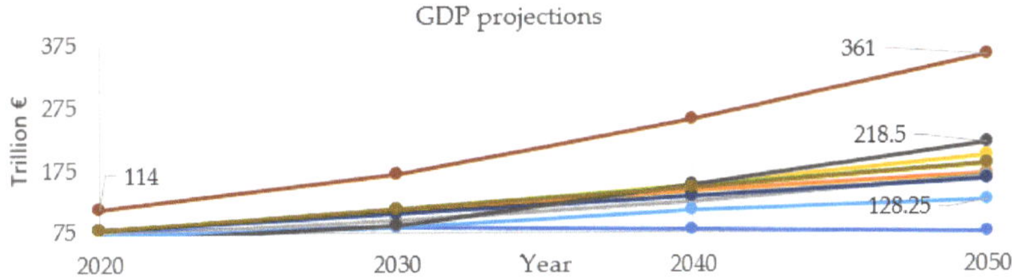

Figure 5. GDP predictions from 2020 to 2050 [36,37,43–49].

Based on the LCA procedure, as shown in [11–13], three main parts (which represent together the whole life cycle) can be identified as Well to Tank (WtT), Tank to Wheel (TtW), and End of Life Cycle (EoL), which include the stages of extraction, manufacturing and assembly (related to WtT or production), and energy generation and distribution (between WtT and TtW), consumption (related to TtW or operation), and recycling (related to EoL). In the production cycle, the main resource indicators are steel, aluminum (Al), lithium (Li), cobalt (Co), nickel (Ni), and platinum (Pt), as presented in [24]. Other materials must be mentioned too, such as iron (Fe), manganese (Mn), copper (Cu), and phosphorous (P) [50]. Due to the high impact on the supply chain, only Li, Co, Ni, and Pt resources will be selected for analysis in the initial cycle, as presented in [14]. Figure 6 is based on the United States Geological Studies (USGS) reports that forecast the identified resources and reserves that are available per year [51].

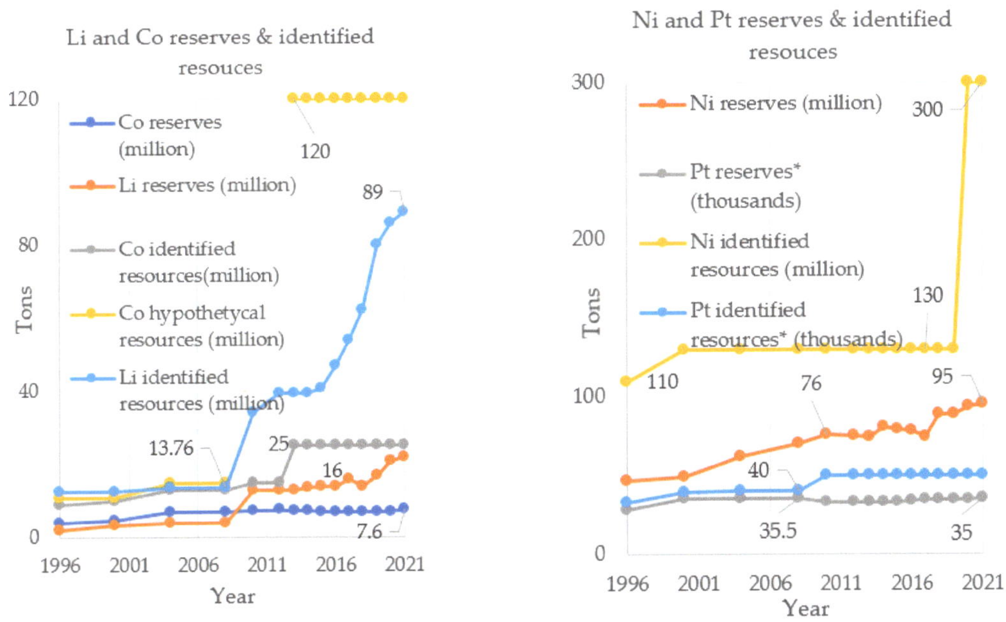

(**a**) (**b**)

Figure 6. Evolution of the identified resources and reserves by USGS for: (**a**) lithium and cobalt; (**b**) nickel and platinum from 1996 to 2021. * Pt reserves and resources represent 50% of the total Platinum-group metals (PGM) reserves and resources in [51].

The vehicle's storage demand is crucial in determining how much of the identified resources or reserves to exploit, and this refers to the vehicle's battery type and capacity. Table 4 synthesizes the storage demand per battery type. Since not all resources can be utilized in the production of LDVs, it is also essential to estimate how much of these resources can be allocated. Both aspects were treated in [14], and in the case of allocated resources, the following considerations were made herein: 40% of the total Li resources, 50% of the total Co resources, 10% of the total Ni resources, and 40% of the total Pt resources can be allocated to the production of LDVs.

Table 4. LDV batteries' Li, Co, Ni, and Pt demand per vehicle.

Battery Type	Lithium (kg/kWh)	Cobalt (kg/kWh)	Nickel (kg/kWh)	Platinum (kg)
For FCV				0.046
For BEV/PHEV/HEV				
NMC/NCM	0.133	0.32	0.435	-
NCA	0.242	0.142	0.79	-
LFP	0.168	-	0.01	-
Li-S	0.412	-	-	-
Li-Air	0.136	-	-	-

By recycling, more resources will be available for manufacturing LDVs in the future, as will be presented at the end of this section, for the final cycle (EoL).

If, in the production cycle, the storage demand dictates the use of materials for producing the required car components, in the operation cycle, the travel demand must be determined in order to find out how many energy resources are necessary to cover the travel mission. Various predictions for the travel demand have been assessed in [52–60]. Figure 7 displays the estimated evolution of travel demand in Passenger km (Pkm) related to LDVs until 2050.

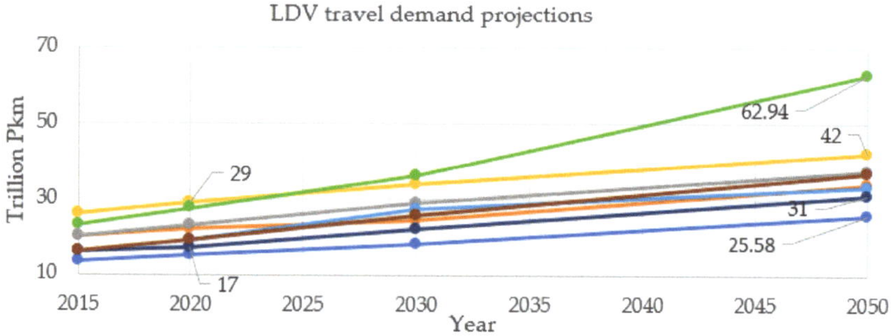

Figure 7. LDVs' travel demand predictions from 2020 to 2050 [52–60].

The correlations between these predictions and the population, GDP, car production, and sales predictions are synthesized in Table 5. As seen in Table 5, the predictions in [58,59] offer both reasonable degrees of correlation.

Table 5. Approximated overall degree of correlation (%) between population, GDP, car production and sales predictions, and predictions for the LDVs' travel demand from 2015 to 2050 (2015, 2020, 2030, 2050).

LN/EXP LDV Production and Sales (Including GDP, Population) Predictions Versus LDV Travel Demand Predictions							
[52]	[53]	[54,55]	[56]	[57]	[58]	[59]	[60]
100/100	99/99	100/99	100/99	96/94	99/100	100/100	100/100

To satisfy the travel demand, the main energy resources, which are needed refer to gasoline reserves and their exploitation in order to fuel ICEVs, HEVs, and some PHEVs and the electricity required for charging PHEVs and BEVs, especially by using RESs, which can lead to the decarbonization of the energy sector [8].

Figure 8 presents the evolution of gasoline reserves in liters gasoline equivalent (LGE) [61] and their exploitation in terms of petroleum production and consumption, also in LGE [5,62,63]. Since not all the gasoline can be used for fueling LDVs [64–66], the estimated gasoline use in cars is assumed at 30%, according to the Baseline predictions in [14]. Based on the same predictions, the evolution of oil consumption per car (ICEV/HEV) is expected to drop from an average of 7 LGE/100 km in 2020 to 4 LGE/100 km in 2050, with a decrement of 1 LGE/100 km per decade.

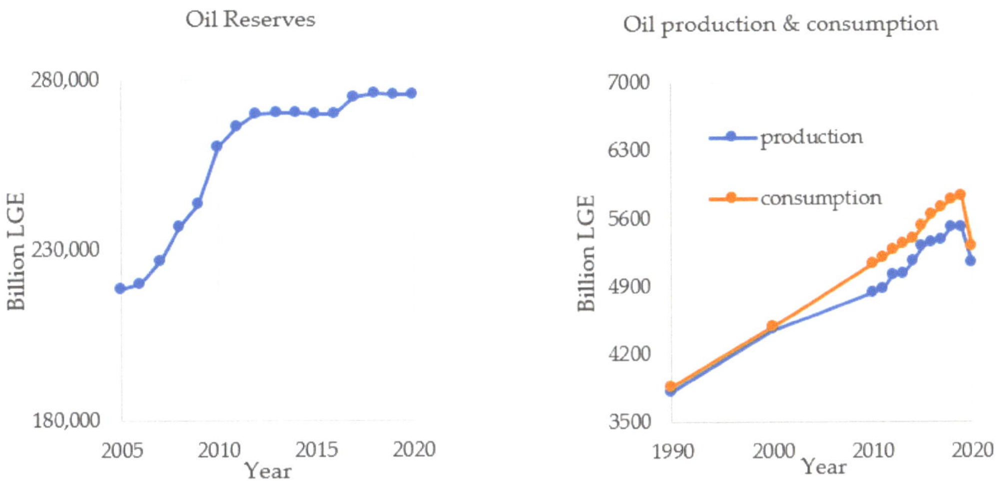

(a) (b)

Figure 8. Evolution of: (**a**) ail reserves, from 2005 to 2020; (**b**) oil exploitation, from 1990 to 2020.

When charging EVs, renewables (RESs) should come as the first option. The quantity of energy that was available until 2018 is depicted in Figure 9.

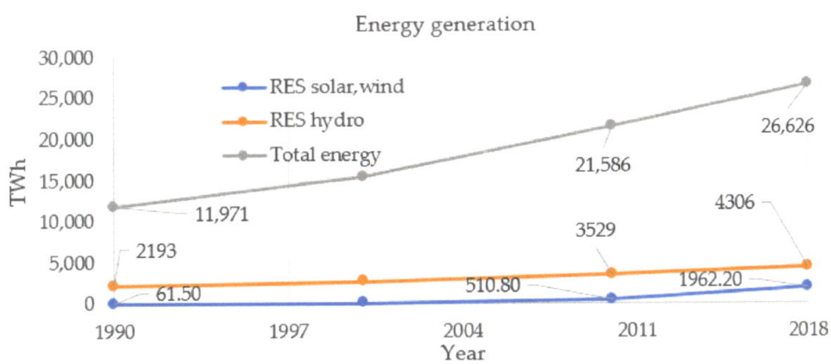

Figure 9. Energy generation from 1990 to 2018.

Based on estimations that reach the year 2050, as presented in the scenarios in [14], solar- and wind-based RESs could increase even by 15–20 times to 40,000 TWh. Table 6 highlights these estimations.

Table 6. Worst-case scenario (WCS) and best-case scenario (BCS) predictions for solar and wind RESs' generation in TWh.

Year	WCS	BCS
2020	2500	3500
2030	8000	16,000
2050	16,500	40,000

The energy demand is directly related to the energy intensity. According to the Baseline scenario in [14], it is estimated at around 20 kWh/100 km in 2020, at 16 kWh/100 km by 2030, at 14 kWh/100 km by 2040, and at 12.5 kWh/100 km by 2050 [67].

The final cycle, which refers to the recycling of car components, can be synthesized in Table 7 when it comes to the raw materials with the most impact on the supply chain, namely, Li, Co, Ni, Pt, as presented in [68].

Table 7. Projected decrease in demand for materials (%) from 2020 to 2050.

Year	Li	Ni	Co	Pt
2020	0.15%	0.13%	0.25%	0.01%
2030	6%	5%	8%	6%
2040	20%	22%	30%	20%
2050	30%	32%	42%	50%

3. Methods

The evolution of the battery capacity for storage-based LDVs: HEV, PHEV, BEV, and FCV is depicted in Figure 10 according to the estimations, predictions, and assumptions in [27,69–81]. Based on the data in Figure 10, the formulas that forecast BEVs' and PHEVs' battery capacity evolutions follow an extrapolation function similar to LN in Figure 3.

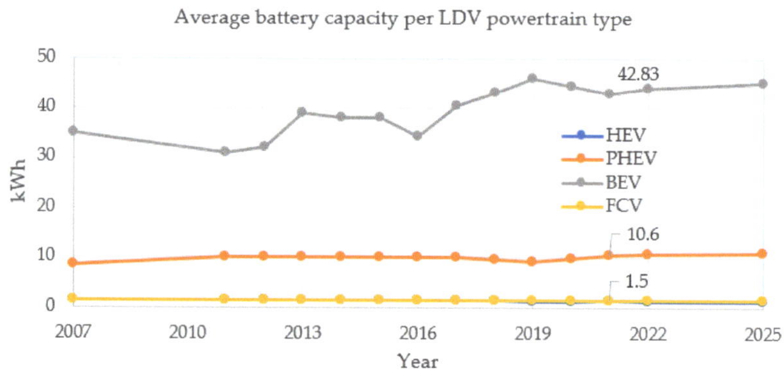

Figure 10. Estimated and predicted evolution of the battery capacity of storage-based LDVs from 2007 to 2025.

Just as important in defining the storage demand as the battery capacity is the adoption rate of the EV's battery chemistry. According to [82], and based on [14] for an optimum ratio between NMC and NCA chemistries, Table 8 highlights the main lithium-based battery technologies that were the most adopted and predicted to be adopted for EVs in the following years.

Table 8. Estimated and projected rate of adoption per battery chemistry in EVs (%), assuming no Li-S or Li-Air market penetration for the 2020–2030 period.

Year	NMC	LFP	NCA
2018 [83]	45%	15%	39%
2019 [83]	53%	3%	43%
2020 [83]	58%	0%	42%
2025 [83]	64%	0%	36%
2030 [14]	72%	0%	28%

Starting from the LDVs' storage and travel demand, plus the decrease in demand of materials, and by taking into account the available resources associated with the production (WtT), operation (TtW), and end of the life cycle of LDVs, it is possible to establish the methodology in Figure 11 which aims to balance the demand over supply ratio (DSR) for various resources available throughout an LDV's whole life cycle.

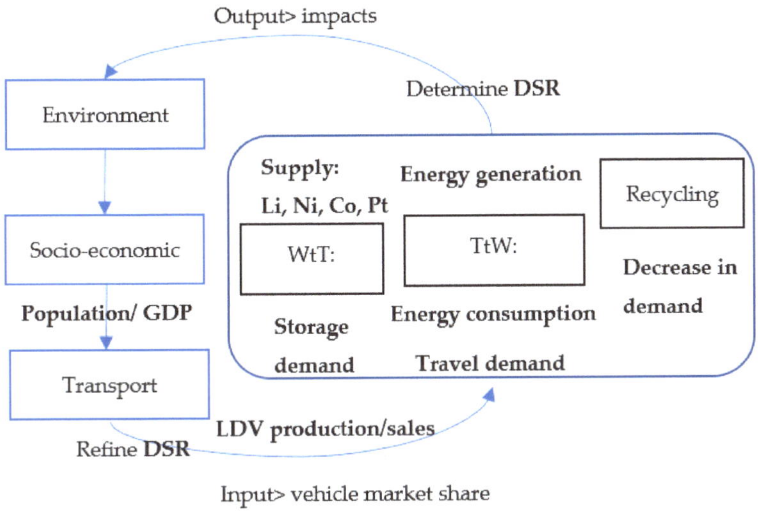

Figure 11. Proposed methodology.

Based on the data in Figure 10, two main scenarios can be formulated for LDV's battery capacity evolution up to 2050 in Table 9: business as usual (BAU), based on tempered storage demand, and natural logarithm-based (LN) predictions.

Table 9. Battery capacity per vehicle (kWh) estimations and predictions for the 2020–2050 period.

Scenario and Year	HEV	PHEV	BEV	FCV
BAU-LN 2020	1.4	10	44.4	1.5
BAU 2030	1.5	11	46	1.5
BAU 2050	1.5	12	50	1.5
LN 2030	1.7	11.2	51	1.7
LN 2050	2	13	67.6	2

The primary inputs for the test methodology addressed in Figure 11, and the vehicle market shares are synthesized in Table 10.

Table 10. Vehicle market share (%) estimations and predictions for the 2020–2050 period.

Scenario and Year	ICEV	HEV	PHEV	BEV	FCV
BAU [23] 2020	91.6	7.2	1.164	0.001	0
BAU [23] 2030	87.6	9.5	2.924	0.001	0
BAU [23] 2050	84	10.8	5.217	0.001	0
S-HEV [23] 2020	80.3	19.2	0.501	0.001	0
S-HEV [23] 2030	53.5	46.1	0.334	0.001	0
S-HEV [23] 2050	0	100	0	0	0
S-PHEV [23] 2020	80.3	4.9	14.786	0	0
S-PHEV [23] 2030	53.5	3.3	43.191	0	0
S-PHEV [23] 2050	0	0	100	0	0
S-EV [23] 2020	80.3	4.9	0.501	14.287	0
S-EV [23] 2030	53.5	3.3	0.334	42.458	0
S-EV [23] 2050	0	0	0	100	0
Base [25] 2020	82.7	12	1.2	4	0.1
Base [25] 2030	33.9	55	3	8	0.1
Base [25] 2050	0.5	10	20	59.5	10
Alt. 1 [25] 2020	82.7	12	1.2	4	0.1
Alt. 1 [25] 2030	17.9	47	15	20	0.1
Alt. 1 [25] 2050	0	0	0	100	0
Alt. 2 [25] 2020	82.7	12	1.2	4	0.1
Alt. 2 [25] 2030	16	47	15	20	2
Alt. 2 [25] 2050	0	0	15	20	65
Scenario [27] 2020	88.5	6.5	3	2	0
Scenario [27] 2030	68	18.5	7.5	5	1
Scenario [27] 2050	45.5	19.5	18	11	6
HER av. NMC [14] 2020	76.613	19.153	0.954	3.18	0.1
HER av. NMC [14] 2030	56	24	1.76	5.865	12.375
HER av. NMC [14] 2050	29.7	29.7	9.362	31.207	0.031
HER av. Li-Air [14] 2020	76.685	19.172	0.933	3.11	0.1
HER av. Li-Air [14] 2030	56	24	1.721	5.735	12.544
HER av. Li-Air [14] 2050	29.7	29.7	9.156	30.519	0.925
LER av. NMC [14] 2020	76.613	19.153	0.954	3.18	0.1
LER av. NMC [14] 2030	49	21	1.496	4.985	23.519
LER av. NMC [14] 2050	15.565	15.565	3.671	12.237	52.962
LER av. Li-Air [14] 2020	76.685	19.172	0.933	3.11	0.1
LER av. Li-Air [14] 2030	49	21	1.463	4.875	23.662
LER av. Li-Air [14] 2050	15.565	15.565	3.81	12.7	52.36

Alt. = alternative case, HER = high exploitation rate, LER = low exploitation rate, av. = average.

Table 11 presents the demand in raw materials sensitive to exploitation (in kg of Ni, Li, and Co and g of Pt per vehicle) for the different LDV powertrains depending on the battery chemistry demand (kg/kWh) data in Table 4 on the scenario type: 72% NMC— 28% NCA, 100% Li-Air, and 100% NMC, based on Table 8, and on capacity evolution scenarios (kWh) BAU and LN in Table 9.

Based on two travel demand scenarios [58] for high demand (WCS) and [59] for low demand (BCS)—, in Figure 7, expressed in Pkm, and by taking into account the estimated consumption in LGE/km, the demand is determined in LGE in Table 12 for the years 2020, 2030, and 2050. Moreover, based on the high and low travel demands, Table 12 includes the available RES solar and wind energy demand in TWh in 2050 if an overall energy intensity of 0.15625 kWh/km is considered, according to the Baseline scenario in [67]. Table 13 synthesizes the available resources for two cases: WCS (low supply) and BCS (high supply) required for battery storage and fueling/charging the vehicle based on low-use scenarios in [14], which considers 0.1 (10%) for Ni, 0.4 (40%) for Li, 0.5 (50%) for Co, and 0.4 (40%) for Pt resources, to which recycling adds more by material recovery, as seen in Table 7.

Table 11. Demand in raw materials (in kg Ni, Li, Co, and g Pt per vehicle) for the BAU and LN battery capacity scenarios based on battery chemistry in 2020, 2030, and 2050.

Scenario and Year	Battery Chemistries	Raw Materials	HEV	PHEV	BEV	FCV
BAU-LN 2020	58% NMC—42% NCA	Ni	0.81774	5.841	25.93404	0.87615
BAU 2030	72% NMC—28% NCA		0.8016	5.8784	24.5824	0.8016
	100% NMC		0.6525	4.785	20.01	0.6525
	100% Li-Air		0	0	0	0
BAU 2050	72% NMC—28% NCA		0.8016	6.4128	26.72	0.8016
	100% NMC		0.6525	5.22	21.75	0.6525
	100% Li-Air		0	0	0	0
LN 2030	72% NMC—28% NCA		0.90848	5.98528	27.2544	0.90848
	100% NMC		0.7395	4.872	22.185	0.7395
	100% Li-Air		0	0	0	0
LN 2050	72% NMC—28% NCA		1.0688	6.9472	36.12544	1.0688
	100% NMC		0.87	5.655	29.406	0.87
	100% Li-Air		0	0	0	0
BAU-LN 2020	58% NMC—42% NCA	Li	0.250292	1.7878	7.937832	0.26817
BAU 2030	72% NMC—28% NCA		0.24528	1.79872	7.52192	0.24528
	100% NMC		0.1995	1.463	6.118	0.1995
	100% Li-Air		0.204	1.496	6.256	0.204
BAU 2050	72% NMC—28% NCA		0.24528	1.96224	8.176	0.24528
	100% NMC		0.1995	1.596	6.65	0.1995
	100% Li-Air		0.204	1.632	6.8	0.204
LN 2030	72% NMC—28% NCA		0.277984	1.831424	8.33952	0.277984
	100% NMC		0.2261	1.4896	6.783	0.2261
	100% Li-Air		0.2312	1.5232	6.936	0.2312
LN 2050	72% NMC—28% NCA		0.32704	2.12576	11.05395	0.32704
	100% NMC		0.266	1.729	8.9908	0.266
	100% Li-Air		0.272	1.768	9.1936	0.272
BAU-LN 2020	58% NMC—42% NCA	Co	0.343336	2.4524	10.88866	0.36786
BAU 2030	72% NMC—28% NCA		0.40524	2.97176	12.42736	0.40524
	100% NMC		0.48	3.52	14.72	0.48
	100% Li-Air		0	0	0	0
BAU 2050	72% NMC—28% NCA		0.40524	3.24192	13.508	0.40524
	100% NMC		0.48	3.84	16	0.48
	100% Li-Air		0	0	0	0
LN 2030	72% NMC—28% NCA		0.459272	3.025792	13.77816	0.459272
	100% NMC		0.544	3.584	16.32	0.544
	100% Li-Air		0	0	0	0
LN 2050	72% NMC—28% NCA		0.54032	3.51208	18.26282	0.54032
	100% NMC		0.64	4.16	21.632	0.64
	100% Li-Air		0	0	0	0
BAU-LN 2020–2050	-	Pt	0	0	0	46

Table 12. Estimated demand in gasoline (billion LGE) in 2020, 2030, and 2050 and in RES energy (TWh) in 2050.

Demand	Year	WCS (High)	BCS (Low)
Gasoline	2020	1923.74	1190
	2030	2168.7	1320
	2050	2517.68	1240
RES energy	2050	9834.688	4843.75

Table 13. Supply (resources) for LDVs' storage and energy (fuel/electricity) available by 2050.

Supply	Unit	Type	WCS (Low)	BCS (High)
Storage for EVs	Million tons	Ni	13.49	42.6
		Li	11.44	46.28
		Co	16.5	79.2
Resources for FCV	Kilotons	Pt	16.5	23.5
Energy for ICEV/HEV	Billion LGE	Gasoline	73,440	
Energy for PHEV/BEV	TWh	RES solar + wind	16,500	40,000

In the case of gasoline and RES resources, the use ratio considered for conventional vehicles (ICEVs/HEVs) is 0.3 (30%) and, respectively, 0.3 (30%) for EVs (PHEVs/BEVs).

Based on the data from Tables 9–13, the following formulas are used in order to determine the demand over supply ratios (DSRs) for LN-WCS (high demand/low supply) and BAU-BCS (low demand/high supply) scenarios, where i = 1:5 in the sum $\sum$ for the 5 LDV powertrains, and the sum $\sum_{year}$ covers the period 2020–2050:

$$DSR\_WCS = \sum_{year} \left(\sum \text{market share}_i \times \text{high demand}_i\right)/\text{Total low supply}, \quad (1)$$

$$DSR\_BCS = \sum_{year} \left(\sum \text{market share}_i \times \text{low demand}_i\right)/\text{Total high supply}, \quad (2)$$

It must be mentioned that, in the case of gasoline resources, in formulas (1) and (2), the Total low supply equals the Total high supply, as seen in Table 13, while, for RES only, the high demand case (WCS) is considered. For simplicity, the sum of ICEVs' and HEVs' market shares is considered altogether in calculations for gasoline resources, and the sum of PHEVs' and BEVs' market shares is the same for RES resources. Based on this, the maximum vehicle market share of EVs (PHEVs and BEVs) that can be charged in WCS (high demand) in 2050 with the available energy (low supply) is around 50.33%.

Table 14 provides the DSR determinations based on the formulas for all scenarios.

As observed in Table 14, most scenarios either exceed the 90–100% range for DSR in WCS or have a very low value of DSR, below 30–40%, in cases mostly related to BCS. While it is unlikely to obtain a one size fits all scenario, which can stay, for example, within the 50–80% range, scenarios with more balanced DSR values can be formulated, such as the two novel scenarios presented in Table 15 together with the calculated DSR values in Table 16.

The two scenarios presented in Table 15 reflect two main battery chemistry evolutions from 2030 to 2050. The first, I Li-Air implies a full transition to Li-Air batteries, which puts the least strain on the supply chain. In the absence of Li-Air battery market penetration, II NMC-NCA proposes an optimum ratio between the two battery chemistries (72% for NMC; 28% for NCA) in terms of least impact on the Ni and Li supply chains. While the first scenario does not consider any restrictions on the Pt dependence of FCVs (46 g Pt per vehicle), the second assumes no Pt is present in FCVs (almost 0 g Pt per vehicle).

Table 14. DSR (%) determinations for the current scenarios for the 2020–2050 period.

Scenario	Nickel WCS/BCS	Lithium WCS/BCS	Cobalt WCS/BCS	Platinum WCS/BCS	Gasoline WCS/BCS	RES Solar Wind WCS
BAU [23]	9.6/2.5	3.5/0.7	4/0.7	0/0	91/52	6.65
S-HEV [23]	19.3/4.3	7/1.2	8/1.2	0/0	94/53.4	0.5
S-PHEV [23]	121/31.8	43.7/9	50/8.6	0/0	38.6/23.6	114
S-EV [23]	601/132	217/37	247/36	0/0	38.6/23.6	113.7
Base [25]	321/67.6	116/19	132/18.3	46.2/27.4	58/35	70
Alt. 1 [25]	529/112	191/31.5	218/30.3	0.64/0.43	43.6/26.7	103
Alt. 2 [25]	200/48	72/13.4	82.2/12.8	306/182	43.6/26.7	54.7
Scenario [27]	95/21.9	34.3/6.2	39/5.9	32.1/19.4	75.5/43.5	33.4
HER NMC [14]	146/31.3	52.7/8.8	86.8/12.1	59.3/39.8	70.9/41	38
HER Li-Air [14]	4.3/1.4	52.7/8.8	1.5/0.3	64.2/42.8	70.9/41	37.1
LER NMC [14]	76.9/17.3	27.7/4.9	45/6.6	354/218	56.7/33.4	18.7
LER Li-Air [14]	4.3/1.4	28.8/5	1.5/0.3	352/217	56.7/33.4	19

Table 15. Vehicle market share (%) for the two novel scenarios for the 2020–2050 period.

Scenario and Year	ICEV	HEV	PHEV	BEV	FCV
I Li-Air 2020	76.6	19.1	1	3.2	0.1
I Li-Air 2030	44.5	20.1	13.5	18.5	3.4
I Li-Air 2050	13.5	20.3	25.4	37	3.8
II NMC-NCA 2020	76.6	19.1	1	3.2	0.1
II NMC-NCA 2030	49	21	9	5	16
II NMC-NCA 2050	15.6	15.6	22	12.2	34.6

Table 16. DSR (%) determinations for the two novel scenarios for the 2020–2050 period.

Scenario	Nickel WCS/BCS	Lithium WCS/BCS	Cobalt WCS/BCS	Platinum WCS/BCS	Gasoline WCS/BCS	RES Solar Wind WCS
I Li-Air	4.43/1.4	80.6/14.2	1.52/0.3	33.7/21.2	55.2/32.4	72.4
II NMC-NCA	111.2/25.7	40.2/7.2	40.1/7	~0/~0 *	56.7/33.4	38

* DSR is almost 0 for no Pt PCVs. If under current conditions (46 g Pt per vehicle), then DSR = 234/145.

4. Results

Figure 12 synthesizes the average DSR values for Li, Ni, Co, and gasoline resources, which represents an average between BCS and WCS values, for the new scenarios: I Li-Air and II NMC-NCA, presented in Table 15, and for S-PHEV [23], Base [25], HER and LER scenarios for NMC and Li-Air in [14], and the Baseline scenario in [27], presented in Table 10. It also includes the DSR values for Pt and RES resources in WCS.

Figure 12 highlights the balanced DSR of the new scenarios, I Li-Air and II No-Pt, except for the case in which Pt will still be used when manufacturing FCVs. It must be mentioned that, in Figure 12f, for the II No-Pt scenario, the value is around 0 if Pt will not be used for FCVs. The value considered is in accordance with the other scenarios which follow the current assumption of a 46 g Pt demand per FCV. Only if the demand in Pt will decrease to 10–20 g per vehicle will the DSR value be acceptable for the II No-Pt scenario. Table 17 assesses the impact on the environment [13,70–72,77,83–90] and, implicitly, on health, including pollution abatement costs [91], and provides the LDV demand profiles of the main operation resources, i.e., gasoline, electricity, and hydrogen [67,70,92].

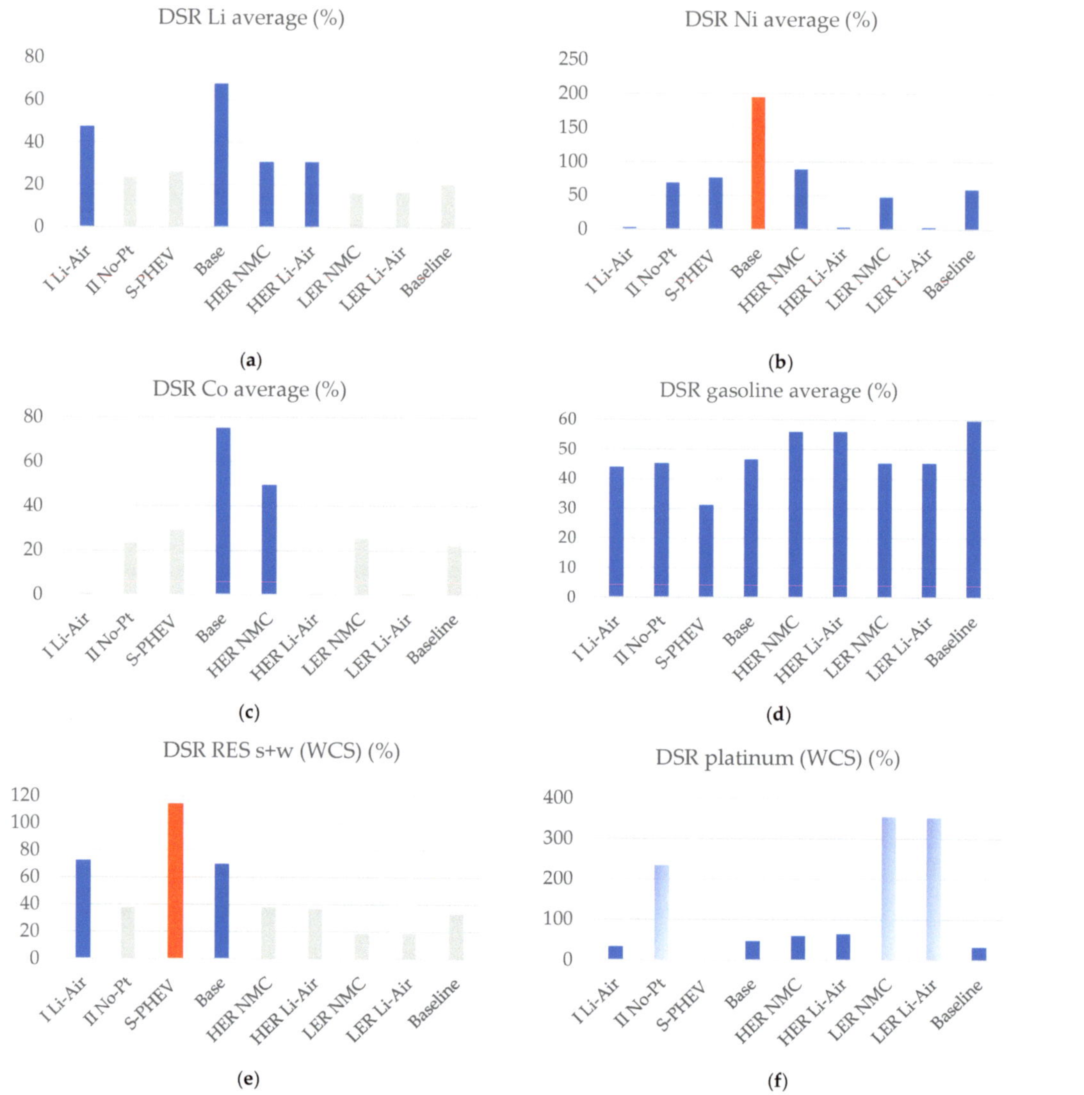

Figure 12. DSR determinations for (**a**) lithium-based storage (average), from 2020 to 2050; (**b**) nickel-based storage (average), from 2005 to 2020; (**c**) cobalt-based storage (average), from 2020 to 2050; (**d**) gasoline resources (average), from 2020 to 2050; (**e**) RES solar and wind resources (WCS), in 2050; (**f**) platinum resources (WCS), from 2005 to 2020, according to nine scenarios.

In order to calculate the total GWP- and PAC-related impacts in Mt CO_2e and Billion EUR- and GEH-related demands in Billion LGE, TWh, and Billion Liters of Hydrogen for the 2020–2050 period (2020, 2030, and 2050); the travel demand in Pkm is added to the formula below, where i = 1:5 in the sum $\sum$ for the five LDV powertrains, and the sum $\sum_{year}$ covers the period 2020–2050:

$$\text{Impact}_{GWP/PAC} \text{ or Demand}_{GEH} = \sum_{year} \left[\sum \text{market share}_i \times (GWP^*_i/PAC_i \text{ or } GEH_i)\right] \times \text{travel demand}_{year}, \quad (3)$$

Table 17. Global Warming Potential (GWP) expressed in kgCO_2 e/km, Pollution Abatement Costs (PAC) in EUR/km, and Gasoline, Electricity, and Hydrogen (GEH) consumption in LGE/km, kWh/km, and, respectively, Liter of Hydrogen/km for various LDV powertrains, based on estimations, forecast, and assumptions, for the 2020–2050 period.

Impact and Year	ICEV	HEV	PHEV	BEV	BEV Clean *	FCV
GWP 2020	0.22	0.2	0.18	0.15	0.04	0.15
GWP 2030	0.19	0.18	0.16	0.13	0.03	0.14
GWP 2050	0.145	0.14	0.12	0.1	0.02	0.13
PAC 2020–2050	0.047	0.041	0.042	0.036	-	0.046
GEH 2020 **	6.5	5.5	21	17	-	7
GEH 2030 **	4.23	3.7632	19	16	-	6
GEH 2050 **	3.0576	2.352	16	12.5	-	5

* Based only on solar and wind RES energy, ** LGE/km for ICEV and HEV, kWh/km for PHEV (electric) and BEV, Liter of Hydrogen/km for FCV.

The travel demand$_{year}$ is calculated as an average between the predicted values in [58] and [59], and is 22.241 trillion Pkm (TPkm) in 2020, 29.072 TPkm in 2030, and, respectively, 46.971 TPkm in 2050. Regarding the GWP calculations, the BEV clean* value in Table 17 is not considered in 2020 due to the lack of meaningful global RES penetration; while, in the 2030–2050 period, only BEV clean* values are considered instead of BEV values due to the reduced impact on GWP, associated with clean energy (solar and wind).

Figure 13 presents the total impacts related to GWP and PAC up to 2050 for the nine scenarios according to the predicted vehicle market share.

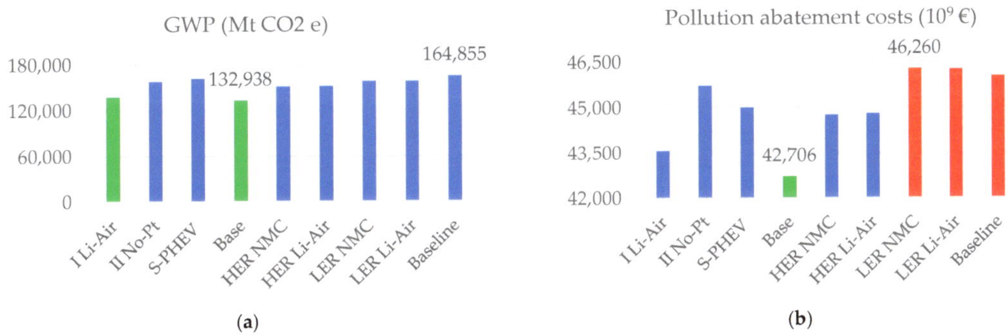

Figure 13. Estimated impacts on: (**a**) GWP; (**b**) PAC, from 2020 to 2050, according to nine scenarios.

Figures 14–16 present the total demand associated with the consumption of gasoline, electricity, and liters of hydrogen required for operating the different LDV powertrains.

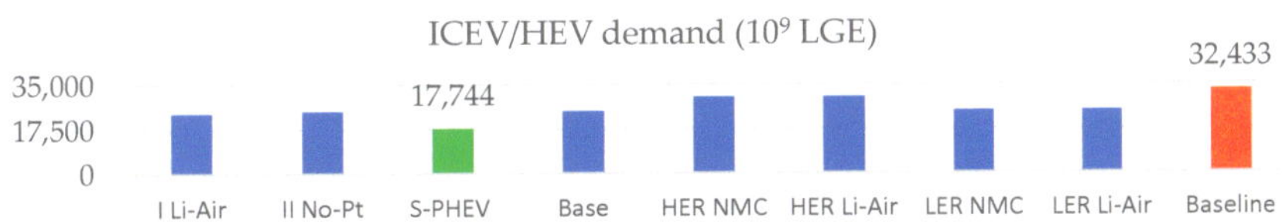

Figure 14. Estimated demand in Billion LGE required for fueling ICEVs and HEVs from 2020 to 2050, according to nine scenarios.

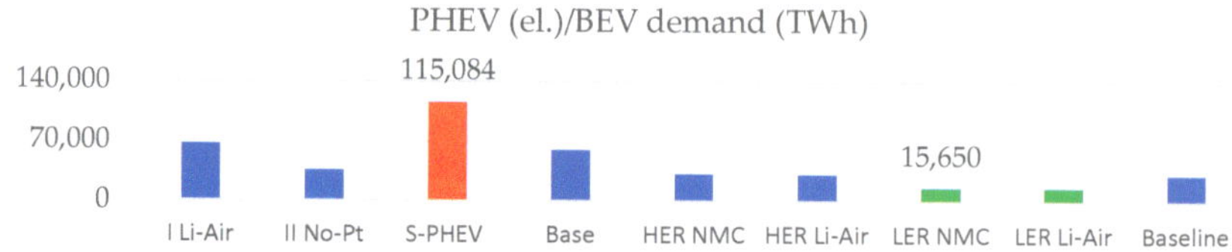

Figure 15. Estimated demand in TWh required for charging PHEVs (electric) and BEVs from 2020 to 2050, according to nine scenarios.

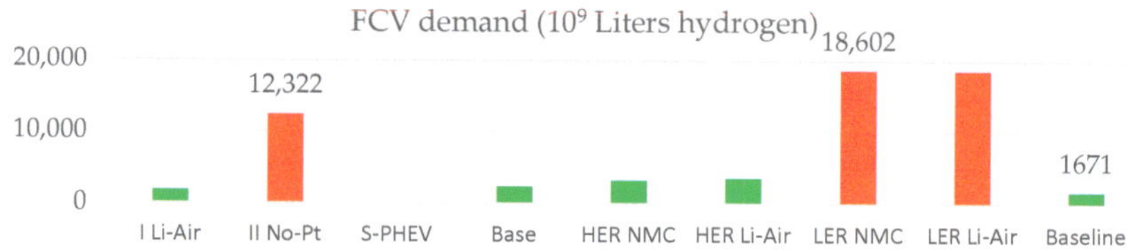

Figure 16. Estimated demand in Billion Liters of Hydrogen required for fueling FEVs from 2020 to 2050, according to nine scenarios.

By analyzing Figures 12–16, it can be concluded that the new scenarios manage not only to balance the DSR in most cases, and thus put less strain on the supply chain, which is related to the available resources' exploitation, but also can prove beneficial in mitigating air pollution, which is mostly related to GHG emissions.

5. Discussion

Due to the uncertain nature of predictions related to the extrapolation functions, which are based on current trends, such as the ones presented in Section 2 and in the beginning of Section 3, it is necessary to provide a sensitivity analysis that can assess the limitations of such projections. Tables 18–22 evoke the projection parameters considered and the type of extrapolations (linear or nonlinear, if known) that modulate the degree of uncertainty, as well as their range when compared to other predictions. The travel demand projections assumed in the methodology depicted in Section 3 are based on a strong correlation with projections that guarantee a 95% confidence interval. Regarding the storage demand projections (modulated by battery's capacity and type), the confidence interval is 80–95% based on the correlations with travel demand and LDV sales. The correlations between GDP, population, and LDVs' travel demand evolutions and LDV sales and production projections are assessed in Tables 3 and 5. The best correlations were those that have shown an almost perfect linear relationship, such as the r coefficient reached values of +0.99 and +1 (+99% to +100% in tables).

In the actual context, of less resources to use for fueling cars, e.g., gasoline, supply chain issues, due to the COVID-19 pandemic, e.g., semiconductor chip shortage and rising prices of gasoline and electricity due to the war in Ukraine, it is more and more important to balance the LDVs' increasing demand of materials and energy with the available resources that must satisfy related travel and storage demands. These can be associated with GDP, population, and emerging technologies' impacts on LDVs' market evolution. The sustainable evolution of the automotive market is reflected also in the air we breathe. Yet, aiming to transition to less polluting cars should be performed without causing disruptions to the supply chain, wherein Li, Ni, Co, Pt, gasoline, hydrogen, and RES resources are at risk of misconduct. One must tackle this issue, as predictions for the vehicle market share are all over the place in terms of market dominance, whether of conventional LDVs (ICE-based) [23,27], of full-electric vehicles (FEVs), mostly BEVs [23,25], or of FCVs [14,25], at the same time almost entirely neglecting the storage demand (battery chemistry and capacity) when predicting the future vehicle market share of various LDV powertrains.

Table 18. Population projections.

Link–Year	Extrapolation *	Parameters	Values	Uncertainty—Confidence Interval
[35]–2004	Nonlinear	Fertility, life expectancy	L to H	Mortality—80–95%
[36]–2015	Almost linear	GDP, growth rate	L	Not assessed
[37]–2006	Unknown	Fertility, growth rate, life expectancy	M	Not assessed
[38]–2020	Unknown	Fertility	M-H	Not assessed
[39]–2017	Nonlinear	Fertility, growth rate, life expectancy	M	Not assessed
[40]–2019	Nonlinear	Fertility-mortality, growth rate-ageing, migration	M to H	Fertility, mortality, migration—95%
[41]–2017	Nonlinear	Fertility, growth rate, life expectancy, education	M	Not assessed
[42]–2021	Nonlinear	Fertility, life expectancy	M	Fertility, ageing, stagnation—0–95%
This study	Nonlinear	Fertility, life expectancy, growth rate, mortality	M to H	Migration, fertility, ageing—90–95%

* Based on current trends, L = low-, M = medium-, H = high-valued predictions.

Table 19. GDP projections.

Link–Year	Extrapolation *	Parameters	Values	Uncertainty—Confidence Interval
[36]–2015	Almost linear	Growth rate, capital, population	L	Population decline—none
[37]–2006	Unknown	Growth rate, production–consumption	L-M	Global public policies—none
[43]–2012	Nonlinear	Growth rate, capital, education, energy–productivity	M	Energy price—none
[44]–2015	Nonlinear	Capital, education, population, energy–productivity	M	Geopolitical context—none
[45]–2018	Unknown	Capital, income, population, and employment data	M	Not assessed
[46]–2022	Unknown	Energy transition	L	Not assessed
[47]–2017	Nonlinear	Investment opportunities	H	Not assessed
[48]–2021	Nonlinear	Growth rate, environment scores	M-H	Not assessed
[49]–2021	Linear	Unknown	M	Unknown
This study	Nonlinear	Capital, population, energy–productivity	L to H	Education, energy price—95%

* Based on current trends, L = low-, M = medium-, H = high-valued predictions by comparison.

Table 20. LDV sales and production projections.

Link–Year	Extrapolation *	Parameters	Values	Uncertainty—Confidence Interval
[14]–2021	Nonlinear	Population, GDP, travel demand	M to H	Vehicle market share—80–95%
[23]–2016	Nonlinear	Population, income, vehicle market share	L-M	Fuel economy, behavior—none
[24]–2019	Nonlinear	Fuel economy **	L to M	Not assessed
[25]–2018	Almost linear	Vehicle market share, purchase probability ***	M	Not assessed
[27]–2012	Nonlinear	Population, GDP	H	Not assessed
This study	Nonlinear	Population, GDP, energy–materials dependency	L to H	Vehicle market share, prices—95%

* Based on current trends, L = low-, M = medium-, H = high-valued predictions when compared to our predictions (this study), ** Based on Chinese and *** U.K. vehicle market reports.

Table 21. LDV travel demand projections.

Link–Year	Extrapolation *	Parameters	Values	Uncertainty—Confidence Interval
[52]–2017	Almost linear	GDP, population, travel preferences	L	Economic growth—none
[53]–2018	Unknown	Unknown	M	Not assessed
[54]–2017	Unknown	GDP, population, travel preferences, technology	M	Not assessed
[55]–2004	Linear	Travel preferences	M	Not assessed
[56]–2012	Nonlinear	Growth rate, travel preferences	M-H	Not assessed
[57]–2015	Nonlinear	Growth rate	M	Fuel prices—none
[58]–2019	Almost linear	GDP, population	H	Not assessed
[59]–2020	Almost linear	Unknown	L-M	Not assessed
[60]–2018	Nonlinear	Population, travel preferences	M	Not assessed
This study	Nonlinear	GDP, population, growth rate, technology	L to H	Economic growth, prices—90–95%

* Based on current trends, L = low, M = medium, H = high-valued predictions by comparison.

Table 22. Battery capacity data estimations and projections.

Link–Year	Study Interval	Vehicle Type (Except ICEV)	Estimation (Studies) or Projection (Uncertainty)
[69]–2021	2012–2019	BEV	Est.: Based on real data/literature (many studies)
[70]–2022	2011–2021	Almost all *	Est.: Based on real data/literature (many studies)
[71]–2020	2019	BEV, PHEV, HEV	Est.: Only one study case
[72]–2021	2021	BEV, HEV	Est.: Only one study case
[73]–2018	2013	BEV, PHEV, HEV	Est.: Only one study case
[74]–2007	2007	BEV, PHEV, HEV	Est.: Based on real data/literature (many studies)
[75]–2011	2007	BEV, PHEV, HEV	Est.: Based on real data/literature (many studies)
[76]–2019	2019	PHEV, BEV	Est.: Based on real data/literature (many studies)
[77]–2022	2013	BEV, PHEV, HEV	Est.: Based on real data/literature (a few studies)
[78]–2018	2017–2025	BEV, PHEV	Pro.: Uncertainty not assessed (based only on 2 years)
[79]-2018	2015–2016	HEV	Est.: Only two study cases
[80]-2019	Unknown	BEV, PHEV, HEV	Est.: Based on real data/literature (many studies)
[81]-2021	2021	BEV, PHEV, HEV	Est.: only one study case
This study	2007–2050	Almost all *	Est.: Based on literature, Pro.: 80–95% confidence

* FCV is neglected for some years, Est. = estimation, Pro. = projection.

In this study, the authors have assessed the impact of the most exploited resources available to LDVs in terms of storage and travel demands. The demand forecast was correlated with the GDP, population, and LDV sales and production for the 2020–2050 period. In order to evaluate the impact on the supply chain, a methodology based on DSR was proposed by the authors for achieving the sustainable exploitation of the resources mentioned above at each stage: WtT (production), TtW (operation), and EoL (recycling), according to an LCA-based analysis. The basis of this methodology is represented by the range defined by the BCS and WCS scenarios that cover the extreme cases of DSR: low demand/high supply (BCS) and high demand/low supply (WCS). Out of the twelve scenarios that forecast the vehicle market share in Table 10, for which the DSR methodology was applied, seven were selected for further analysis, since their DSR values have remained within acceptable bounds. Moreover, based on a balanced DSR, two novel scenarios were proposed herein. These scenarios refer to the future battery chemistry evolutions that can cause less strain on the Ni, Li, Co, and Pt supply chains I Li-Air and II No Pt (72% NMC, 28% NCA), as presented in Figure 12. In terms of air pollution abatement, which aims to decarbonize the energy sector [8], the new scenarios with balanced DSR presented benefits in mitigating GHG emissions (GWP) and reducing related costs (PAC) when compared to the seven selected scenarios, as seen in Figure 13. Moreover, in terms of gasoline, hydrogen, and RES (solar and wind) demand, Figures 14–16 validate that the novel scenarios stay within acceptable limits.

Other aspects that are just as important when making the transition to EVs or FCVs, that were not discussed or detailed in this article, since they are beyond its scope, refer to social and psychological factors, environmental regulations and associated health impact, infrastructure investments and various vehicle costs, and technology transition impacts, to mention a few.

As the system complexity of the future vehicles is expected to increase, wireless sensor networks (WSNs), connected to the cloud, perhaps in a mesh configuration [93–95], will operate beside the Controller Area Network (CAN) bus, and maybe even replace it altogether in the not so far away future, since cables take up a lot of space and require a considerable amount of copper. Moreover, the large deployment of Internet of Things (IoT) and Vehicle to Everything (V2X) permits both the improvement of the extensive traffic management and security and paves the way for autonomous vehicles [96]. Since Industry 4.0 is already here, it is possible that these vehicles will penetrate the automotive market [7]. However, an important aspect that was highlighted in this paper, and that should always work in conjunction with such smart networks based on IoT, is the necessity to charge and store the energy in an effective way. The dependence on energy and raw materials, which was addressed herein, will become more and more demanding in the evolution of LDVs. The energetic challenge also refers to finding new energy management solutions that include the development of communication protocols. The parameters considered are related to the energy and power capabilities of vehicles. Strictly related to the storage of energy, new features such as intelligent power control will facilitate the implementation of heterogeneous solutions such as Hybrid Energy Storage Systems (HESS). Because no single storage device can perfectly fit the applications' requirements, electric hybrid storage solutions were developed in [97,98]. This implies the use of, at minimum, two different storage cells or storage device-like supercapacitors, batteries, and fuel cells [99,100]. Such solutions will increase the overall lifespan of the vehicles' energetic sources, improving not only the energy efficiency, but also allowing a more efficient usage of raw materials for developing the new transport means. Regarding the materials sensitive to exploitation, e.g., Li and Ni, it is possible that new battery chemistries such as NaS and LFP 4680 could reduce this dependence should they become technologically mature. The novel methodology proposed in this study can be adapted accordingly in this sense.

The authors consider that, when addressing the future challenges related to LDV powertrains' adoption, both the demand and supply of energy and raw materials play an important role in defining their market penetration. In this regard, the improvement of on-board electric storage technologies will represent a genuine catalyzer for the new LDVs.

Author Contributions: Conceptualization, M.M.-P.; methodology, M.M.-P.; software, M.M.-P. and P.N.B.; validation, M.M.-P. and P.N.B.; formal analysis, M.M.-P. and P.N.B.; investigation, M.M.-P. and P.N.B.; resources, M.M.-P.; data curation, M.M.-P. and P.N.B.; writing—original draft preparation, M.M.-P. and P.N.B.; writing—review and editing, M.M.-P. and P.N.B.; visualization, M.M.-P. and P.N.B.; supervision, M.M.-P. and P.N.B.; project administration, M.M.-P. and P.N.B.; funding acquisition, M.M.-P. and P.N.B. All authors have read and agreed to the published version of the manuscript.

Funding: This research received no external funding.

Institutional Review Board Statement: Not applicable.

Informed Consent Statement: Not applicable.

Data Availability Statement: The data presented in this study are available on request from the corresponding author.

Conflicts of Interest: The authors declare no conflict of interest.

References

1. Dostál, I.; Adamec, V. Transport and its Role in the Society. *Trans. Transp. Sci.* **2011**, *4*, 43–56. [CrossRef]
2. International Energy Agency. *Energy Technology Perspectives 2012: Pathways to a Clean Energy System*; International Energy Agency: Paris, France, 2012.
3. Ager-Wick Ellingsen, L.; Hung, C.R. Research for TRAN committee—Battery-powered electric vehicles: Market development and lifecycle emissions. In-depth analysis. In *Policy Department B: Structural and Cohesion Policies*; Directorate-general for Internal Policies; European Parliament's Committee on Transport and Tourism: Luxembourg, 2018. [CrossRef]
4. Hayes, J.G.; Goodarzi, G.A. *Electric Powertrain*; John Wiley & Sons Ltd.: Hoboken, NJ, USA, 2017. [CrossRef]
5. Mi, C.; Masrur, M.A.; Gao, D.W. *Hybrid Electric Vehicles*; John Wiley & Sons Ltd.: Hoboken, NJ, USA, 2011. [CrossRef]
6. European Environment Agency Air Quality in Europe—2019 Report. 2019. Available online: https://www.eea.europa.eu/publications/air-quality-in-europe-2019 (accessed on 17 May 2022).
7. Rafael, S.; Correia, L.P.; Lopes, D.; Bandeira, J.; Coelho, M.C.; Andrade, M.; Miranda, A.I. Autonomous vehicles opportunities for cities air quality. *Sci. Total Environ.* **2020**, *712*, 136546. [CrossRef]
8. World Equestrian Center. World Equestrian Center. World Energy Trilemma 2016 Defining Measures to Accelerate the Energy Transition. In *World Energy Council Report*; World Equestrian Center: Ocala, FL, USA, 2016.
9. Ehsani, M.; Gao, Y.; Emadi, A. Modern Electric, Hybrid Electric, and Fuel Cell Vehicles. In *Modern Electric, Hybrid Electric, and Fuel Cell Vehicles*; CRC Press: Boca Raton, FL, USA, 2017. [CrossRef]
10. Heshmati, A. A review of the circular economy and its implementation. *Int. J. Green Econ.* **2017**, *11*, 251. [CrossRef]
11. Machedon-Pisu, M.; Borza, P.N. Are Personal Electric Vehicles Sustainable? A Hybrid E-Bike Case Study. *Sustainability* **2019**, *12*, 32. [CrossRef]
12. Requia, W.J.; Mohamed, M.; Higgins, C.D.; Arain, A.; Ferguson, M. How clean are electric vehicles? Evidence-based review of the effects of electric mobility on air pollutants, greenhouse gas emissions and human health. *Atmos. Environ.* **2018**, *185*, 64–77. [CrossRef]

13. Hooftman, N.; Oliveira, L.; Messagie, M.; Coosemans, T.; Van Mierlo, J. Environmental analysis of petrol, diesel and electric passenger cars in a Belgian urban setting. *Energies* **2016**, *9*, 84. [CrossRef]
14. Machedon-Pisu, M.; Borza, P.N. A methodological approach to assess the impact of energy and raw materials constraints on the sustainable deployment of light-duty vehicles by 2050. *Sustainability* **2021**, *13*, 11826. [CrossRef]
15. Longo, M.; Yaïci, W.; Zaninelli, D. "Team play" between renewable energy sources and vehicle fleet to decrease air pollution. *Sustainability* **2015**, *8*, 27. [CrossRef]
16. Keramidas, K.; Kitous, A.; Despres, J.; Schmitz, A. POLES-JRC model documentation. In *JRC Technical Reports*; Publications Office of the European Union: Luxembourg, 2017. [CrossRef]
17. Capros, P.; van Regemorter, D.; Paroussos, L.; Karkatsoulis, P.; Fragkiadakis, C.; Tsani, S.; Revesz, T. GEM-E3 Model Documentation. In *JRC Technical Reports*; Publications Office of the European Union: Luxembourg, 2013. [CrossRef]
18. Vilchez, J.J.G.; Julea, A.; Peduzzi, E.; Pisoni, E.; Krause, J.; Siskos, P.; Thiel, C. Modelling the impacts of EU countries' electric car deployment plans on atmospheric emissions and concentrations. *Eur. Transp. Res. Rev.* **2019**, *11*, 40. [CrossRef]
19. EU Commission. the PRIMES Energy Model. 2014. Available online: https://ec.europa.eu/clima/sites/clima/files/strategies/analysis/models/docs/primes_model_2013-2014_en.pdf (accessed on 9 July 2021).
20. EU Commission. The TREMOVE Model and Baseline Description. 2006. Available online: https://ec.europa.eu/clima/sites/clima/files/transport/vehicles/docs/2006_tremove_en.pdf (accessed on 11 July 2021).
21. Lopez-Arboleda, E.; Sarmiento, A.T.; Cardenas, L.M. Systematic Review of Integrated Sustainable Transportation Models for Electric Passenger Vehicle Diffusion. *Sustainability* **2019**, *11*, 2513. [CrossRef]
22. Kieckhäfer, K.; Wachter, K.; Spengler, T.S. Analyzing manufacturers' impact on green products' market diffusion—The case of electric vehicles. *J. Clean. Prod.* **2017**, *162*, S11–S25. [CrossRef]
23. Onat, N.C.; Kucukvar, M.; Tatari, O.; Egilmez, G. Integration of system dynamics approach toward deepening and broadening the life cycle sustainability assessment framework: A case for electric vehicles. *Int. J. Life Cycle Assess.* **2016**, *21*, 1009–1034. [CrossRef]
24. Hao, H.; Liu, F.; Sun, X.; Liu, Z.; Zhao, F. Quantifying the Energy, Environmental, Economic, Resource Co-Benefits and Risks of GHG Emissions Abatement: The Case of Passenger Vehicles in China. *Sustainability* **2019**, *11*, 1344. [CrossRef]
25. Mazur, C.; Offer, G.; Contestabile, M.; Brandon, N. Comparing the Effects of Vehicle Automation, Policy-Making and Changed User Preferences on the Uptake of Electric Cars and Emissions from Transport. *Sustainability* **2018**, *10*, 676. [CrossRef]
26. Onat, N.C.; Aboushaqrah, N.N.; Kucukvar, M.; Tarlochan, F.; Hamouda, A.M. From sustainability assessment to sustainability management for policy development: The case for electric vehicles. *Energy Convers. Manag.* **2020**, *216*, 112937. [CrossRef]
27. Tartakovsky, L.; Gutman, M.; Mosyak, A. Energy efficiency of road vehicles—Trends and challenges. In *Energy Efficiency: Methods, Limitations and Challenges*; Nova Science Publishers: New York, NY, USA, 2012.
28. Enerdata. *Global Energy Statistical Yearbook 2019: Total Energy Consumption*; Enerdata: Grenoble, France, 2019.
29. Enerdata. *Costs and Benefits to EU Member States of 2030 Climate and Energy Targets*; Enerdata: Grenoble, France, 2014.
30. UNCTAD. *Handbook of Statistics 2021*; United Nations Conference on Trade and Development: Geneva, Switzerland, 2022. Available online: https://unctad.org/webflyer/handbook-statistics-2021 (accessed on 1 May 2022).
31. World Bank Group. GDP (Current US$) 1960–2020. 2020. Available online: https://data.worldbank.org/indicator/NY.GDP.MKTP.CD (accessed on 20 May 2022).
32. Statista. *Estimated Worldwide Motor Vehicle Production from 2000 to 2021*; Statista: Hamburg, Germany, 2021.
33. Statista. *Global Car Sales 1990–2019*; Statista: Hamburg, Germany, 2020.
34. International Energy Agency. *Passenger Car Sales by Key Region, 2010–2020e*; International Energy Agency: Paris, France, 2020.
35. United Nations. *World Population to 2300: United Nations*; Department of Economic and Social Affairs/Population Division: New York, NY, USA, 2004.
36. Economist Intelligence Unit. *Long-Term Macroeconomic Forecasts: Key Trends to 2050*; The Economist: London, UK, 2015.
37. World Bank Group. The Road to 2050; Sustainable Development for the 21st Century. 2006. Available online: https://documents1.worldbank.org/curated/en/192421468341095824/pdf/360210rev0The0Road0to0205001PUBLIC1.pdf (accessed on 21 May 2022).
38. IISD. World Population to Reach 9.9 Billion by 2050. 2020. Available online: https://sdg.iisd.org/news/world-population-to-reach-9-9-billion-by-2050/ (accessed on 21 May 2022).
39. United Nations. *World Population Prospects: The 2017 Revision, Custom Data Acquired via Website*; Department of Economic and Social Affairs, Population Division: New York, NY, USA, 2017. Available online: https://www.un.org/en/desa/world-population-projected-reach-98-billion-2050-and-112-billion-2100 (accessed on 3 May 2022).
40. UN DESAPD. *World Population Prospects 2019*; Department of Economic and Social Affairs: New York, NY, USA, 2017. Available online: https://population.un.org/wpp/publications/files/wpp2019_highlights.pdf (accessed on 3 May 2022).
41. Roser, M. Future Population Growth. 2020. Available online: https://ourworldindata.org/future-population-growth (accessed on 5 May 2022).
42. Adam, D. How far will global population rise? Researchers can't agree. *Nature* **2021**, *597*, 462–465. [CrossRef]
43. Foure, J.; Bénassy-Quéré, A.; Fontagne, L. *The Great Shift: Macroeconomic Projections for the World Economy at the 2050 Horizon*; CEPII Working Papers, 2012-03; CEPII: Paris, France, 2012. [CrossRef]
44. Hawksworth, J.; Chan, D. The World in 2050: Will the shift in global economic power continue? *PwC Anal.* **2015**. [CrossRef]
45. OECD. Real GDP Long-Term Forecast. 2018. Available online: https://data.oecd.org/gdp/gdp-long-term-forecast.htm (accessed on 21 May 2022).

46. Wood Mackenzie. Energy Transition May Shave 2% off Global GDP by 2050. 2022. Available online: https://www.woodmac.com/press-releases/the-economic-impact-of-an-accelerated-energy-transition/ (accessed on 21 May 2022).
47. Buiter, W.; Rahbari, E. Global Growth Generators: Moving beyond Emerging Markets and BRICs. Citi Economics Forum 2011. Available online: https://www.willembuiter.com/3G.pdf (accessed on 24 May 2022).
48. Sachs, G. The N-11: More Than an Acronym. In Global Economics Paper. 2007. Available online: https://www.goldmansachs.com/insights/archive/archive-pdfs/brics-book/brics-chap-11.pdf (accessed on 25 May 2022).
49. European Environment Agency. Historic and Projected GDP in the EU, the US, the BRIICS Countries and Other Countries. 2017. Available online: https://www.eea.europa.eu/data-and-maps/figures/past-and-projected-global-economic-output-1 (accessed on 21 May 2022).
50. Pistoia, G.; Liaw, B. *Behaviour of Lithium-Ion. Batteries in Electric Vehicles. Battery Health, Performance, Safety, and Cost*; Springer International Publishing: Cham, Switzerland, 2018; ISSN 1865-3529. [CrossRef]
51. U.S. Geological Survey. *Mineral. Commodity Summaries 2021*; U.S. Geological Survey: Reston, VI, USA, 2021.
52. ITF International Transport Forum. *Transport Outlook 2017*; OECD Publishing: Paris, France, 2017; Available online: https://www.oecd-ilibrary.org/transport/itf-transport-outlook-2017_9789282108000-en (accessed on 26 May 2022).
53. Statista. *Urban Passenger Mobility Demand Worldwide between 2010 and 2050*; Statista: Hamburg, Germany, 2018.
54. Kristalina, G.; Hart, S.; Laura, T.; Elif, K.; Nicholas, B. Global Mobility Report 2017. Tracking Sector Performance. 2017. Available online: https://sustainabledevelopment.un.org/content/documents/2643Global_Mobility_Report_2017.pdf (accessed on 26 May 2022).
55. Moriarty, P.; Honnery, D. Forecasting world transport in the year 2050. *Int. J. Veh. Des.* **2004**, *35*, 151. [CrossRef]
56. International Energy Agency. *Global Transport Outlook to 2050. Targets and Scenarios for a Low-Carbon Transport Sector*; International Energy Agency: Paris, France, 2012.
57. Seixas, J.; Simões, S.; Dias, L.; Kanudia, A.; Fortes, P.; Gargiulo, M. Assessing the cost-effectiveness of electric vehicles in European countries using integrated modeling. *Energy Policy* **2015**, *80*, 165–176. [CrossRef]
58. Khalili, S.; Rantanen, E.; Bogdanov, D.; Breyer, C. Global transportation demand development with impacts on the energy demand and greenhouse gas emissions in a climate-constrained world. *Energies* **2019**, *12*, 3870. [CrossRef]
59. European Union Road Federation. Passenger Transport 2020. 2020. Available online: https://erf.be/statistics/passenger-transport-2020/ (accessed on 22 May 2022).
60. Koster, A.; Kuhnert, F.; Stürmer, C. Five trends transforming the Automotive Industry. PwC, 2018. Available online: https://www.pwc.com/gx/en/industries/automotive/assets/pwc-five-trends-transforming-the-automotive-industry.pdf (accessed on 27 May 2022).
61. Statista. *Proved Crude Oil Reserves Worldwide from 1990 to 2020*; Statista: Hamburg, Germany, 2021.
62. Statista. *Daily Global Crude Oil Demand 2006–2026*; Statista: Hamburg, Germany, 2021.
63. BP Energy outlook. *Statistical Review of World Energy*, 70th ed.; BP Energy Outlook: London, UK, 2021.
64. Eurostat. *OIL-Final-Energy-Consumption-by-Sector-EU28-1990–2014: Statistics Explained*; Eurostat: Luxembourg, 2016.
65. Statista. *Leading Oil Demanding Sectors in the OECD 2019*; Statista: Hamburg, Germany, 2021.
66. Lopez, G.; Independent Commodity Intelligence Services. Petchems Demand for Crude Oil Set to Boom Despite Rising Recycling Rates—IEA. 2019. Available online: https://www.icis.com/explore/resources/news/2019/11/13/10443245/petchems-demand-for-crude-oil-set-to-boom-despite-rising-recycling-rates-iea (accessed on 15 January 2021).
67. Hou, F.; Chen, X.; Chen, X.; Yang, F.; Ma, Z.; Zhang, S.; Guo, F. Comprehensive analysis method of determining global long-term GHG mitigation potential of passenger battery electric vehicles. *J. Clean. Prod.* **2021**, *289*, 125137. [CrossRef]
68. Xu, C.; Dai, Q.; Gaines, L.; Hu, M.; Tukker, A.; Steubing, B. Future material demand for automotive lithium-based batteries. *Commun. Mater.* **2020**, *1*, 99. [CrossRef]
69. The International Council on Clean Transportation. Race to Electrify Light-Duty Vehicles in China, The United States, and Europe: A Comparison of Key EV Market Development Indicators. 2021. Available online: https://theicct.org/publication/race-to-electrify-light-duty-vehicles-in-china-the-united-states-and-europe-a-comparison-of-key-ev-market-development-indicators/ (accessed on 24 May 2022).
70. Wang, N.; Tang, G. A Review on Environmental Efficiency Evaluation of New Energy Vehicles Using Life Cycle Analysis. *Sustainability* **2022**, *14*, 3371. [CrossRef]
71. Bouter, A.; Hache, E.; Ternel, C.; Beauchet, S. Comparative environmental life cycle assessment of several powertrain types for cars and buses in France for two driving cycles: "Worldwide harmonized light vehicle test procedure" cycle and urban cycle. *Int. J. Life Cycle Assess.* **2020**, *25*, 1545–1565. [CrossRef]
72. Puig-Samper Naranjo, G.; Bolonio, D.; Ortega, M.F.; García-Martínez, M.J. Comparative life cycle assessment of conventional, electric and hybrid passenger vehicles in Spain. *J. Clean. Prod.* **2021**, *291*, 125883. [CrossRef]
73. Ziemann, S.; Müller, D.B.; Schebek, L.; Weil, M. Modeling the potential impact of lithium recycling from EV batteries on lithium demand: A dynamic MFA approach. *Resour. Conserv. Recycl.* **2018**, *133*, 76–85. [CrossRef]
74. Pesaran, A.A.; Markel, T.; Tataria, H.S.; Howell, D. Battery requirements for plug-in hybrid electric vehicles—Analysis and rationale. In Proceedings of the 23rd International Electric Vehicle Symposium, Anaheim, CA, USA, 2–5 December 2007; McMaster University: Hamilton, ON, Canada, 2007.

75. The National Academies. *Assessment of Fuel Economy Technologies for Light-Duty Vehicles*; The National Academies: Washington, DC, USA, 2011. [CrossRef]
76. Vermeulen, I.; Helmus, J.R.; Lees, M.; van den Hoed, R. Simulation of future electric vehicle charging behavior-effects of transition from PHEV to FEV. *World Electr. Veh. J.* **2019**, *10*, 42. [CrossRef]
77. Hirz, M.; Nguyen, T.T. Life-Cycle CO_2-Equivalent Emissions of Cars Driven by Conventional and Electric Propulsion Systems. *World Electr. Veh. J.* **2022**, *13*, 61. [CrossRef]
78. Statista. *Estimated Average Battery Capacity in Electric Vehicles Worldwide from 2017 to 2025, by Type of Vehicle*; Statista: Hamburg, Germany, 2018.
79. Jung, H.; Li, C. Emissions from Plug-in Hybrid Electric Vehicle (PHEV) During Real World Driving Under Various Weather Conditions. 2018. Available online: https://rosap.ntl.bts.gov/view/dot/35123. (accessed on 24 May 2022).
80. Green Car Reports. Analysis Finds Hybrids Make Better Use of-Scarce Batteries than Pure EVs. 2019. Available online: https://www.greencarreports.com/news/1123815_analysis-finds-hybrids-make-better-use-of-scarce-batteries-than-pure-evs (accessed on 24 May 2022).
81. Car and Driver. Battery Taxonomy: The Differences between Hybrid and EV Batteries. 2021. Available online: https://www.caranddriver.com/news/a15345397/battery-taxonomy-the-differences-between-hybrid-and-ev-batteries/ (accessed on 24 May 2022).
82. Salgado, R.M.; Danzi, F.; Oliveira, J.E.; El-Azab, A.; Camanho, P.P.; Braga, M.H. The latest trends in electric vehicles batteries. *Molecules* **2021**, *26*, 3188. [CrossRef]
83. Granovskii, M.; Dincer, I.; Rosen, M.A. Economic and environmental comparison of conventional, hybrid, electric and hydrogen fuel cell vehicles. *J. Power Sources* **2006**, *159*, 1186–1193. [CrossRef]
84. Sharma, I.; Chandel, M.K. Will electric vehicles (EVs) be less polluting than conventional automobiles under Indian city conditions? *Case Stud. Transp. Policy* **2020**, *8*, 1489–1503. [CrossRef]
85. Wu, Z.; Wang, M.; Zheng, J.; Sun, X.; Zhao, M.; Wang, X. Life cycle greenhouse gas emission reduction potential of battery electric vehicle. *J. Clean. Prod.* **2018**, *190*, 462–470. [CrossRef]
86. Congress, G.C. Green Car Congress. 2018. Available online: https://www.greencarcongress.com/2017/09/20170918-diesellca.html (accessed on 30 May 2022).
87. Zeng, D.; Dong, Y.; Cao, H.; Li, Y.; Wang, J.; Li, Z.; Hauschild, M.Z. Are the electric vehicles more sustainable than the conventional ones? Influences of the assumptions and modeling approaches in the case of typical cars in China. *Resour. Conserv. Recycl.* **2021**, *167*, 105210. [CrossRef]
88. Van Mierlo, J.; Messagie, M.; Rangaraju, S. Comparative environmental assessment of alternative fueled vehicles using a life cycle assessment. *Transp. Res. Procedia* **2017**, *25*, 3435–3445. [CrossRef]
89. Hooftman, N.; Messagie, M.; Joint, F.; Segard, J.B.; Coosemans, T. In-life range modularity for electric vehicles: The environmental impact of a range-extender trailer system. *Appl. Sci.* **2018**, *8*, 1016. [CrossRef]
90. Shi, S.; Zhang, H.; Yang, W.; Zhang, Q.; Wang, X. A life-cycle assessment of battery electric and internal combustion engine vehicles: A case in Hebei Province, China. *J. Clean. Prod.* **2019**, *228*, 606–618. [CrossRef]
91. Ahmadi, P. Environmental impacts and behavioral drivers of deep decarbonization for transportation through electric vehicles. *J. Clean. Prod.* **2019**, *225*, 1209–1219. [CrossRef]
92. Greene, D.L.; Park, S.; Liu, C. Public policy and the transition to electric drive vehicles in the U.S.: The role of the zero emission vehicles mandates. *Energy Strategy Rev.* **2014**, *5*, 66–77. [CrossRef]
93. Machedon-Pisu, M.; Nedelcu, A.V.; Szekely, I.; Morariu, G.; Miron, M.; Kertész, C.Z. Energy-efficient tracking for wireless sensor networks. In Proceedings of the IEEE International Workshop on Robotic and Sensors Environments, ROSE, Lecco, Italy, 15–16 October 2010. 2009. [CrossRef]
94. Nedelcu, A.V.; Talaba, D.; Stoianovici, V.C.; Machedon-Pisu, M.; Szekely, I. Conceptual integration of wireless sensor networks with 3D virtual environments. In Proceedings of the IEEE International Conference on Wireless Communications, Networking and Information Security (WCNIS), Beijing, China, 25–27 June 2010; Volume 2. [CrossRef]
95. Machedon-Pisu, M. The impact of propagation media and radio interference on the performance of wireless sensor networks with MicaZ motes. In Proceedings of the International Conference on Optimization of Electrical and Electronic Equipment, OPTIM, Bran, Romania, 22–24 May 2014. [CrossRef]
96. Wang, J.; Shao, Y.; Ge, Y.; Yu, R. A survey of vehicle to everything (V2X) testing. *Sensors* **2019**, *19*, 334. [CrossRef]
97. Grama, A.; Petreus, D.; Borza, P.; Grama, L. Experimental determination of equivalent series resistance of a supercapacitor. In Proceedings of the 32nd International Spring Seminar on Electronics Technology, Brno, Czech Republic, 13–17 May 2009; pp. 1–4. [CrossRef]
98. Puşcaş, A.M.; Carp, M.C.; Kertész, C.Z.; Borza, P.N.; Coquery, G. Thermal and voltage testing and characterization of supercapacitors and batteries. In Proceedings of the International Conference on Optimisation of Electrical and Electronic Equipment, OPTIM, Brasov, Romania, 20–22 May 2010; pp. 125–132. [CrossRef]

99. Stanca, A.C.; Borza, P.N.; Romanca, M.; Paun, R.; Zamfir, S. Model of supercapacitor-starter assembly used for internal combustion engines starting. In Proceedings of the International Conference on Optimisation of Electrical and Electronic Equipment, OPTIM, Brasov, Romania, 20–22 May 2010; pp. 943–948. [CrossRef]
100. Carp, M.C.; Puşcaş, A.M.; Borza, P.N. Energy Management System and Controlling Methods for a LDH1250HP Diesel Locomotive Based on Supercapacitors. In *Camarinha-Matos, Technological Innovation for Sustainability. DoCEIS*; Camarinha-Matos, L.M., Ed.; IFIP Advances in Information and Communication Technology; Springer: Berlin, Germany, 2011; Volume 349. [CrossRef]

sustainability

MDPI

Review

Steel Slag and Recycled Concrete Aggregates: Replacing Quarries to Supply Sustainable Materials for the Asphalt Paving Industry

Carlos D. A. Loureiro [1], Caroline F. N. Moura [1], Mafalda Rodrigues [2], Fernando C. G. Martinho [3], Hugo M. R. D. Silva [1] and Joel R. M. Oliveira [1,*]

1 ISISE—Institute for Sustainability and Innovation in Structural Engineering, Department of Civil Engineering, University of Minho, 4800-058 Guimaraes, Portugal; id9629@alunos.uminho.pt (C.D.A.L.); id8972@alunos.uminho.pt (C.F.N.M.); hugo@civil.uminho.pt (H.M.R.D.S.)

2 DST—Domingos da Silva Teixeira, 4700-727 Braga, Portugal; mafalda.rodrigues@dstsgps.com

3 CERENA—Centro de Recursos Naturais e Ambiente, Instituto Superior Técnico, Lisbon University, 1049-001 Lisbon, Portugal; fernando.martinho@tecnico.ulisboa.pt

* Correspondence: joliveira@civil.uminho.pt; Tel.: +351-253-510200

Abstract: Various researchers are developing efforts to integrate waste and by-products as alternative materials in road construction and maintenance, reducing environmental impacts and promoting a circular economy. Among the alternative materials that several authors have studied regarding their use as partial or total substitutes for natural aggregates in the asphalt paving industry, the steel slag aggregate (SSA) and recycled concrete aggregate (RCA) from construction demolition waste (CDW) stand out. This paper reviews and discusses the characteristics and performance of these materials when used as aggregates in asphalt mixtures. Based on the various studies analyzed, it was possible to conclude that incorporating SSA or RCA in asphalt mixtures for road pavements has functional, mechanical, and environmental advantages. However, it is essential to consider some possible drawbacks of these aggregates that are discussed in this paper, to define the acceptable uses of SSA and RCA as sustainable feedstocks for road paving works.

Keywords: steel slag aggregate; recycled concrete aggregates; waste products; asphalt mixture; sustainable materials

check for updates

Citation: Loureiro, C.D.A.; Moura, C.F.N.; Rodrigues, M.; Martinho, F.C.G.; Silva, H.M.R.D.; Oliveira, J.R.M. Steel Slag and Recycled Concrete Aggregates: Replacing Quarries to Supply Sustainable Materials for the Asphalt Paving Industry. *Sustainability* **2022**, *14*, 5022. https://doi.org/10.3390/su14095022

Academic Editor: Rui Micaelo

Received: 18 March 2022
Accepted: 20 April 2022
Published: 22 April 2022

1. Introduction

The European Community aims to contribute to a radical increase in the construction sector sustainability through a paradigm shift that implements the circular economy principles [1].

A pavement's sustainability is highly influenced by the proper selection of the materials used [2]. The material circularity principle for a cleaner production of asphalt mixtures aims to completely or partially replace traditional aggregates and fillers with wastes or by-products, which can also be used on asphalt binder modifications. Adapting asphalt mixtures to incorporate waste materials reduces the consumption of naturally mined materials, minimizing the carbon footprint and pavement industry's impact on the environment. Furthermore, the use of waste materials minimizes landfills and deposition in piles and limits the disposal of enormous volumes of waste produced from various sources [3–5].

Previous research works have demonstrated the advantages of applying new recycled materials in the granular or asphalt layers of road pavements. Steel slags and construction and demolition waste are among the alternative materials that several authors have studied as partial or total substitutes for natural aggregates in asphalt mixtures [2,6–12].

Formerly considered waste material, steel slag has become a valuable by-product used as a raw material for many industries and is almost fully utilized in some countries [13]. Steel slag aggregates (SSA) can replace natural aggregates in asphalt and cement-based road construction, lowering the environmental impact by reducing the consumption of natural and non-renewable aggregates and the quantity of steel slag deposited on landfill sites [14].

The rapid development of the construction engineering sector has led to a massive and unsustainable production with an excessive consumption of natural resources. Moreover, Menegaki and Damigos state that around 35% of the global construction and demolition waste (CDW) is disposed of directly in landfills without treatment [15]. These landfills are detrimental to a significant land area and bring major environmental concerns, such as global warming and soil and water pollution [16]. Excess construction and demolition waste puts a massive strain on the environment. Therefore, recovering CDW as aggregates for the partial substitution of natural aggregates in asphalt or concrete cement roads minimizes this effect and the consumption of new materials.

Concrete is the primary waste source in CWD. This concrete can be reused as recycled concrete aggregate (RCA), prepared through the crushing and grading of waste concrete, which can be wholly or partially used as a raw material in new concrete mixtures [17]. Because the high volume of RCA produced exceeds the amount placed back into the construction cycle, finding alternative applications for RCA is essential to maintain the construction industry as globally sustainable. Hence, utilizing recycled aggregates for highway pavement construction offers tangible environmental and economic benefits [18].

Several studies and reviews [13,15,19–25] have recently focused on the technical and economic feasibility of using recycled concrete and steel slag aggregates to produce new asphalt concrete mixtures. This topic became essential for the current and future development of the road paving industry.

Therefore, this review paper thoroughly identifies and evaluates innovative, efficient, and sustainable alternatives for the current uses of SSA and RCA in asphalt mixtures, thus promoting the circularity of road paving materials. It also aligns with the sustainable development goal twelve defined by the United Nations, i.e., responsible consumption and production [26]. This paper shows that SSA and RCA are the two main by-products capable of substituting significant amounts of natural aggregates in producing asphalt mixtures, either separately or together, as "alternative quarries".

After a brief description of steel slag and recycled concrete from construction and demolition wastes, the legal and environmental aspects of using these wastes or by-products are discussed using different approaches. Then, the mechanical properties of steel slag and the main treatments of recycled concrete were the main points focused on for each alternative aggregate. Finally, the current research on the various applications of these wastes as road paving materials is presented, demonstrating their potential and efficient use as new aggregates for asphalt mixtures.

A greater demand for environmental and legal requirements in the construction industry is expected soon, and this article aims to stimulate the use of eco-efficient building materials in road construction.

2. Steel Slag Aggregates Applications in the Asphalt Paving Industry

2.1. Steel Slag Properties and Their Application in Road Pavements

2.1.1. Steel Slag Types and Main Properties

Steel slag is a synthetic aggregate produced as a by-product in the steel production process, recycled in many countries worldwide for decades. This by-product is one of the most used materials, and it can be considered a green resource because it has a great potential to substitute natural aggregates and reduce the environmental impact of virgin resources quarrying [27–29].

There are many possibilities for using steel slag in building and transport infrastructure construction [30]. Material researchers and civil engineers studied its use in construction applications. They concluded that this material could be used in broad areas, such as a raw material for blended cement manufacturing, aggregates for asphalt mixture production, or granular material for a road pavement base or subbase courses [19,31].

As mentioned by Skaf et al. [20], there are currently five main types of metallurgical slags: the iron-making slag of a Blast Furnace (BF), steel slag divided into Basic Oxygen Furnace (BOF) slag (sometimes referred to as Linz–Donawitz slag), the Electric Arc Furnace (EAF) slag, and secondary refining slag that are the Ladle Furnace (LD) and the stainless steel slag.

The BOF and EAF steel slags from different sources within Europe are generally comparable, but the properties may differ based on the steel produced and the furnace. Compared to EAF, the main problem with BOF is the excess quantity of its free lime and magnesia contents [32,33]. Although EAF steel slag aggregate (Figure 1) is the most common type used in road construction [34,35], some researchers [36–46] have developed research on applications of BOF steel slag aggregates in asphalt mixtures.

Figure 1. Steel slag aggregates from electric arc furnaces (EAF SSA).

The EAF steelmaking process is essentially a steel scrap recycling process. Therefore, the chemical composition of EAF slag depends significantly on the source and properties of recycled steel, and its chemical composition can vary from one batch to another. The main chemical components of EAF steel slag are lime (CaO), magnesium oxide (MgO), iron oxides (FeO, Fe_2O_3), silicon dioxide (SiO_2), and aluminum oxide (Al_2O_3). Therefore, the CaO–MgO–SiO_2–FeO–Al_2O_3 system can represent the steel slag. The FeO, CaO, SiO_2, Al_2O_3, and MgO contents of EAF slags are typically 10–40%, 22–60%, 6–34%, 3–14%, and 3–13%, respectively. EAF slags also contain free CaO and MgO and other complex minerals and solid solutions of CaO, FeO, and MgO [20,31,47,48].

2.1.2. Characteristics of Steel Slag Aggregates Relevant for Road Pavements

Zumrawi and Khalill [49] state that particular attention has been directed at investigating the possible use of SSA as a substitute for natural mineral aggregates to produce asphalt mixtures. The steel slag has been utilized as a coarse or a fine aggregate in asphalt pavements and had a reliable response [50]. Numerous researchers have found that the angular shape, the rough-textured surface, and the high specific gravity of steel slag provides a high skid resistance, mechanical interlocking, better stability, a high-temperature deformation resistance, and a low-temperature cracking resistance [47,51,52]. Moreover, steel slag aggregates (SSA) have shown excellent polishing resistance, meaning surface courses with SSA will keep their friction properties over time. This result was confirmed in field studies reporting that sections made with SSA have equal or better skid resistance and macrotexture than sections made with conventional asphalt mixtures produced with natural aggregates [53].

The mechanical performance of EAF slags when they replace natural coarse aggregates in asphalt mixtures has been evaluated in-depth, obtaining excellent results. Steel slag's mechanical and physical properties productively meet high-class material requirements, making it an alternative high-quality aggregate compared to the natural aggregate. A comparison of some properties of steel slag with natural aggregate is shown in Table 1. Moreover, according to several authors, the asphalt mixtures with SSA have excellent workability, durability, permeability, stability at high temperatures, fatigue and thermal cracking resistance, abrasion resistance, and increased stiffness [30,32,47,54,55].

Table 1. Technical properties of EAF steel slag against granite natural aggregates adapted from [2,13].

Characteristics	EAF Steel Slags	Granite
Bulk density (g/cm^3)	3.4–3.5	2.5–2.7
Shape—thin and elongated pieces (%)	<10	<10
Impact value (wt.%)	18	12
Crushing value (wt.%)	13	17
Micro-Deval Coefficient (wt.%)	10	15
Polished stone value, PSV (%)	51–61	48–52
Water absorption (wt.%)	0.7–1.3	0.5–0.7
Resistance to freeze-thaw (wt.%)	<0.5	<0.5
Binder affinity (%)	50–65	10–15

Steel slag has also been reported to retain heat considerably longer than natural aggregates. The heat retention characteristics of SSA can be beneficial for hot mix asphalt (HMA) production, as less energy can be used to heat aggregates in the asphalt plant being maintained for more extended periods during the asphalt paving works [20,52].

Another property that makes steel slag an excellent aggregate for asphalt mixtures is its strong affinity with bitumen. EAF slag aggregates are tough and dense, but they have excellent adhesion with bitumen due to their high alkali character. This property of steel slag helps to resist bitumen stripping and minimizes potential moisture damage to asphalt mixtures [20,32,56,57].

In general, using EAF steel slag to substitute part of the natural aggregates improves the mechanical performance of the resulting asphalt mixtures. This tendency is stated in the literature for skid resistance, Marshall stability, indirect tensile strength, stiffness, resistance to permanent deformation, fatigue, and low-temperature cracking resistance [29,34,58,59]. Table 2 summarizes previous research on SSA incorporation in different HMA and warm mix asphalt (WMA) mixtures and the corresponding advantages.

Table 2. Research work developed with the incorporation of SSA in different asphalt mixtures.

Reference	Type of Mixture	Improved Properties with SSA
Kavussi and Qazizadeh [59]	HMA	Fatigue cracking resistance
Maharaj et al. [47]	HMA	Marshall stability and surface characteristics
Pasetto and Baldo [30]	HMA	Stiffness modulus, fatigue and rutting resistance, and indirect tensile strength
Abd Alhay and Jassim [28]	HMA	Marshall stability and temperature susceptibility
Shiha et al. [29]	HMA	Marshall stability and fatigue cracking resistance
Masoudi et al. [34]	WMA	Marshall stability, stiffness, resilient modulus, and indirect tensile strength
Ameri et al. [58]	WMA	Marshall stability, tensile strength, resilient modulus, moisture resistance, and rutting resistance
Ziaee and Behnia [60]	WMA	Indirect tensile strength, resilient modulus, and dynamic creep
Keymanesh et al. [61]	Microsurfacing	Abrasion resistance, curing time, bleeding, and vertical displacement

However, the incorporation of SSA in asphalt mixtures can have some restrictions due to its chemical and physical characteristics. According to Swathi et al. [62], three of the most discussed issues associated with the use of SSA in asphalt mixtures are failures due to volume instability, increased bitumen demand due to its texture and porous structure, high air void contents, and voids in mineral aggregate (VMA) of the corresponding asphalt mixture. However, the lack of volume stability of steel slag can be improved with a pre-treatment or aging treatment [63].

Although many factors influence the stability of the EAF steel slag, there seems to be an agreement that the hydration reactions of free lime and periclase of SSA are responsible for its expansive behavior [20]. In contact with water, free MgO and CaO in steel slag will react to hydroxides. Depending on the free lime or free MgO rate, this reaction causes a slag's volume increase, mostly combined with a loss of strength and partial disintegration of the slag pieces [33]. In the free lime hydration, there may be an increase in the volume of 99%, and in the case of hydration of free magnesium oxide, the volume may increase by 119% [64,65].

The bitumen and air voids contents are the two most crucial mix design control indexes [66]. Asphalt mixtures with high steel slag aggregate contents tend to have more voids, and the optimum bitumen content can be excessive and lead to binder drainage [20]. The higher air voids content in 100% SSA asphalt mixtures can be attributed to the difference in the specific gravity of the coarse and fine fractions of SSA. Adopting the weight-based grading envelopes for asphalt mix design without considering the difference in the specific gravity, for a unit volume of the asphalt mix, few coarse aggregates may be obtained than the mixture with natural aggregates with similar specific gravity. The consequence is that the asphalt mixture with SSA tends to have more fine aggregates than the corresponding mixture with natural aggregates of a similar specific gravity, resulting in a higher VMA [62].

He et al. [63] stated that using steel slag in an asphalt mixture benefits the environment and saves cost. Nonetheless, the critical points of mix design of asphalt mixtures with SSA include the ratio of NA substitution, the gradation correction, the determination of effective relative density, and the optimum binder–aggregate ratio. The same authors mentioned that the main limiting factors (i.e., large density, poor volume stability, and the increasing binder content) had not been fundamentally resolved, and the long-term performance and quality control system of pavements with SSA should be further researched.

2.2. Legislation on the Use of Steel Slag

Despite the general use of steel slag in construction, there has been an ongoing legal argument about classifying steel slags as waste or by-products. The classification of steel slags is not uniform within the European member states. Some steel slag types are considered by-products in several countries and waste in others [67].

According to the European Waste Directive 2008/98/CE, later amended by Directive (UE) 2018/851, steel slag is initially classified as waste, leaving this condition after processing, meeting technical criteria, and proving no risks to health and the environment. Thus, the SSA is a by-product obtained from various stages of steel slag beneficiation, has a well-defined market and demand, and meets the legislation, technical standards, and criteria so that its intended applications do not cause adverse impacts on the environment or human health. Therefore, it aligns with Article 5 of the European Waste Directive concepts and can be legally framed as a by-product by meeting the following conditions:

(a) further use of SSA is certain;
(b) SSA can be used directly without any further processing other than regular industrial practice;
(c) SSA is produced as an integral part of a production process; and
(d) further use of SSA is lawful, fulfilling all the requirements for its specific use without causing adverse environmental or human health impacts.

The same waste framework directive (WFD) also introduced the end of waste status in Article 6 for wastes undergoing recycling or recovery operations but falling outside the definition of by-products according to Article 5. Therefore, the waste must be applied with specific objectives and assure a market or demand. Furthermore, it should also fulfill the technical requirements and applicable legislation without compromising the environment or human health.

Figure 2 shows the different classifications of steel slag according to Articles 5 and 6 of the European Waste Directive 2008/98/EC, later amended by Directive (UE) 2018/851.

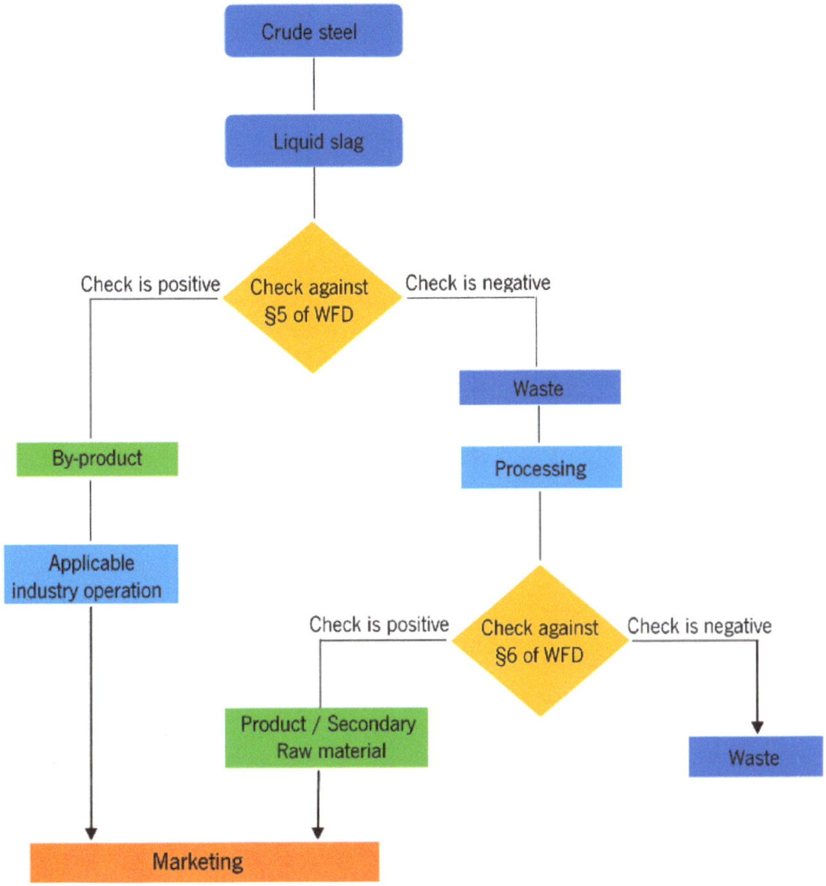

Figure 2. Different classifications of steel slag according to Articles 5 and 6 (§5 and §6) of the European Waste Directive 2008/98/EC, amended by Directive (UE) 2018/851 adapted from [68].

The same Directive mentions a list of waste (LoW) in Article 7, created to identify wastes from various activities, called the European list of waste. The list consists of 20 chapters, numbered from 01 to 20, which groups waste that concerns a specific activity area that generates waste. The term steel slag has two references in the LoW, being considered a waste from thermal processes in the steel industry, identified by code 10 02 01 (wastes from the processing of slag) and 10 02 02 (unprocessed slag). However, the existence of several stages for the processing of steel slag automatically excludes them from item 10 02 02, once it is processed, becoming a new material with economic value. Moreover, such materials cannot be classified under code 10 02 01 since they are classified as by-products according to Articles 5 and 6. Thus, it can be inferred that the European waste list does not cover the steel slag aggregates that have undergone further processing steps and, therefore, should not be classified as waste [69].

2.3. Environmental Evaluation of Steel Slag Aggregates

Replacing natural aggregates with wastes and by-products has several environmental benefits. On the one hand, it reduces the production of raw materials, the consumption of water, electricity, diesel, and the emission of noise and dust. On the other hand, waste disposal in landfills is avoided, extending the landfill's life, and reducing emissions. Besides, this solution represents considerable cost reductions for both generators and consumers since the industries, in most cases, do not intend to make profits through the commercialization of steel slag aggregates precisely because they are not their main product [69,70].

The mechanical performance of asphalt mixtures incorporating EAF steel slags to replace natural coarse aggregates has been evaluated in-depth, obtaining excellent workability, stiffness, fatigue and rutting resistances, and moisture damage performance. However, the environmental aspects of this solution have not been studied in such detail [70].

According to Mombelli et al. [71], the use of EAF SSA in construction is limited by some polluting chemical elements in their composition, such as chromium [72], barium, and vanadium, which can be dangerous to humans and the environment. Nevertheless, Motz and Geiseler [33] state that the assessment of aggregates' environmental compatibility as a building material is not determined by the content of environmentally relevant elements in the solid material but by the potential leaching behavior.

Gan et al. [73] studied the risk of leaching of heavy metal elements during the steel slag stockpiling process and after paving asphalt mixtures. They concluded that the leaching concentration of heavy metals (particularly Cu) from steel slag increased rapidly due to the acid rain effect. Increasing the specific surface area and decreasing pH value will increase the leaching risk of heavy metals in SSA. The leaching risk of SSA was significantly reduced after being introduced into asphalt mixes. Another work [74] focused on the leaching behavior study of EAF slag to assess its suitability as a construction material because it contains toxic metals that can leach and affect the eco-system. It was found that toxic metals were not leached beyond the permissible limits for pH values between 4.0 and 8.5. By applying different leaching test methods, these authors concluded that the leaching of hazardous heavy metals from EAF slag is negligible or within permissible limits, confirming it can be used as an eco-friendly construction material.

Moreover, Li et al. [75] stated that steel slag is alkaline in an aqueous solution due to its basic oxide components. They studied the pH value of the short-term and long-term deposited slag leaching solution and concluded that it was below the specifications and, hence, was not considered as hazardous waste. However, the specification limit was exceeded for ground steel slag, highlighting the importance of adequate storage of SSA.

A few studies quantified the environmental performance of incorporating EAF steel slag as an aggregate in asphalt mixtures using the Life Cycle Assessment (LCA) methodology. Esther et al. [70] analyzed the environmental impact of replacing high-quality coarse aggregates with EAF SSA. The researchers performed an LCA on three asphalt mixtures containing different aggregates, one ophite, and two steel slag types. The results showed the importance of the aggregate absorption rate and the humidity, and steel slags can replace natural aggregates even when located up to 144 km from the asphalt plant.

Another study analyzed the environmental impact of using EAF SSA in road pavements compared with the natural materials used in road construction. The authors verified that some of the most relevant environmental impacts, such as carbon footprint, ozone layer depletion, abiotic depletion, and photochemical oxidation, depend highly on the road construction technologies used. Moreover, they concluded that significant environmental benefits could be obtained from using SSA in road construction [14].

Mladenovič et al. [76] carried out an LCA to compare the environmental impacts of asphalt surface course construction in two scenarios: a conventional scenario (using siliceous aggregates) and an alternative scenario (using SSA). The Life Cycle Assessment results (based on consequential modeling) showed that the alternative scenario is more sustainable than the conventional one if the following impact categories are considered: acidification, eutrophication, photochemical ozone creation, and human toxicity. The impacts are reduced by approximately 80% in the above indicators compared to the conventional scenario. The authors found that the alternative scenario is more sustainable even when considering long SSA delivery distances (around 100 km) and minimal delivery distances of the siliceous aggregate. Nevertheless, LCA results changed when the delivery distance exceeded 160 km due to the high SSA density.

2.4. Case Studies of Steel Slag Aggregate Incorporation in Asphalt Mixtures or Road Pavements

2.4.1. Incorporation of Steel Slag Aggregate in Conventional HMA

The initial steel slag applications occurred in the late 1960s when Canada and the United States constructed the first test roads using steel slag to evaluate the pavements' stability. From the 1970s to the middle 1980s, Baltimore City, in the United States, paved many sidewalks with steel slag asphalt mixtures. From 1990 to 1995, the total steel slag asphalt mixtures applied in New York reached about 250 thousand tons. Furthermore, it is reported that steel slag aggregates have been produced and used successfully as a road construction material in various European countries due to their advantageous technical properties [77].

Motevalizadeh et al. [78] successfully employed different EAF steel slag types containing coarse and fine aggregates to substitute mineral aggregates in asphalt mixtures. Rohde [79] also studied the possibility of using steel slag from electric arc furnaces in different layers of road pavements. The study results showed that EAF SSA could be used as an aggregate in asphalt mixtures, presenting no risks to the environment and public health when adequately treated. The same author found that a 40 cm layer of graduated basalt gravel has elastic deformations equivalent to a 25 cm base layer with EAF SSA. Thus, the economic advantage of reducing the granular base layer by 15 cm when using steel slag aggregates was demonstrated while maintaining the mechanical characteristics.

Ahmedzade and Sengoz [80] determined the influences of steel slag utilization as a coarse aggregate on the HMA properties. Four different asphalt mixtures containing two binders and coarse aggregates (i.e., limestone, steel slag) were produced to obtain Marshall specimens and determine the optimum bitumen content. The mechanical characteristics of all mixtures were evaluated by Marshall stability, indirect tensile stiffness modulus, creep stiffness, and indirect tensile strength tests. The researchers observed that steel slag used as a coarse aggregate improved the mechanical properties of asphalt mixtures.

A more comprehensive study on this topic also evaluated the effectiveness of using SSA to improve the engineering properties of asphalt mixtures. For this purpose, thirteen mixtures were tested containing steel slag in a portion of fine or coarse aggregates or all the portions. The effectiveness of the SSA was judged by the improvement in Marshall stability, indirect tensile strength, resilient modulus, and fatigue life of the asphalt mixtures samples. It was found that replacing 50% of the limestones coarse or fine aggregates with SSA improved the mechanical properties of the mixtures [50].

Kavussi and Qazizadeh [59] evaluated the fatigue behavior of asphalt mixtures containing EAF SSA. Asphalt mixture samples were produced, replacing various coarse limestone aggregate fractions (greater than 2.36 mm) with EAF SSA. Fatigue testing was performed in a controlled stress mode at various stress levels to characterize the fatigue behavior of the asphalt mixes. The results showed that incorporating EAF SSA in asphalt mixtures considerably improved the fatigue life of tested samples.

Pasetto and Baldo [30] verified the possibility of using EAF steel slags to substitute the natural aggregates (NA) in the composition of surface course asphalt mixtures for flexible pavements. The researchers evaluated the asphalt mixtures' performance through gyratory compaction, permanent deformation, stiffness modulus tests at various temperatures, fatigue, and indirect tensile strength tests. The asphalt mixtures with EAF SSA have generally presented better mechanical characteristics than those of the corresponding mixtures produced with natural aggregates.

Kim et al. [81] compared the behavior of HMA containing SSA or natural aggregates. They observed that the dynamic modulus of mixtures with SSA remained higher than that containing natural aggregates because of a favorable SSA interlock due to its high strength and the suitable grain shape. Sorlini et al. [54] and Zumrawi and Khalill [49] also evaluated the use of SSA as a substitute for NA in the production of HMA and observed that the incorporation of SSA significantly improves the properties of HMA.

A group of researchers from Germany [82] carried out a comprehensive investigation of the performance-based properties of different asphalt mixtures produced with 100% Linz–Donawitz slag aggregate compared to conventional mixtures with NA. They concluded that, in most cases, mixtures produced with this type of SSA perform better than conventional ones, making them suitable for flexible pavement construction.

2.4.2. Incorporation of Steel Slag Aggregate in WMA Mixtures

Several researchers also recommend using EAF SSA to produce WMA mixtures. Ameri et al. [58], Goli et al. [83], and Masoudi et al. [34] developed laboratory studies using coarse EAF SSA aggregates in WMA mixtures. In all studies, the authors verified that using coarse SSA in WMA mixtures enhances the mechanical performance of the mixture, including Marshall stability, tensile strength, resistance to moisture damage, resilient modulus, and resistance to permanent deformation. Furthermore, WMA mixtures containing SSA as the coarse portion of aggregates have a better fatigue performance than HMA mixtures. The authors concluded that WMA mixtures containing EAF steel slag aggregates are recommended as eco-friendly, economical, and suitable mixtures for road paving industries.

Ziaee and Behnia [60] investigated the use of EAF steel slag as a substitute for coarse aggregates in WMA mixtures, replacing 0%, 25%, 50%, and 75% of natural coarse limestone with EAF SSA. The authors concluded that replacing 25% to 50% of the natural coarse aggregates with EAF steel slag can optimize the WMA mechanical properties.

2.4.3. Evaluation of Different Incorporation Ratios of Steel Slag Aggregate

The literature review shows no standard rule for the proportion of steel slag used as a natural aggregate substitute, but satisfactory results are found for replacement rates between 20% and 100% [32]. Several studies investigated the ideal percentage of natural aggregate substitution for steel slag aggregate.

Asi et al. [84] studied the replacement of 0%, 25%, 50%, 75%, and 100% of the coarse limestone aggregate by SSA in asphalt concrete mixtures. The authors verified an improved mechanical performance of the asphalt mixtures when SSA replaced up to 75% of the coarse limestone aggregates. Kasaf and Prastyanto [85] studied HMA mixtures with steel slag ratios of 0%, 50%, 80%, and 100% as coarse aggregates. The best Marshall stability results were obtained using 80% steel slag as coarse aggregates.

Abd Alhay and Jassim [28] applied SSA with percentages of 10–40% of the weight of the asphalt concrete mixtures produced. The results revealed improved mechanical properties mainly when adding 30–40% SSA percentages, increasing Marshal stability from 8 to 15 kN. Behiry [57] studied the effect of using different ratios of steel slag combined with limestone aggregates to improve the mechanical properties and the resistance to failure factors of unbound layer mixtures. The author verified that the highest density and strength of the subbase layer were obtained for a blend of 70% SSA to 30% limestone aggregates. Moreover, the resistance to the deformation of the mixture increased proportionally to the steel slag content.

Studies were also developed to evaluate the possibility of producing asphalt mixtures incorporating 75% SSA. Rodrigues [86] concluded that steel slag could replace up to 75% of the natural aggregates used in asphalt mixtures but reduced the fatigue cracking resistance, implying a slight increase in the pavement thickness. Nascimento et al. [87] and Moura et al. [2] evaluated the mechanical performance of asphalt mixtures with 75% SSA. They concluded that mixtures incorporating steel slag might present slightly higher air void contents due to their lower workability. Nevertheless, those mixtures showed higher moisture and rutting resistance than conventional ones with natural aggregates, confirming the feasibility of using SSA as an alternative aggregate for asphalt mixtures.

2.4.4. Combined Use of Steel Slag Aggregate with Other Waste or By-Products

Some researchers have also combined SSA with other waste or by-products to develop more sustainable asphalt mixtures following a circular production system. Chen et al. [88] stated that it is worth further studying the addition of two types of recycled materials or by-products in asphalt mixtures to meet the mix design requirements (e.g., substituting coarse or fine aggregates or filler). Oluwasola et al. [35] investigated the rutting potential and skid resistance of HMA incorporating EAF SSA and copper mine tailings. Four different mixtures containing different proportions of copper mine waste and EAF steel slag were investigated. The results showed that the mixture with 20% copper mine tailings and 80% EAF steel slag had the highest skid number, intermediate texture depth, and the lowest rut depth.

Viana [89] studied the feasibility of producing asphalt mixtures using high amounts of SSA and reclaimed asphalt pavement (RAP). The excellent performance of those asphalt mixtures, assessed with laboratory tests (water sensitivity, wheel tracking, fatigue), revealed that the joint incorporation of SSA and RAP is a viable and successful solution for producing sustainable asphalt mixtures. Pasetto and Baldo [90] analyzed the rutting and moisture susceptibility of asphalt concrete mixtures incorporating RAP and EAF SSA, using these materials to partially substitute limestone aggregates at different proportions (up to 70% of their weight). The experimental results were highly satisfactory for all the asphalt mixtures produced with SSA and RAP regarding fulfilling technical acceptance requisites (i.e., air voids, ITS) for road paving applications. Furthermore, these mixtures were characterized by low water damage and a high permanent deformation resistance, demonstrating their good durability.

Fakhri and Ahmadi [53] investigated the effects of using SSA and RAP in warm asphalt mixtures. Therefore, six WMA asphalt mixtures with two contents of coarse steel slag aggregates (0% and 40%) and three contents of fine RAP material (0%, 20%, and 40%) were produced. Overall, the authors concluded that simultaneous use of SSA and RAP materials in WMA mixtures was an environmentally and economically friendly option with a comparable or even better performance than conventional WMA mixtures produced with natural aggregates.

Crisman et al. [72] researched the influence of using recycled crumb rubber (CR) when producing asphalt mixtures with SSA. The asphalt mixture was modified using the "dry" method for CR incorporation. The results indicated a significant stiffness increase at high temperatures (up to 30%), a slight reduction at low temperatures (up to 8%), and a reduction in permanent deformation under cyclic loads after incorporating CR in the mixtures with SSA. The authors stated that SSA has a higher bulk specific gravity than natural aggregates and a highly porous surface that permits a distinct interaction with the bitumen and the crumb rubber compared to natural aggregates. These factors justify the improved performance of the asphalt mixtures with SSA when modified with CR.

2.4.5. Incorporation of Steel Slag Aggregate in Other Types of Asphalt Mixtures

Steel slag has also been used as aggregates in other asphalt mixtures. Some authors [66,72,91,92] explored the feasibility of using steel slag as aggregates in Stone Mastic Asphalt (SMA) and concluded that SSA improves the characteristics of SMA mixtures. The authors state the performance is superior to that of SMA mixtures only with natural aggregates, highlighting the high stiffness, excellent friction resistance, better volumetric characteristics, and dynamic stability in situ with twice the value of a mixture with natural aggregates. SMA mixtures with steel slag used in an in-service pavement also presented excellent surface characteristics, including roughness and British Pendulum Number.

Another study was conducted using two open-graded aggregate gradations to investigate the use of SSA in porous asphalt (PA) mixtures. Those mixtures were tested for the resilient modulus, the rutting resistance, and permeability and were later compared with similar mixtures with natural aggregates. The resilient modulus and the rutting test results were significantly different, with the PA mixture with steel slag aggregate performing better. However, in this study, the porous asphalt mixture produced with conventional aggregates had higher permeability values than the PA mixture with steel slag [52].

Keymanesh et al. [61] studied the feasibility of using steel slag rather than conventional filler materials for a microsurfacing mix design to improve the mixture's ultimate performance. Microsurfacing mixtures were prepared with five compositions containing EAF steel slag filler at 0%, 25%, 50%, 75%, and 100% as the replacement filler (passing the 0.075 mm sieve). The tests showed that the mixtures with EAF steel slag satisfied the standard requirements and were compatible with the bitumen emulsion, thus obtaining the desirable mechanical, chemical, and physical properties. Furthermore, the authors verified that the mixture containing 50% EAF steel slag filler obtained the best performance among all studied mixtures.

Several authors [93–97] recently evaluated the incorporation of SSA in dense or porous asphalt mixtures to promote induced crack healing per induction or microwave radiation heating. These works with SSA demonstrated the importance of taking advantage of specific features of this by-product (e.g., electromagnetic capacity) to obtain new multifunctional products. In these solutions, steel slag will work as aggregate and healing/heating promoter in the asphalt mixtures.

3. Recycled Concrete Aggregates Applications in the Asphalt Paving Industry

3.1. Characteristics of Construction and Demolition Waste

The construction sector is associated with almost 36% of waste production volume in Europe and close to 67% in the United States [21], with the corresponding potential environmental impacts. This reality represented the production of around 924 million tons of CDW in the European Union in 2016 and 2.36 billion tons only in China in 2018 [98].

In addition to the very significant quantities of CDW produced, these materials have other characteristics that make their management and recycling difficult, as typified in Figure 3, primarily due to:

- A heterogeneous constitution with fractions of several size gradings and levels of hazard;
- Scattered origins in terms of geography;
- Occasional or temporary production at each place of origin considering the construction works' temporary nature.

Figure 3. Some of the difficulties that CDW recycling has faced.

These problems lead to problematic quantification, frequent uncontrolled deposition, and systems supported by end-of-line treatments promoted by the low exploitation of mineral resources and landfill costs. This situation is challenging to revert and has very high environmental, social, and economic impacts, such as the environmental and visual impacts of illegal disposal and the costs to eliminate them [99]. However, some advantages and benefits of possible CDW recycling, such as those mentioned in Table 3, will contribute to the increased use of more sustainable materials and a cleaner environment.

Table 3. Advantages/benefits and drawbacks of CDW recycling adapted from [22].

Advantages	Benefits
It prevents the depletion of natural resources	Protection of natural habitats
Minimizing dependency on raw materials	Minimization of consumption of natural resources Prioritize ready-to-use materials
Reduction in necessary financial resources	Reduction in energy costs used in the extraction, processing, and transport of natural materials
Elimination of waste deposits	Minimization of greenhouse gas emissions Reductions in water and air pollution
Minimization of disposal expenses	Decrease in transport and disposal costs
Environmental protection	Contribution to climate change mitigation
Drawbacks	
Heterogeneity in the CDW composition Difficulties in pre-screening CDWs It may have contamination Uncertainties about the standards to be met Processing and crushing equipment may not be suitable Lack of incentives from some public entities	

3.2. Properties and Treatments of Recycled Concrete Aggregates Applied in Road Pavements

3.2.1. Properties of Recycled Concrete Aggregates Applied in Road Pavements

Concrete is globally the second most widely used material, after water, with an annual global production estimated at 25 billion tons in 2009 [100,101]. This situation generates significant impacts on the environment and society in general. One of the problems is cement production, which accounts for 4.4% of global annual industry emissions. Another problem is the end-of-life of concrete materials and structures since these materials are landfilled after demolition, generating large amounts of CDW—more than 850 million tons/year in the European Union [22]. Therefore, one way to minimize this problem is to incorporate RCA into paving materials to produce sustainable asphalt mixtures [102].

Several researchers have been pointing out some ways to process and incorporate recycled concrete aggregates (Figure 4) in different types of asphalt mixtures during the last decade, namely in hot mix asphalt [103–106], warm mix asphalt [107,108], stone mastic asphalt [103], foamed asphalt mixtures [109], and emulsified asphalt mixtures [110].

Nowadays, the main concepts involved in the comprehensive waste management process should prioritize strategies favoring implementing the most recent 7 R's rule: Rethink, Redesign, Reuse, Repair, Remanufacturing, Recycling and Recover [111]. These principles may also explain why incorporating recycled concrete aggregates in asphalt mixtures or other layers for flexible pavements, or even in other types of work, has also been recently described in several articles and technical documents [112].

Figure 4. Recycled concrete aggregate (RCA) resulting from CDW processing.

A suitable way to accomplish the goals mentioned above is to reuse RCA in asphalt mixtures. However, recycled concrete aggregates usually have a high porosity and water absorption, which has been reported as the main reason for increasing bitumen content in asphalt mixtures. Meanwhile, some recycled aggregate fractions usually have many angular and rough-textured particles. Thus Marshall stability tends to increase, and Marshall flow tends to decrease [113].

Considering the general requirements listed in the relevant standards, the adequacy of these recycled concrete aggregates resulting from construction and demolition waste processing shall be evaluated to be used in asphalt mixtures. The selection and assessment of RCA quality must be made for each type of asphalt mixture and layer to offer the best mechanical and functional performances; namely, regarding the quality of asphalt, the resistance to raveling and stripping, the resistance to weathering, the resistance to freezing and thawing, the bitumen absorption, the skid resistance, the compaction, the strength and stability, and the resistance to fragmentation [114]. In order to ensure adequate performance levels in these tests, Martinho et al. [115] reported that different natural, artificial and recycled aggregates could be combined and used in more sustainable asphalt mixtures.

Table 4 shows some of the relevant properties of recycled concrete aggregates for their incorporation in asphalt mixtures [109,116–120].

Table 4. Some RCA properties reported by several authors.

Properties	References						
	[116]	[109]	[108]	[119]	[117]	[118]	[120]
Flakiness index, FI (%)	3.4	-	-	6.0	-	4.5	34.0
Sand equivalent, SE (%)	-	-	62	30	-	77	67
Los Angeles fragmentation, LA (%)	19	28	-	43	-	38	34
Micro-Deval abrasion loss, M_{DE} (%)	-	-	-	-	24	-	-
Bulk specific gravity, G (Mg/m^3)	2.52	2.28	2.50	2.30	2.30	2.64	2.63
Water absorption, WA_{24} (%)	4.8	5.8	1.0	6.1	5.9	7.0	6.1
Flat and elongated particles (%)	-	-	-	-	2.9	-	-
Porosity, ϕ (%)	-	-	-	-	13.6	-	-

Đokić et al. [121] analyzed the mineralogical–petrographic and physical–mechanical properties of RCA, the natural aggregate (dolerite), and their combination (RCA rates between 15% and 60%). They confirmed that the granular mixtures presented good wearing and fragmentation resistance (M_{DE} 14–15%, LA 22–27%) and an acceptable Polishing Stone

Value (PSV) of 55–57. They concluded that RCA could be used in asphalt mixtures or flexible pavement layers and several traffic loads.

Moreover, as Galan et al. [104] reported, among other differences in using RCA in HMA, a 50 % higher duration on the mixing time is required until a 100% bitumen coating is reached, and the HMA compaction is more difficult due to greater internal friction on the aggregate. However, these authors also concluded that using RCA rates up to 60% improves the mechanical performance of the HMA.

The recycled concrete aggregates also tend to improve the rutting resistance of asphalt mixtures, but some works analyzed by Pasandín and Pérez [122] mention that this parameter can decrease when high percentages of RCA are used.

Regarding the resistance to fatigue cracking, using these recycled concrete aggregates usually slightly reduces the performance of HMA mixtures [123], even though this tendency was not observed in some studies mentioned by Pasandín and Pérez [122].

3.2.2. Treatments of Recycled Concrete Aggregates for Road Pavement Application

The RCA is included in the European Waste Catalogue (EWC) under code number 17 01 07 and, in general, has lower quality when compared with other natural aggregates. Therefore, many treatments [22,106,122,124] have been considered to avoid the consequences of using RCA material with inferior properties on asphalt mixtures. Table 5 presents some of the treatments studied and proposed by different researchers and their effects.

Table 5. Some treatments for RCA and their effects adapted from [22,122].

Treatments	Effects	Ref.
Double coating with cement slag paste and "Sika Tite-BE"	Decrease in water absorption and a marked increase in stiffness and moisture resistance	[125]
Microbial carbonate precipitation	Compressive strength increases (up to 40%), and water absorption decreases (up to 27%)	[126]
Pre-soaking with hydrochloric acid, nitric acid, and sulfuric acid	Increase in the compressive strength	[127]
Coating with bitumen emulsion (5%)	Improvement in stripping resistance	[128]
Coating with waste plastic bottles	Reduces water absorption and improves its mechanical behavior	[124]
Activation by organic silicon resin	Improvement in the dynamic stability of asphalt treated base	[129]
Curing at 170 °C in the oven	Improvement in moisture resistance	[130]
Modification by calcium carbonate bio deposition	A decline in water absorption	[131]
Modification with liquid silicone resin	Improved low-temperature flexibility and higher moisture and rutting resistance	[132]
Precoating with cement slag paste	Resulting in high pore contents, absorption of water, and asphalt content	[133]
Calcination process	Transform RCA calcium carbonate into lime	[134]
Silica fume solution and ultrasonic cleaning	Increase in compressive strength	[135]
Carbonation and hydrochloric acid	Significantly reduced RCA porosity	[136]

3.3. Legislation on the Use of Recycled Aggregates from CDW

The first European Directive (2008/98/EC) regarding the use of construction and demolition waste already defined a minimum threshold increase to 70%, by 2020, in respect of preparation for reuse, recycling, and other forms of material valorization for non-hazardous CDW (except natural materials as defined in the category 17 05 04 of the EWC).

In this regard, the situation in Europe remains very heterogeneous, with the CDW recovery rate ranging from less than 10% to more than 90% [137]. Therefore, these authors claim that it is essential to create conditions for secondary materials generated from the recycling of CDW to be effectively integrated into the market and used in high added value applications, dynamizing this market and promoting a circular economy in the construction sector.

Thus, the European Commission prepared a protocol in 2016 [138] to strengthen the confidence in the CDW management and the quality of recycled materials obtained. The Commission thought that this objective would be achieved as follows:

- Improving the identification, separation of the origin, and collection of waste;
- The improvement of waste logistics;
- The advance in waste processing;
- Quality management;
- The appropriate policy and framework conditions.

In particular, the importance of improving the identification, separation of the origin, and collection of CDW have been highlighted by several authors as one of the most relevant features affecting the general acceptance and use of these alternative materials [139–143].

In 2018, the European Directive 2018/851 amended the Waste Framework Directive 2008/98/EC. Among other changes, this new document replaced the concept of a "European recycling society" with that of a "European circular economy", and a new disposition was introduced to promote selective demolition. Consequently, hazardous substances must be safely removed, and selective reuse and high-quality recycling be facilitated by establishing a sorting system for CDW, including mineral fractions (e.g., concrete, bricks, tiles, ceramics, and stones). This objective's fulfillment is expected to contribute to obtaining RCA with better quality in the future.

Nowadays, recycled aggregates obtained from the CDW can also have the CE mark under the Regulation EU No-305/2011 of the European Parliament and Council because they can be used in different civil construction activities, replacing natural aggregates extracted from quarries.

Nonetheless, there is still an urgent necessity to implement measures to promote new applications of these recycled materials by all private and public owners that should contemplate the mandatory incorporation of a minimum percentage of recycled aggregates in the works' technical specifications. Thus, national environmental agencies have prepared guides that allow municipalities and companies to improve CDW streams' management and effectively fulfill their legal obligations [144–146].

3.4. Environmental Evaluation of Recycled Concrete Aggregates from CDW

Sun et al. [147] studied the workability and fracture toughness of natural aggregate and recycled concrete aggregate combined with a blowing agent through an environmental study. The acceptability of these materials was investigated and discussed based on their mechanical, micro-mechanical, and ecological performance evaluation through compressive strength, flexural strength, X-ray diffraction (XRD), scanning electron microscope (SEM), and CO_2 emission tests. The results indicated that natural aggregates (NA) have better compressive strength performance, while recycled concrete aggregates have improved flexural strength. Finally, the CO_2 emission per unit of NA was higher than RCA, indicating that using recycled concrete aggregate over other conventional resources will reduce energy consumption and meet the goal of being environmentally friendly.

Another approach to evaluate the viability of using recycled aggregates in asphalt mixtures consists of their LCA [148,149]. The work of Nwakaire et al. [149] on the use of RCA for sustainable highway pavement applications included a complete diagram with the life cycle phases to be considered for their use in road pavements (Figure 5). The authors concluded that RCA could be fully used in pavement lower layers and is a sustainable substitute for natural aggregates for highway pavements.

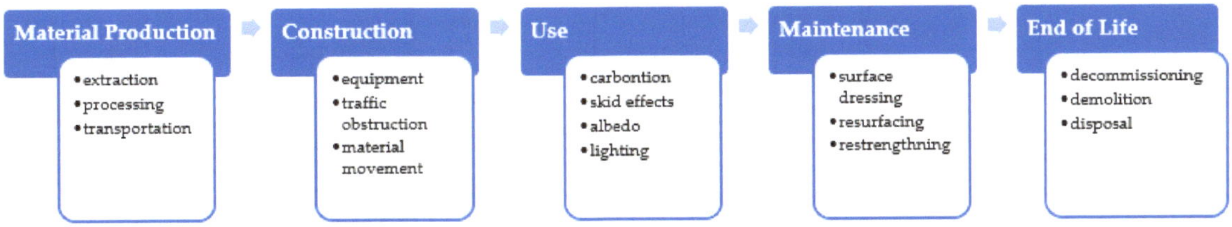

Figure 5. Life cycle phases for road pavements and factors/processes considered (adapted from Nwakaire et al. [149]).

Another example considered a cradle-to-laid LCA method to evaluate the prospective environmental impacts related to the use of RCA in an HMA for a binder course [148]. This study identifies the main processes and the system boundaries, taking into account the following four life cycle phases:

- The aggregates production and transportation to the asphalt plant;
- The production at the asphalt plant;
- The asphalt mixture transportation to the construction site;
- The pavement construction.

These authors defined three percentages of RCA replacements, ranging from 15% to 45%, and used data collected from Colombian contractors to model the foreground system. Next, they used the SimaPro software for modeling the processes, and all the life cycle inputs and outputs related to the functional unit were considered for potential impacts studied with the TRACI v.2.1 impact assessment methodology. According to their conclusions, the HMA with 15% and 30% RCA can be considered eco-friendly alternatives to the HMA with NA. However, the HMA with 45% RCA presented a lower environmental performance. Nevertheless, this work demonstrated the advantages of using recycled concrete aggregates processed from CDW in new asphalt mixtures [148].

3.5. Case studies of Recycled Concrete Aggregates Incorporation in Asphalt Mixtures or Road Pavements

3.5.1. Incorporation of Recycled Concrete Aggregate in Conventional HMA

As already well recognized, one possible approach for using recycled concrete aggregates processed from CDW is to use them as an alternative to natural aggregates, particularly in hot asphalt mixtures.

As concluded in the final report of the IRCOW project [150], the main reasons that have led many European public authorities to pay more attention to the CDW problem were the following:

- CDW is an essential source of waste in the EU and can be reused or recycled;
- Directive 2008/98/EC, from 2008, amended by Directive (UE) 2018/851, already indicated a target of 70% for the reuse of these materials by 2020;
- Recycling and reusing CDW saves natural resources and energy and can be cheaper than natural aggregates.

Another worthy effort to encourage the reuse of recycled aggregates was given by the DIRECT-MAT project, developed between 2009 and 2011 and involved twenty partners from fifteen European countries [151].

Many projects and research studies have followed this CDW recycling or reuse principle, resulting in numerous reports and technical articles over the last two decades. Since one of the materials being addressed in this review paper is recycled concrete aggregate, the main conclusions of some of the most recently released studies conducted by different authors on diverse RCA applications are presented below.

Sánchez-Cotte et al. [17,152] evaluated the incorporation of RCA of two different origins (CDW of a building; CDW of a rigid pavement) on HMA mixtures, assessing the RCA's physical and chemical properties and comparing the performance of the mixtures produced with both RCA and NA aggregates. They used different techniques to evaluate these aggregates and their influence on HMA: X-ray fluorescence, XRD, UV spectroscopy, and atomic absorption spectrometry. They replaced NA with RCA at 15%, 30%, and 45% for each coarse fraction. The mineralogical tests showed that the potential reactions in asphalt mixtures by nitration, sulfonation, amination of organic compounds, and reactions by alkaline activation in the aggregates could be neglected. These authors concluded that coarse RCA could be used in asphalt mixtures without affecting their chemical stability.

Furthermore, the laboratory test results showed that the HMA with and without RCA offered a similar behavior. However, RCA promotes more significant environmental benefits and potential cost savings. Nevertheless, the authors also stated that the HMA performance is strongly associated with the RCA source and dosage. Finally, similarly to other authors' conclusions, they note that HMA with RCA induces a higher optimum binder content (OBC) than NA.

Pasandín and Pérez [122] published an extensive and comprehensive review of the properties of HMA with RCA. These authors found that the laboratory results had significant variations, probably due to RCA's different origins. These aggregates include cement mortar, promoting a higher water/bitumen absorption, and a lower mixture quality. Most of the analyzed studies also reported a high tendency for stripping on mixtures with RCA. As previously mentioned in Section 3.2, some treatments were also identified which minimize this problematic characteristic and increase the moisture resistance of mixtures. However, other aspects need to be analyzed, including their economic and environmental impacts and practical feasibility at the asphalt mixing plant. These authors also documented the costs of using RCA in asphalt mixtures (e.g., with a higher bitumen consumption, lower density, and lower environmental impacts) and the lack of technical specifications as two critical aspects to take into account.

Regarding the specifications, the Marshall method can underestimate HMA properties because compaction can break some RCA particles and some countries use the same limits for natural aggregates in this test. Pasandín and Pérez [122] also mentioned that some studies point out that RCA can be used on the pavements of low-traffic roads and favor sustainable growth. Thus, they have concluded that new specifications must frame RCA's use in asphalt mixtures to increase the application success in trial sections, defining the type of roads and heavy traffic categories suitable to each use.

Tahmoorian and Samali [114] highlighted that asphalt mixtures include more than 65% coarse aggregates, and the large quantities of RCA available in different construction sites turn their use in asphalt mixtures to almost mandatory. In their laboratory work, they assessed the suitability of RCA and basalt to be incorporated in asphalt mixtures as coarse aggregates. In the tests performed, these authors [114] confirmed that, compared to basalt, the RCA provides better workability, compaction, and permanent deformation resistance to the asphalt mixtures. The test results also revealed that the RCA exhibits more absorption and wet/dry resistance variation than conventional aggregates and that the RCA still complies with aggregate requirements for asphalt mixtures in the remaining tests.

Gopalam et al. [153] also developed an experimental study to evaluate the influence of the binder type on the performance of dense graded asphalt mixtures that included RCA replacing NA in the HMA coarse fraction. They produced different Marshall samples of dense asphalt macadam (DBM) mixtures under the Indian technical specifications, using three binders, conventional VG 30 and VG 40 bitumen's, and a crumb rubber modified binder (CRMB), all of them with RCA or NA. Then, they evaluated the Marshall characteristics, indirect tensile strength, moisture sensitivity, resilient modulus, and permanent deformation resistance. Based on the results, they confirmed that, in general, all the mixtures fulfilled the requirements for the Marshall and moisture susceptibility characteristics. The CRMB and the VG40 binders offered the best performance with RCA or NA, although the latter showed slightly better results.

The need for recycling CDW (as RCA) for use in road pavement construction was also studied by Kanoungo et al. [154]. They compared some of RCA's available treatment methods; namely, acid treatment, thermal treatment, and asphalt emulsion, to define what would be the best method. Using these treatments, they produced DBM mixtures with RCA and analyzed the Marshall characteristics and moisture sensitivity. The results pointed to the asphalt emulsion treatment of RCA as the most suitable method.

The possible use of fine and coarse RCA as NA substitutes in nine asphalt mixtures for base course layers was also studied by Radević et al. [155]. These researchers assessed the influence of the RCA rate (up to 45 wt.%) on the physical and mechanical properties of an AC 22 base 50/70 mixture and compared the results with those obtained in a reference mixture produced with NA. They found that RCA's addition needs a higher binder content (up to 1%), which resulted in lower stiffness and higher fatigue cracking resistance of the asphalt mixtures, while their low-temperature resistance was slightly inferior. In conclusion, it is possible to use up to 45% RCA in asphalt mixtures without significantly reducing the mechanical performance.

Nwakaire et al. [149] claimed that more unanimous standard guidelines should be developed to guarantee excellence and sustainability when using RCA in HMA. These authors stated that more studies are needed on the use of RCA for porous and SMA mixtures and rigid pavements. Moreover, other essential research areas are assessing the actual field performance of in-situ RCA pavements, surveying the professional's perspectives on the challenges of using RCA, and identifying all feasible utilization potentials for RCA based on different circumstances and scenarios using LCA methods. Regarding the mix design of RCA mixtures, it is necessary to harmonize the binder requirement for RCA mixtures, develop an innovative asphalt binder with an improved affinity with RCA, adjust the particle size requirements of the conventional HMA for RCA mixtures, and adjust the solution to the sources and nature of RCA used.

3.5.2. Incorporation of Recycled Concrete Aggregate in WMA Mixtures

The use of RCA in WMA was evaluated by other authors [107,108,119,156] to assess the influence of these alternative aggregates (processed from CDW) on the performance of WMAs. Martinho et al. [108] selected three WMA mixtures with RCA considering different laboratory results, optimizing them before being produced in an asphalt plant, and then compacted under real conditions in several pavement trial sections. A dense-graded AC 20 base mixture was used as a reference mixture. Using up to 60% RCA was possible because its grading curve (Figure 6) matches the selected AC 20 base mixture gradation. The RCA characteristics were also evaluated according to EN 933-11, mainly comprising concrete (84%), unbound natural aggregates (9%), masonry elements of clay materials (5%), and traces of asphalt material and other CDWs (respectively, 0.7% and 0.6%).

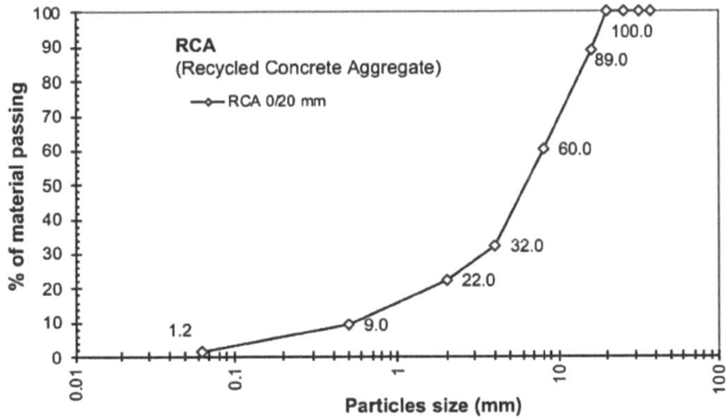

Figure 6. Grading curve of the RCA used in a WMA, adapted from Martinho et al. [108].

The results obtained in that study, supported by the literature (up to that date), led to the conclusion that the general performance of WMA with 60% of RCA was good; namely, when using an organic wax as the WMA additive. The use of a chemical additive reduced the rutting resistance compared to the wax. The alternative aggregates slightly reduced the stiffness modulus and water sensitivity results of the WMA mixtures, although the results are still satisfactory for a conventional base/binder layer. The fatigue life of WMA with RCA was adequate and close to the performance observed for the HMA and WMA mixtures used as references. Although conventional equipment was used to build the experimental pavement sections, no specific problems were observed in the field for the WMA mixtures with RCA in any of the necessary operations: mixing, transport, laying, and compaction.

Abass and Albayati [157] investigated the possibility of using RCA in WMA, testing five replacement rates (ranging from 0% to 100%) for the coarse fraction of natural aggregates, using untreated RCA and RCA treated with hydrated lime slurry or hydrochloric acid. The binder contents were obtained with the Marshall mix design method, and the WMA performance was assessed by moisture damage, resilient modulus, and permanent deformation tests. The untreated RCA mixes presented a higher binder content than treated RCA mixes. The moisture susceptibility of treated RCA was improved by nearly 10% compared to untreated RCA. However, the resilient modulus and rutting resistance for mixtures with 100% RCA were lower than those obtained with natural aggregates.

Another innovative paving technique toward a circular economy model was promoted by Neves et al. [158], studying the properties of an AC 20 WMA mixture composed of 60% RCA and a modified binder (4.5%). After assessing the aggregate-bitumen affinity, Marshall properties, moisture sensitivity, stiffness, rutting, and fatigue resistance, the authors concluded that the WMA with RCA shows adequate performance for base course layer use, comparable to equivalent WMA or HMA mixtures with natural aggregates.

In several case studies, Polo-Mendoza et al. [156] have examined WMA mixtures with various RCA contents. This study developed a trade-off methodology based on the LCA technique and statistical analysis to determine the maximum incorporation rate of RCA in WMA without generating increasing environmental impacts due to the higher binder content of those mixtures.

In a different type of WMA mixture, Zou et al. [109] reported that using RCA in foamed asphalt mixtures (FAM) does not affect the binder curing time, but the mixture performance is influenced when the processed aggregates include redbrick. These authors also concluded that FAM fulfills technical requirements for the strength and moisture resistance of base and subbase course with 100% RCA. The inclusion of redbrick in RCA weakens the FAM performance (a redbrick content limit of 5% is suggested in FAM with RCA), but the addition of 0.5% to 1.5% cement content increases the indirect tensile strength (ITS) and wet/dry indirect tensile strength ratio (ITSR) of FAM. These authors also recommended using harder bitumen's in foaming, as they include more asphaltenes and offer a higher viscosity.

Monu et al. [159] investigated the optimum proportion of hydrophobic RAP and hydrophilic RCA (separately and combined) for the production of FAM mixtures. The incorporation of RCA affected FAM performance but delayed the hydration of residual cement grains, which could enhance the performance of FAM. FAM mixtures with 20% RCA, 40% RAP, or a blend of 30% RCA and 10% RAP satisfied the specified requirements for pavement applications.

3.5.3. Evaluation of Different Incorporation Ratios of Recycled Concrete Aggregate

Different amounts of fine and coarse RCA (from 20% to 60%) were integrated by Daquan et al. [160] in an AC 20 HMA to evaluate their effect on mechanical performance. The authors concluded that the OBC increased, and the bulk density of mixtures decreased for higher amounts of RCA. Besides, they observed that fine RCA had higher OBC than coarse RCA, and mixtures with 50% coarse RCA or more had reduced moisture resistance.

Furthermore, all mixtures presented good low-temperature properties, but the rutting resistance of asphalt mixtures with RCA was lower than that of HMA with natural aggregates. The decrease in RCA content led to an increased resilient modulus and extended fatigue life (particularly when reducing coarse RCA). The mixtures with 20–40% fine and coarse RCA generally showed the best performance.

Galan et al. [161] studied the influence of RCA percentage on binder content, curing time, and temperature by studying the stiffness of HMA samples. The asphalt mixtures were produced with different percentages of RCA (0%, 5%, 10%, 20%, and 30%) and bitumen (3.5%, 4.0%, and 4.5%). The samples were cured for 0 h, 2 h, or 4 h before being tested at different temperatures (0 °C, 10 °C, and 20 °C). This study concluded that temperature was the most influential factor in decreasing the stiffness modulus. Conversely, the percentage of RCA was not very relevant to changing the stiffness modulus.

Zhang et al. [162] produced a dense-gradated AC 16 HMA incorporating RCA processed from low strength concrete [163] as a partial substitute for NA at different rates: 30%, 50%, and 75% by weight of NA. Compared with NA, this RCA presented higher wearing and fragmentation values, a lower apparent relative density, a higher water absorption, and a more inadequate bitumen adhesion. The water sensitivity of the mixture decreased up to 50% with the increase in RCA content. Nevertheless, the authors concluded that HMA incorporating up to 50% RCA could be satisfactorily used in road construction.

Finally, after reviewing the available literature on RCA uses in sustainable flexible and rigid pavement applications, Nwakaire et al. [149] presented a summary of several studies where different rates of RCA (15% to 100%) were incorporated in HMA (Table 6).

Table 6. RCA replacement levels in asphalt mixtures, adapted from Nwakaire et al. [149].

References	Type of Mixture	%RCA Included	Conclusions
Paranavithana and Mohajerani [164]	HMA	100	Volumetric properties and stability similar to other mixtures
Lee et al. [133]	HMA	100	Satisfactory mechanical performance, including the rutting resistance and moisture sensitivity
Mills-Beale and You [165]	Asphalt mixtures	75	Satisfactory mechanical performance, including the rutting resistance and moisture sensitivity
Zulkati et al. [166]	HMA	60	Satisfactory rutting resistance
Al-Bayati et al. [117]	HMA	60	Above 60%, the requirements for volumetric properties and stability were not met
Rafi et al. [167]	HMA	50	Above 50%, it did not meet the Marshall requirements (stability and flow)
Zhang et al. [168]	HMA	50	Shows a considerable reduction in the flexural tensile strength and the stiffness modulus
Wong et al. [134]	HMA	45	Adequate performance based on creep resistance and stiffness modulus
Pérez et al. [120]	HMA	40	Satisfactory rutting resistance
Pasandín and Pérez [130]	HMA	30	Satisfactory fatigue life and rutting resistance
Pasandín and Pérez [122]	HMA	30	Satisfactory water sensitivity, fatigue life, and rutting resistance
Dhir et al. [169]	Asphalt mixtures	30	Satisfactory stiffness modulus, rutting resistance, and fatigue life
Qasrawi and Asi [170]	HMA	25	Did not meet the requirements for volumetric properties above 25%
Ossa et al. [99]	Surface HMA	20	Above 20% caused moisture damage
Kowalski et al. [171]	Asphalt mixtures	15	Above 15% caused moisture damage

3.5.4. Combined Use of Recycled Concrete Aggregate with Other Waste or By-Products

Most studies combining RCA with other waste or byproducts to produce new road paving materials include RAP [9,172–174]. Nevertheless, it should be noted that some examples of the simultaneous use of RCA and SSA are presented in Section 3.6.

Abedalqader et al. [172] studied the temperature influence on the performance of asphalt mixtures with RAP and coarse RCA, concluding that the mechanical properties of these asphalt mixtures decreased as RAP and RCA incorporation levels increased for the same test temperature.

Another work was developed to obtain a 5–10 mm RCA and RAP aggregate fraction to substitute natural aggregates in hot mix asphalt [173]. The results show the possible combination of both wastes in RCA/RAP ratios equal to 25/75 or 50/50 to obtain a coarse aggregate fraction meeting the specifications and great environmental benefit due to the reduced use of natural resources.

Coban et al. [174] investigated recycled aggregate base layers for road pavements with two different RCA materials with different gradations and a blend of RCA and RAP materials, compared to the conventional solution with natural aggregates. After performing lab tests (resilient modulus) and field tests (falling weight deflectometer), they concluded that all the recycled aggregate base layers had satisfactory performance.

The fatigue cracking and moisture resistance of HMA mixtures produced with 0%, 35%, and 42% RCA and 10% waste tire rubber modified bitumen were evaluated by other authors [175,176]. This investigation showed the beneficial effect of simultaneously using RCA and crumb rubber on fatigue life, although crumb rubber increased the water sensitivity of the RCA mixture. Nevertheless, these solutions have adequate properties for medium-traffic roads.

Giri et al. [177] explored combining waste materials such as coarse RCA and waste polyethylene in asphalt paving mixtures. They observed that all the developed mixtures satisfy the Marshall and moisture susceptibility specified requirements. The use of waste PE chiefly improves the engineering properties at higher temperatures.

3.5.5. Incorporation of Recycled Concrete Aggregate in Other Types of Asphalt Mixtures

Nwakaire et al. [116] studied the performance of an SMA 14 mixture after replacing 20% to 100% of natural coarse aggregates with RCA. They also used a control SMA with 100% granite. The effect of RCA replacement on SMA quality was assessed through volumetric properties, Marshal stability, indirect tensile strength (ITS), moisture susceptibility, resilient modulus, fatigue and rutting performance, abrasion, and skid resistance. The SMA with RCA performed worse than the control SMA in the ITS and resilient modulus tests, contrary to the remaining tests. Nevertheless, the authors recommended an optimum replacement of 40% RCA because the SMA with RCA requires higher binder contents for optimum performance. The skid resistance of all SMA mixtures was satisfactory, and the rutting resistance of the SMA with RCA was lower than the control SMA at the initial cycles but was better at the end of the test.

The incorporation of RCA in cold asphalt mixtures was studied by Zou et al. [110] by investigating the feasibility of using RCA from CDW to replace natural aggregates in emulsified asphalt mixtures (EAM). Their work was based on the laboratory's assessment of the optimum moisture and emulsified asphalt contents of some EAM samples that included different RCA rates. They also evaluated the RCA EAM in-service performance after adding cement, including the moisture sensitivity and high and low-temperature performance. This work found that RCA increased the high-temperature performance and reduced the low-temperature performance and moisture damage resistance of EAM mixtures. Moreover, the addition of cement enhances the in-service performance of EAM so that RCA can substitute NA in EAM mixtures when combined with cement.

Chen and Wong [178] evaluated mechanically and functionally PA mixtures made of 100% RCA. Drain down, Cantabro, Marshall, permeability, and aging tests were used to assess the performance of three PA designs: 100% RCA; 100% RCA with enhanced asphalt binder; and control PA with granite aggregates. The results for PA made of 100% RCA assured the essential drainage function, although it is necessary to use enhanced binders to fulfill the Marshall criteria for regular highway applications. The results suggest the possible application of 100% RCA in PA.

Another study incorporated several RCA fractions in semi-dense asphalt (SDA) mixtures. Mikhailenko et al. [18] replaced 100% and 50% of the natural aggregates with three fractions of RCA (coarse, sand, and filler) and evaluated the mixtures' volumetric properties, water sensitivity, ITS, fracture energy, and rutting resistance. The results confirmed that coarse RCA absorbed higher amounts of binder and reduced the workability. The RCA mixtures presented increased ITS results and brittleness, reducing crack resistance. Higher aggregate replacements significantly affected the moisture susceptibility of the mixtures and decreased the fracture energy (mainly when replacing sand fraction). The incorporation of recycled concrete aggregates enhanced the SDA's rutting resistance, especially when replacing the coarse fraction. In general, RCA's use in SDA can be incorporated in limited amounts, and replacement by volume is recommended.

3.6. Case Studies of SSA and RCA Simultaneous Incorporation in Asphalt Mixtures

This paper presents a comprehensive literature review on two alternative by-products or secondary materials (i.e., SSA and recycled concrete aggregate) arising among the principal substitutes of natural aggregates in asphalt mixtures to fulfill the circular economy model. Various studies evaluated the potential use of these two alternative materials in asphalt mixtures separately. However, only a few studies assessed asphalt mixtures with the simultaneous incorporation of SSA and RCA, being presented in this section.

Martinho et al. [108] compared the mechanical performance of several warm asphalt mixtures with recycled concrete aggregate, EAF steel slag, or both by-products as partial substitutes for the natural aggregates. Initially developed in the laboratory, this study selected asphalt mixtures later applied in road pavement trials. Conventional HMA and WMA mixtures without by-products were used as references. The research evaluated aggregate substitution rates of 60% RCA, 30% EAF SSA, or a blend of 40% RCA and 35% EAF SSA. The authors concluded that using EAF SSA or RCA in WMA mixtures increases Marshall stability and could increase or decrease the rutting resistance. The results also showed that the water sensitivity and the stiffness modulus are slightly reduced, and the fatigue resistance does not change significantly. The overall performance of WMA mixtures with RCA or SSA was satisfactory, and the best results were obtained with 60% RCA.

Arabani and Azarhoosh [179] developed a study to determine the mechanical properties of asphalt mixtures produced simultaneously with recycled concrete and SSA. Six different asphalt mixtures containing three types of aggregate (dacite, recycled concrete, and steel slag) were produced to obtain Marshall specimens and determine the optimum binder content. Marshall stability, indirect tensile resilient modulus, dynamic creep, and indirect tensile fatigue tests evaluated the mechanical characteristics of the asphalt mixtures. The results indicated that the asphalt mixture with the best performance contains steel slag coarse aggregates and recycled concrete fine aggregates.

Roque et al. [180] evaluated the concurrent incorporation of RCA processed from construction and demolition waste and a steel slag aggregate in granular drainage layers of road pavements. Considering the high durability and permeability of the granular materials prepared with these by-products, the authors concluded that these materials could be used together in the drainage base or sub-base layers of transport infrastructures.

After presenting a comprehensive literature review on SSA and RCA aggregates' use in asphalt mixtures, including their combined use described in this last section, these alternative aggregates' main advantages and drawbacks are summarized in Table 7.

Table 7. Advantages and drawbacks of SSA and RCA in asphalt mixtures.

Aggregate	Advantages	Drawbacks
SSA	• Minimizes depletion of natural resources • Reduces consumption of natural aggregates • May reduce the production costs • Reduces waste landfill • Presents excellent wearing and polishing resistances • Can replace high amounts of NA • Increases the mechanical performance of asphalt mixtures regarding water sensitivity and rutting resistance • May be used as a self-healing promoter	• Increases the transportation costs due to its higher density • Reduces the mixture workability and increases air void and binder contents due to its rough and porous surface • Untreated SSA may leach heavy metals, increase the eluates' pH values, and present long-term expansion problems • Shows higher variability depending on steel slag origin and treatment • Demands new specifications • Increases equipment wearing and production complexity
RCA	• Minimizes depletion of natural resources • Reduces consumption of natural aggregates • Shows a mechanical performance similar to natural aggregates when the mixture is adequately designed • Minimizes the impacts of material transportation when reused on-site • May reduce the production costs • Reduces environmental liabilities	• Increases the heterogeneity due to different concrete origins • Shows higher processing variability • May present low fragmentation resistance and high water absorption • Demands new and specifically developed standards • Lacks trust from public authorities • Unknown long-term evolution of mixtures with these aggregates

4. Conclusions

This review paper showed that it is possible to use steel slag aggregates and recycled concrete aggregates from CDW to substitute significant amounts of natural resources used to produce asphalt mixtures. This substitution is especially relevant in the current situation of very high energy costs and shortages of natural aggregates. Moreover, these alternative aggregates improve the mixtures' mechanical performance, durability, and long-term sustainability in most cases.

The mechanical properties of SSA, such as roughness, shape, angularity, hardness, polishing resistance, and wear resistance, make it suitable for use as aggregates in different asphalt mixtures. Consequently, the mechanical behavior of mixtures with steel slag aggregate was generally better than that of mixtures with natural aggregates.

Different rates of SSA incorporation were observed in the literature, with values between 20% and 100%, without compromising the performance of asphalt mixtures. Nevertheless, several authors have noted that the best results were obtained when replacing up to 75% of natural aggregates with SSA.

However, some studies concluded that mixtures with SSA require a higher bitumen content than natural aggregates. In addition, some authors report difficulties in assuring the specified air voids contents due to the lower workability of these mixtures. It is also essential to pay special attention to the high density of SSA during the mix design phase, highlighting the need to use a volumetric approach.

Regarding the recycled concrete aggregates processed from CDW, different solutions have been presented for their use in asphalt mixtures as a natural aggregates replacement. Nevertheless, since RCA properties have limitations (e.g., lower wearing resistance and higher water absorption), several treatment methods were mentioned to improve its quality, and different binders can be incorporated. Different asphalt mixtures have also been produced with a wide range of RCA incorporation rates (between 15% and 100%), substituting different fractions (coarse, medium, or fine) of the natural aggregates.

Both alternative aggregates addressed in this paper allow various combinations (e.g., individual or simultaneous use of different byproducts) in replacing natural aggregates in several asphalt mixtures (e.g., HMA, WMA, SMA, foamed mixtures, cold mixtures, and recycled mixtures). This range of solutions demonstrates the importance of appropriately selecting, in each actual case, the composition that can maximize the performance and sustainability of the alternative asphalt mixtures.

Thus, researchers and engineers should continue to investigate this type of sustainable asphalt mixture to understand and evaluate its long-term mechanical and environmental performance.

The sustainability of replacing natural aggregates with SSA or RCA in asphalt mixtures is the inspiration of several research studies mentioned in this review, and it was proven using LCA techniques in some of those studies. Nevertheless, the environmental advantages depend on the amount of SSA or RCA replacement and the scenarios and boundaries considered.

In conclusion, this comprehensive literature review on incorporating steel slag and recycled concrete aggregates in asphalt mixtures for road pavements showed an adequate mechanical and environmental performance. A few restrictions of these solutions were identified, which can be corrected with an appropriate mix design. Therefore, these byproducts can be widely used to replace quarries in supplying alternative and sustainable aggregates for the asphalt paving industry.

Author Contributions: Conceptualization, C.D.A.L., C.F.N.M., M.R., F.C.G.M., H.M.R.D.S. and J.R.M.O.; methodology, C.D.A.L., C.F.N.M., F.C.G.M., H.M.R.D.S. and J.R.M.O.; validation, F.C.G.M., H.M.R.D.S. and J.R.M.O.; formal analysis, F.C.G.M., H.M.R.D.S. and J.R.M.O.; investigation, C.D.A.L., C.F.N.M., M.R., F.C.G.M., H.M.R.D.S. and J.R.M.O.; writing—original draft preparation, C.D.A.L., C.F.N.M., M.R., F.C.G.M., H.M.R.D.S. and J.R.M.O.; writing—review and editing, F.C.G.M., H.M.R.D.S. and J.R.M.O.; supervision, M.R., H.M.R.D.S. and J.R.M.O. All authors have read and agreed to the published version of the manuscript.

Funding: This research was funded by the "Environment, Climate Change and Low Carbon Economy Programme—Environment Programme" (EEA financial mechanism 2014–2021) through the Funding Mechanism Commission established by Iceland, Liechtenstein, Norway, and Portugal, under the scope of project CirMat—CIRcular aggregates for sustainable road and building MATerials. This study was also supported by Fundação para a Ciência e a Tecnologia through the Ph.D. grants number 2021.06428.BD and 2021.08004.BD.

Institutional Review Board Statement: Not applicable.

Informed Consent Statement: Not applicable.

Data Availability Statement: Not applicable.

Acknowledgments: The authors would like to thank all the CirMat project members who contributed to achieving this study's objectives through the motivating discussions during several project meetings.

Conflicts of Interest: The authors declare no conflict of interest. The funders had no role in the study's design, collection, analyses, or interpretation of data, writing of the manuscript, or decision to publish the results.

Abbreviations

The following abbreviations are used in this manuscript:

BF	Iron-making slag of Blast Furnace
BOF	Basic Oxygen Furnace
CDW	Construction Demolition Waste
CR	Crumb Rubber
CRMB	Crumb Rubber Modified Binder
DBM	Dense asphalt macadam
EAF	Slag and Electric Arc Furnace
EAM	Emulsified Asphalt Mixtures
EWC	European Waste Catalogue

FAM	Foamed Asphalt Mixtures
HMA	Hot Mix Asphalt
ITS	Indirect Tensile Strength
LD	Ladle Furnace
LCA	Life Cycle Assessment
LoW	List of Waste
NA	Natural Aggregates
OBC	Optimum Binder Content
PA	Porous Asphalt
PSV	Polishing Stone Value
RAP	Reclaimed Asphalt Pavement
RCA	Recycled Concrete Aggregate
SDA	Semi-Dense Asphalt mixtures
SEM	Scanning Electron Microscope
SMA	Stone Mastic Asphalt
SSA	Steel Slag Aggregate
VMA	Voids in Mineral Aggregate
WFD	Waste Framework Directive
WMA	Warm Mix Asphalt
ITSR	Wet/dry Indirect Tensile Strength Ratio
XRD	X-ray diffraction

References

1. European Commission. *A New Circular Economy Action Plan for a Cleaner and More Competitive Europe*; European Commission: Brussels, Belgium, 2020.
2. Moura, C.; Nascimento, L.; Loureiro, C.; Rodrigues, M.; Oliveira, J.; Silva, H. Viability of Using High Amounts of Steel Slag Aggregates to Improve the Circularity and Performance of Asphalt Mixtures. *Appl. Sci.* **2022**, *12*, 490. [CrossRef]
3. Choudhary, J.; Kumar, B.; Gupta, A. Utilization of solid waste materials as alternative fillers in asphalt mixes: A review. *Constr. Build. Mater.* **2020**, *234*, 117271. [CrossRef]
4. Woszuk, A.; Bandura, L.; Franus, W. Fly ash as low cost and environmentally friendly filler and its effect on the properties of mix asphalt. *J. Clean. Prod.* **2019**, *235*, 493–502. [CrossRef]
5. Plati, C. Sustainability factors in pavement materials, design, and preservation strategies: A literature review. *Constr. Build. Mater.* **2019**, *211*, 539–555. [CrossRef]
6. Topini, D.; Toraldo, E.; Andena, L.; Mariani, E. Use of recycled fillers in bituminous mixtures for road pavements. *Constr. Build. Mater.* **2018**, *159*, 189–197. [CrossRef]
7. Goli, A. The study of the feasibility of using recycled steel slag aggregate in hot mix asphalt. *Case Stud. Constr. Mater.* **2022**, *16*, e00861. [CrossRef]
8. Zhao, X.; Sheng, Y.; Lv, H.; Jia, H.; Liu, Q.; Ji, X.; Xiong, R.; Meng, J. Laboratory investigation on road performances of asphalt mixtures using steel slag and granite as aggregate. *Constr. Build. Mater.* **2022**, *315*, 125655. [CrossRef]
9. Purohit, S.; Panda, M.; Chattaraj, U. Use of Reclaimed Asphalt Pavement and Recycled Concrete Aggregate for Bituminous Paving Mixes: A Simple Approach. *J. Mater. Civ. Eng.* **2021**, *33*, 04020395. [CrossRef]
10. Chen, W.; Wei, J.; Xu, X.; Zhang, X.; Han, W.; Yan, X.; Hu, G.; Lu, Z. Study on the optimum steel slag content of sma-13 asphalt mixes based on road performance. *Coatings* **2021**, *11*, 1436. [CrossRef]
11. Jitsangiam, P.; Nusit, K.; Nikraz, H.; Leng, Z.; Prommarin, J.; Chindaprasirt, P. Dense-Graded Hot Mix Asphalt with 100% Recycled Concrete Aggregate Based on Thermal-Mechanical Surface Treatment. *J. Mater. Civ. Eng.* **2021**, *33*, 04021156. [CrossRef]
12. Nwakaire, C.M.; Onn, C.C.; Yap, S.P.; Yuen, C.W.; Koting, S.; Mo, K.H.; Othman, F. The strength and environmental performance of asphalt mixtures with recycled concrete aggregates. *Transp. Res. Part D Transp. Environ.* **2021**, *100*, 103065. [CrossRef]
13. Dhoble, Y.N.; Ahmed, S. Review on the innovative uses of steel slag for waste minimization. *J. Mater. Cycles Waste Manag.* **2018**, *20*, 1373–1382. [CrossRef]
14. Ferreira, V.J.; Sáez-De-Guinoa Vilaplana, A.; García-Armingol, T.; Aranda-Usón, A.; Lausín-González, C.; López-Sabirón, A.M.; Ferreira, G. Evaluation of the steel slag incorporation as coarse aggregate for road construction: Technical requirements and environmental impact assessment. *J. Clean. Prod.* **2016**, *130*, 175–186. [CrossRef]
15. Menegaki, M.; Damigos, D. A review on current situation and challenges of construction and demolition waste management. *Curr. Opin. Green Sustain. Chem.* **2018**, *13*, 8–15. [CrossRef]
16. Fatemi, S.; Imaninasab, R. Performance evaluation of recycled asphalt mixtures by construction and demolition waste materials. *Constr. Build. Mater.* **2016**, *120*, 450–456. [CrossRef]

17. Sánchez-Cotte, E.H.; Pacheco-Bustos, C.A.; Fonseca, A.; Triana, Y.P.; Mercado, R.; Yepes-Martínez, J.; Lagares Espinoza, R.G. The Chemical-Mineralogical Characterization of Recycled Concrete Aggregates from Different Sources and Their Potential Reactions in Asphalt Mixtures. *Materials* **2020**, *13*, 5592. [CrossRef]
18. Mikhailenko, P.; Kakar, M.R.; Piao, Z.Y.; Bueno, M.; Poulikakos, L. Incorporation of recycled concrete aggregate (RCA) fractions in semi-dense asphalt (SDA) pavements: Volumetrics, durability and mechanical properties. *Constr. Build. Mater.* **2020**, *264*, 120166. [CrossRef]
19. Kumar, H.; Varma, S. A review on utilization of steel slag in hot mix asphalt. *Int. J. Pavement Res. Technol.* **2021**, *14*, 232–242. [CrossRef]
20. Skaf, M.; Manso, J.M.; Aragon, A.; Fuente-Alonso, J.A.; Ortega-Lopez, V. EAF slag in asphalt mixes: A brief review of its possible re-use. *Resour. Conserv. Recycl.* **2017**, *120*, 176–185. [CrossRef]
21. Ruiz, L.A.L.; Ramón, X.R.; Domingo, S.G. The circular economy in the construction and demolition waste sector–A review and an integrative model approach. *J. Clean. Prod.* **2020**, *248*, 119238. [CrossRef]
22. Gedik, A. A review on the evaluation of the potential utilization of construction and demolition waste in hot mix asphalt pavements. *Resour. Conserv. Recycl.* **2020**, *161*, 104956. [CrossRef]
23. Prasad, D.; Singh, B.; Suman, S.K. Utilization of recycled concrete aggregate in bituminous mixtures: A comprehensive review. *Constr. Build. Mater.* **2022**, *326*, 126859. [CrossRef]
24. Dos Reis, G.S.; Quattrone, M.; Ambrós, W.M.; Cazacliu, B.G.; Sampaio, C.H. Current applications of recycled aggregates from construction and demolition: A review. *Mater.* **2021**, *14*, 1700. [CrossRef] [PubMed]
25. Xu, X.; Luo, Y.; Sreeram, A.; Wu, Q.; Chen, G.; Cheng, S.; Chen, Z.; Chen, X. Potential use of recycled concrete aggregate (RCA) for sustainable asphalt pavements of the future: A state-of-the-art review. *J. Clean. Prod.* **2022**, *344*, 130893. [CrossRef]
26. Lee, K.-H.; Noh, J.; Khim, J.S. The Blue Economy and the United Nations' sustainable development goals: Challenges and opportunities. *Environ. Int.* **2020**, *137*, 105528. [CrossRef]
27. Jiang, Y.; Ling, T.C.; Shi, C.J.; Pan, S.Y. Characteristics of steel slags and their use in cement and concrete-A review. *Resour. Conserv. Recycl.* **2018**, *136*, 187–197. [CrossRef]
28. Abd Alhay, B.A.; Jassim, A.K. Steel Slag Waste Applied to Modify Road Pavement. In Proceedings of the 1st International Conference on Pure Science (ISCPS-2020), Najaf, Iraq, 20 September 2020.
29. Shiha, M.; El-Badawy, S.; Gabr, A. Modeling and performance evaluation of asphalt mixtures and aggregate bases containing steel slag. *Constr. Build. Mater.* **2020**, *248*, 118710. [CrossRef]
30. Pasetto, M.; Baldo, N. Mix design and performance analysis of asphalt concretes with electric arc furnace slag. *Constr. Build. Mater.* **2011**, *25*, 3458–3468. [CrossRef]
31. Wang, G.; Wang, Y.; Gao, Z. Use of steel slag as a granular material: Volume expansion prediction and usability criteria. *J. Hazard. Mater.* **2010**, *184*, 555–560. [CrossRef]
32. Hainin, M.R.; Aziz, M.M.A.; Ali, Z.; Jaya, R.P.; El-Sergany, M.M.; Yaacob, H. Steel Slag as A Road Construction Material. *J. Teknol.* **2015**, *73*, 33–38. [CrossRef]
33. Motz, H.; Geiseler, J. Products of steel slags an opportunity to save natural resources. *Waste Manag.* **2001**, *21*, 285–293. [CrossRef]
34. Masoudi, S.; Abtahi, S.M.; Goli, A. Evaluation of electric arc furnace steel slag coarse aggregate in warm mix asphalt subjected to long-term aging. *Constr. Build. Mater.* **2017**, *135*, 260–266. [CrossRef]
35. Oluwasola, E.A.; Hainin, M.R.; Aziz, M.M.A. Evaluation of rutting potential and skid resistance of hot mix asphalt incorporating electric arc furnace steel slag and copper mine tailing. *Indian J. Eng. Mater. Sci.* **2015**, *22*, 550–558.
36. Kong, D.; Chen, M.; Xie, J.; Zhao, M.; Yang, C. Geometric characteristics of BOF slag coarse aggregate and its influence on asphalt concrete. *Materials* **2019**, *12*, 741. [CrossRef] [PubMed]
37. Li, S.; Xiong, R.; Zhai, J.; Zhang, K.; Jiang, W.; Yang, F.; Yang, X.; Zhao, H. Research progress on skid resistance of basic oxygen furnace (BOF) slag asphalt mixtures. *Materials* **2020**, *13*, 2169. [CrossRef]
38. Xie, J.; Wang, Z.; Wang, F.; Wu, S.; Chen, Z.; Yang, C. The life cycle energy consumption and emissions of asphalt pavement incorporating basic oxygen furnace slag by comparative study. *Sustainability* **2021**, *13*, 4540. [CrossRef]
39. Ye, Y.; Wu, S.; Li, C.; Kong, D.; Shu, B. Morphological discrepancy of various basic oxygen furnace steel slags and road performance of corresponding asphalt mixtures. *Materials* **2019**, *12*, 2322. [CrossRef]
40. Shen, D.H.; Wu, C.M.; Du, J.C. Laboratory investigation of basic oxygen furnace slag for substitution of aggregate in porous asphalt mixture. *Constr. Build. Mater.* **2009**, *23*, 453–461. [CrossRef]
41. Xue, Y.; Wu, S.; Hou, H.; Zha, J. Experimental investigation of basic oxygen furnace slag used as aggregate in asphalt mixture. *J. Hazard. Mater.* **2006**, *138*, 261–268. [CrossRef]
42. Chen, Z.; Wu, S.; Xiao, Y.; Zeng, W.; Yi, M.; Wan, J. Effect of hydration and silicone resin on Basic Oxygen Furnace slag and its asphalt mixture. *J. Clean. Prod.* **2016**, *112*, 392–400. [CrossRef]
43. Roe, P. *Basic Oxygen Steel Slag as Surface Course Aggregate: An Investigation of Skid Resistance*; 0968-4107; Transport Research Laboratory: Wokingham, UK, 2003.
44. Chen, J.S.; Wei, S.H. Engineering properties and performance of asphalt mixtures incorporating steel slag. *Constr. Build. Mater.* **2016**, *128*, 148–153. [CrossRef]
45. Cui, P.D.; Wu, S.P.; Xiao, Y.; Yang, C.; Wang, F. Enhancement mechanism of skid resistance in preventive maintenance of asphalt pavement by steel slag based on micro-surfacing. *Constr. Build. Mater.* **2020**, *239*, 117870. [CrossRef]

46. Zalnezhad, M.; Hesami, E. Effect of steel slag aggregate and bitumen emulsion types on the performance of microsurfacing mixture. *J. Traffic Transp. Eng. Engl. Ed.* **2020**, *7*, 215–226. [CrossRef]
47. Maharaj, C.; White, D.; Maharaj, R.; Morin, C. Re-use of steel slag as an aggregate to asphaltic road pavement surface. *Cogent Eng.* **2017**, *4*, 1416889. [CrossRef]
48. Yildirim, I.Z.; Prezzi, M. Chemical, Mineralogical, and Morphological Properties of Steel Slag. *Adv. Civ. Eng.* **2011**, *2011*, 463638. [CrossRef]
49. Zumrawi, M.M.; Khalill, F.O. Experimental study of steel slag used as aggregate in asphalt mixture. *Int. J. Civ. Environ. Eng.* **2015**, *9*, 753–758.
50. Ziari, H.; Nowbakht, S.; Rezaei, S.; Mahboob, A. Laboratory Investigation of Fatigue Characteristics of Asphalt Mixtures with Steel Slag Aggregates. *Adv. Mater. Sci. Eng.* **2015**, *2015*, 623245. [CrossRef]
51. Chen, Z.W.; Gong, Z.L.; Jiao, Y.Y.; Wang, Y.; Shi, K.; Wu, J.C. Moisture stability improvement of asphalt mixture considering the surface characteristics of steel slag coarse aggregate. *Constr. Build. Mater.* **2020**, *251*, 118987. [CrossRef]
52. Hainin, M.R.; Rusbintardjo, G.; Hameed, M.A.S.; Hassan, N.A.; Yusoff, N.I.M. Utilisation of Steel Slag as an Aggregate Replacement in Porous Asphalt Mixtures. *J. Teknol.* **2014**, *69*, 67–73. [CrossRef]
53. Fakhri, M.; Ahmadi, A. Recycling of RAP and steel slag aggregates into the warm mix asphalt: A performance evaluation. *Constr. Build. Mater.* **2017**, *147*, 630–638. [CrossRef]
54. Sorlini, S.; Sanzeni, A.; Rondi, L. Reuse of steel slag in bituminous paving mixtures. *J. Hazard. Mater.* **2012**, *209–210*, 84–91. [CrossRef] [PubMed]
55. Yi, H.; Xu, G.P.; Cheng, H.G.; Wang, J.S.; Wan, Y.F.; Chen, H. An overview of utilization of steel slag. *Procedia Environ. Sci.* **2012**, *16*, 791–801. [CrossRef]
56. Barra, M.; Aponte, D.; Vazquez, E.; Mendez, B.; Miro, R.; Valls, S. Experimental study of the effect of the thermal conductivity of EAF slag aggregates used in asphaltic concrete of wearing courses on the durability of road pavements. In Proceedings of the Fourth International Conference on Sustainable Construction Materials and Technologies, Las Vegas, NV, USA, 7–11 August 2016.
57. Behiry, A.E.A.E.M. Evaluation of steel slag and crushed limestone mixtures as subbase material in flexible pavement. *Ain Shams Eng. J.* **2013**, *4*, 43–53. [CrossRef]
58. Ameri, M.; Hesami, S.; Goli, H. Laboratory evaluation of warm mix asphalt mixtures containing electric arc furnace (EAF) steel slag. *Constr. Build. Mater.* **2013**, *49*, 611–617. [CrossRef]
59. Kavussi, A.; Qazizadeh, M.J. Fatigue characterization of asphalt mixes containing electric arc furnace (EAF) steel slag subjected to long term aging. *Constr. Build. Mater.* **2014**, *72*, 158–166. [CrossRef]
60. Ziaee, S.A.; Behnia, K. Evaluating the effect of electric arc furnace steel slag on dynamic and static mechanical behavior of warm mix asphalt mixtures. *J. Clean. Prod.* **2020**, *274*, 123092. [CrossRef]
61. Keymanesh, M.R.; Ziari, H.; Zalnezhad, H.; Zalnezhad, M. Mix design and performance evaluation of microsurfacing containing electric arc furnace (EAF) steel slag filler. *Constr. Build. Mater.* **2021**, *269*, 121336. [CrossRef]
62. Swathi, M.; Andiyappan, T.; Guduru, G.; Reddy, M.A.; Kuna, K.K. Design of asphalt mixes with steel slag aggregates using the Bailey method of gradation selection. *Constr. Build. Mater.* **2021**, *279*, 122426. [CrossRef]
63. He, L.; Zhan, C.Y.; Lyu, S.T.; Grenfell, J.; Gao, J.; Kowalski, K.J.; Valentin, J.; Xie, J.; Ržek, L.; Ling, T.Q. Application status of steel slag asphalt mixture. *Jiaotong Yunshu Gongcheng Xuebao* **2020**, *20*, 15–33. [CrossRef]
64. Machado, A.T.; John, V.M. Estudo Comparativo dos Métodos de Ensaio para Avaliação da Expansibilidade das Escórias de Aciaria. Master Thesis, Universidade de São Paulo, São Paulo, Brazil, 2000.
65. Ferreira, S.R. Comportamento Mecânico e Ambiental de Materiais Granulares: Aplicação às Escórias de Aciaria Nacionais. Ph.D. Thesis, Universidade do Minho, Guimarães, Portugal, 2010.
66. Shao-Peng, W.; Wen-Feng, Y.; Yong-Jie, X.; Zhen-Hua, L. Design and preparation of steel slag SMA. *J. Wuhan Univ. Technol. Mater. Sci. Ed.* **2003**, *18*, 86–88. [CrossRef]
67. Euroslag. Legislation. Available online: https://www.euroslag.com/status-of-slag/legislation/ (accessed on 22 November 2021).
68. Euroslag; Eurofer. *Position Paper on the Status of Ferrous Slag Complying with the Waste Framework Directive 2008/98/CE (Articles 5/6) and the REACH Regulation*; The European Slag Association: Duisburg, Germany, 2012.
69. Borges, P.H.A. Uso da escória de aciaria como agregado siderúrgico: Uma discussão sobre os aspectos ambientais e legais no Brasil. Master Thesis, Universidade Estadual de Campinas, Limeira, Brazil, 2020.
70. Esther, L.-A.; Pedro, L.-G.; Irune, I.-V.; Gerardo, F. Comprehensive analysis of the environmental impact of electric arc furnace steel slag on asphalt mixtures. *J. Clean. Prod.* **2020**, *275*, 123121. [CrossRef]
71. Mombelli, D.; Mapelli, C.; Barella, S.; Di Cecca, C.; Le Saout, G.; Garcia-Diaz, E. The effect of chemical composition on the leaching behaviour of electric [illegible] furnace (EAF) carbon steel slag during a standard leaching test. *J. Environ. Chem. Eng.* **2016**, *4*, 1050–1060. [CrossRef]
72. Crisman, B.; [illegible]; Lorenzi, L.D.; Bevilacqua, P.; Roberti, R. A Laboratory Assessment of the Influence of Crumb Rubber in Hot Mix Asph[illegible] with Recycled Steel Slag. *Sustainability* **2020**, *12*, 8045. [CrossRef]
73. Gan, Y.; Li, C.; Zou, J.; Wan[illegible], W.; Yu, T. Evaluation of the Impact Factors on the Leaching Risk of Steel Slag and its Asphalt Mixture. *Case Stud. Constr. Mater.* **2022**, *16*, e01067. [CrossRef]
74. Singh, S.K.; Vashistha, P.; Chandra, R.; Rai, A.K. Study on leaching of electric arc furnace (EAF) slag for its sustainable applications as construction material. *Process Saf. Environ. Prot.* **2021**, *148*, 1315–1326. [CrossRef]

75. Li, Y.; Ni, W.; Gao, W.; Zhang, Y.; Yan, Q.; Zhang, S. Corrosion evaluation of steel slag based on a leaching solution test. *Energy Sources Recovery Util. Environ. Eff.* **2019**, *41*, 790–801. [CrossRef]
76. Mladenovič, A.; Turk, J.; Kovač, J.; Mauko, A.; Cotič, Z. Environmental evaluation of two scenarios for the selection of materials for asphalt wearing courses. *J. Clean. Prod.* **2015**, *87*, 683–691. [CrossRef]
77. Ziari, H.; Khabiri, M.M. Preventive maintenance of flexible pavement and mechanical properties of steel slag asphalt. *J. Environ. Eng. Landsc. Manag.* **2007**, *15*, 188–192. [CrossRef]
78. Motevalizadeh, S.M.; Sedghi, R.; Rooholamini, H. Fracture properties of asphalt mixtures containing electric arc furnace slag at low and intermediate temperatures. *Constr. Build. Mater.* **2020**, *240*, 117965. [CrossRef]
79. Rohde, L. Escória de Aciaria Elétrica em Camadas Granulares de Pavimentos: Estudo laboratorial. Master Thesis, Universidade Federal do Rio Grande do Sul, Porto Alegre, Brazil, 2002.
80. Ahmedzade, P.; Sengoz, B. Evaluation of steel slag coarse aggregate in hot mix asphalt concrete. *J. Hazard. Mater.* **2009**, *165*, 300–305. [CrossRef] [PubMed]
81. Kim, K.; Jo, S.H.; Kim, N.; Kim, H. Characteristics of hot mix asphalt containing steel slag aggregate according to temperature and void percentage. *Constr. Build. Mater.* **2018**, *188*, 1128–1136. [CrossRef]
82. Groenniger, J.; Cannone Falchetto, A.; Isailović, I.; Wang, D.; Wistuba, M.P. Experimental investigation of asphalt mixture containing Linz-Donawitz steel slag. *J. Traffic Transp. Eng.* **2017**, *4*, 372–379. [CrossRef]
83. Goli, H.; Hesami, S.; Ameri, M. Laboratory Evaluation of Damage Behavior of Warm Mix Asphalt Containing Steel Slag Aggregates. *J. Mater. Civ. Eng.* **2017**, *29*, 04017009. [CrossRef]
84. Asi, I.M.; Qasrawi, H.Y.; Shalabi, F.I. Use of steel slag aggregate in asphalt concrete mixes. *Can. J. Civ. Eng.* **2007**, *34*, 902–911. [CrossRef]
85. Kasaf, M.; Prastyanto, C.A. Analysis the use of steel slag as a replacement of natural aggregate in the asphalt concrete binder course (AC-BC) mixture. In Proceedings of the IOP Conference Series: Materials Science and Engineering, Surabaya, Indonesia, 22–23 July 2020; p. 012064.
86. Rodrigues, M.F. Estudo da Utilização de Escórias de Aciaria em Misturas Betuminosas. Master Thesis, Universidade do Minho, Guimarães, Portugal, 2019.
87. Nascimento, L.P.; Silva, H.M.R.D.; Oliveira, J.R.M.; Vilarinho, C. Valorisation of steel slag as aggregates for asphalt mixtures. In Proceedings of the 5th International Conference Wastes: Solutions, Treatments and Opportunities, Costa da Caparica, Portugal, 4–6 September 2019; pp. 489–494.
88. Chen, S.H.; Smimine, H.A.; Tsai, W.L.; Hung, C.T.; Kuo, M.H.; Zen, C.L. Performance Evaluation of Hybrid EAF Slag and RAP in Pavement. In *Lecture Notes in Civil Engineering*; Springer: Berlin/Heidelberg, Germany, 2020; Volume 76, pp. 375–384.
89. Viana, B.M.A. Soluções Integradas e Complementares para Otimizar a Utilização de Subprodutos Industriais em Pavimentação. Master Thesis, Universidade do Minho, Guimarães, Portugal, 2019.
90. Pasetto, M.; Baldo, N. Resistance to permanent deformation of base courses asphalt concretes made with RAP aggregate and steel slag. In Proceedings of the 12th ISAP Conference on Asphalt Pavements, Raleigh, NC, USA, 1–5 June 2014; pp. 1199–1208.
91. Wu, S.P.; Xue, Y.J.; Ye, Q.S.; Chen, Y.C. Utilization of steel slag as aggregates for stone mastic asphalt (SMA) mixtures. *Build. Environ.* **2007**, *42*, 2580–2585. [CrossRef]
92. Pasetto, M.; Baldo, N. Fatigue performance and stiffness properties of Stone Mastic Asphalts with steel slag and coal ash. In Proceedings of the 12th ISAP Conference on Asphalt Pavements, Raleigh, NC, USA, 1–5 June 2014; pp. 881–889.
93. Vila-Cortavitarte, M.; Jato-Espino, D.; Tabakovic, A.; Castro-Fresno, D. Optimizing the valorization of industrial by-products for the induction healing of asphalt mixtures. *Constr. Build. Mater.* **2019**, *228*, 116715. [CrossRef]
94. Xu, S.; García, A.; Su, J.; Liu, Q.; Tabaković, A.; Schlangen, E. Self-Healing Asphalt Review: From Idea to Practice. *Adv. Mater. Interfaces* **2018**, *5*, 1800536. [CrossRef]
95. Xu, S.; Liu, X.; Tabaković, A.; Schlangen, E. A novel self-healing system: Towards a sustainable porous asphalt. *J. Clean. Prod.* **2020**, *259*, 120815. [CrossRef]
96. Phan, T.M.; Park, D.W.; Le, T.H.M. Crack healing performance of hot mix asphalt containing steel slag by microwaves heating. *Constr. Build. Mater.* **2018**, *180*, 503–511. [CrossRef]
97. Loureiro, C. Use of Steel by-Products as an Enhancer Element for Pavement Repair Solutions (in Portuguese). Master Thesis, University of Minho, Guimarães, Portugal, 2021.
98. Ginga, C.P.; Ongpeng, J.M.C.; Daly, M.; Klarissa, M. Circular economy on construction and demolition waste: A literature review on material recovery and production. *Materials* **2020**, *13*, 2970. [CrossRef]
99. Ossa, A.; García, J.; Botero-Jaramillo, E. Use of recycled construction and demolition waste (CDW) aggregates: A sustainable alternative for the pavement construction industry. *J. Clean. Prod.* **2016**, *135*, 379–386. [CrossRef]
100. Paulraj, S.; Balasundaram, N.; Sates Kumar, K.; Dharshna Devi, M. Experimental studies on stren[illegible] [illegible]C characteristics of basalt fiber reinforced concrete. *Int.J. Civ. Eng. Technol.* **2017**, *8*, 704–711.
101. Klee, H. *The Cement Sustainability Initiative: Recycling Concrete*; World Business Council for Sustainable [illegible]elopment (WBCSD): Geneva, Switzerland, 2009.
102. Silva, H.M.R.D.; Oliveira, J.R.M. May recycled concrete be used as an alternative material for asphalt mixtures? In Proceedings of the 5th International Conference Wastes: Solutions, Treatments and Opportunities, Costa da Caparica, Portugal, 4–6 September 2019; pp. 482–488.

103. Pourtahmasb, M.S.; Karim, M.R.; Shamshirband, S. Resilient modulus prediction of asphalt mixtures containing recycled concrete aggregate using an adaptive neuro-fuzzy methodology. *Constr. Build. Mater.* **2015**, *82*, 257–263. [CrossRef]
104. Galan, J.J.; Silva, L.M.; Pérez, I.; Pasandín, A.R. Mechanical behavior of hot-mix asphalt made with recycled concrete aggregates from construction and demolition waste: A design of experiments approach. *Sustainability* **2019**, *11*, 3730. [CrossRef]
105. Pasandín, A.R.; Pérez, I.; Oliveira, J.R.M.; Silva, H.M.R.D.; Pereira, P.A.A. Influence of ageing on the properties of bitumen from asphalt mixtures with recycled concrete aggregates. *J. Clean. Prod.* **2015**, *101*, 165–173. [CrossRef]
106. Makul, N. A review on methods to improve the quality of recycled concrete aggregates. *J. Sustain. Cem. Based Mater.* **2021**, *10*, 65–91. [CrossRef]
107. Martinho, F.; Picado-Santos, L.; Capitão, S. Mechanical properties of warm-mix asphalt concrete containing different additives and recycled asphalt as constituents applied in real production conditions. *Constr. Build. Mater.* **2017**, *131*, 78–89. [CrossRef]
108. Martinho, F.C.G.; Picado-Santos, L.G.; Capitao, S.D. Influence of recycled concrete and steel slag aggregates on warm-mix asphalt properties. *Constr. Build. Mater.* **2018**, *185*, 684–696. [CrossRef]
109. Zou, G.; Sun, X.; Liu, X.; Zhang, J. Influence factors on using recycled concrete aggregate in foamed asphalt mixtures based on tensile strength and moisture resistance. *Constr. Build. Mater.* **2020**, *265*, 120363. [CrossRef]
110. Zou, G.; Zhang, J.; Liu, X.; Lin, Y.; Yu, H. Design and performance of emulsified asphalt mixtures containing construction and demolition waste. *Constr. Build. Mater.* **2020**, *239*, 117846. [CrossRef]
111. Rood, T.; Hanemaaijer, A. *Opportunities for a Circular Economy*; PBL Netherlands Environmental Assessment Agency: The Hague, The Netherlands, 2017.
112. Anastasiou, E.; Liapis, A.; Papayianni, I. Comparative life cycle assessment of concrete road pavements using industrial by-products as alternative materials. *Resour. Conserv. Recycl.* **2015**, *101*, 1–8. [CrossRef]
113. Fernandes, G.; Capitão, S.; Picado-Santos, L.G. Use of construction and demolition waste in road pavements. In Proceedings of the 7° Congresso Rodoviário Português, Novos Desafios para a Atividade Rodoviária, Lisboa, Portugal, 10–12 April 2013; pp. 10–12. (In Portuguese).
114. Tahmoorian, F.; Samali, B. Experimental and correlational study on the utilisation of RCA as an alternative coarse aggregate in asphalt mixtures. *Aust. J. Civ. Eng.* **2017**, *15*, 80–92. [CrossRef]
115. Martinho, F.; Picado-Santos, L.; Capitão, S. Feasibility assessment of the use of recycled aggregates for asphalt mixtures. *Sustainability* **2018**, *10*, 1737. [CrossRef]
116. Nwakaire, C.M.; Yap, S.P.; Yuen, C.W.; Onn, C.C.; Koting, S.; Babalghaith, A.M. Laboratory study on recycled concrete aggregate based asphalt mixtures for sustainable flexible pavement surfacing. *J. Clean. Prod.* **2020**, *262*, 121462. [CrossRef]
117. Al-Bayati, H.K.A.; Tighe, S.L.; Achebe, J. Influence of recycled concrete aggregate on volumetric properties of hot mix asphalt. *Resour. Conserv. Recycl.* **2018**, *130*, 200–214. [CrossRef]
118. Gómez-Meijide, B.; Pérez, I.; Pasandín, A.R. Recycled construction and demolition waste in Cold Asphalt Mixtures: Evolutionary properties. *J. Clean. Prod.* **2016**, *112*, 588–598. [CrossRef]
119. Martinho, F.C.; Picado-Santos, L.G.; Capitão, S.D. Assessment of warm-mix asphalt concrete containing sub-products as part of aggregate blend. *Int. J. Pavement Eng.* **2018**, *21*, 1213–1222. [CrossRef]
120. Pérez, I.; Pasandín, A.R.; Gallego, J. Stripping in hot mix asphalt produced by aggregates from construction and demolition waste. *Waste Manag. Res.* **2012**, *30*, 3–11. [CrossRef]
121. Đokić, O.; Radević, A.; Zakić, D.; Đokić, B. Potential of Natural and Recycled Concrete Aggregate Mixtures for Use in Pavement Structures. *Minerals* **2020**, *10*, 744. [CrossRef]
122. Pasandín, A.; Pérez, I. Overview of bituminous mixtures made with recycled concrete aggregates. *Constr. Build. Mater.* **2015**, *74*, 151–161. [CrossRef]
123. Wu, S.; Muhunthan, B.; Wen, H. Investigation of effectiveness of prediction of fatigue life for hot mix asphalt blended with recycled concrete aggregate using monotonic fracture testing. *Constr. Build. Mater.* **2017**, *131*, 50–56. [CrossRef]
124. Azarhoosh, A.; Koohmishi, M.; Hamedi, G.H. Rutting Resistance of Hot Mix Asphalt Containing Coarse Recycled Concrete Aggregates Coated with Waste Plastic Bottles. *Adv. Civ. Eng.* **2021**, *2021*, 9558241. [CrossRef]
125. Kareem, A.I.; Nikraz, H.; Asadi, H. Evaluation of the double coated recycled concrete aggregates for hot mix asphalt. *Constr. Build. Mater.* **2018**, *172*, 544–552. [CrossRef]
126. Wang, J.; Vandevyvere, B.; Vanhessche, S.; Schoon, J.; Boon, N.; De Belie, N. Microbial carbonate precipitation for the improvement of quality of recycled aggregates. *J. Clean. Prod.* **2017**, *156*, 355–366. [CrossRef]
127. Palaniraj, S.; Abhiram, K.; Manoj, B. Properties of treated recycled aggregates and its influence on concrete strength characteristics. *Constr. Build. Mater.* **2016**, *111*, 611–617. [CrossRef]
128. Pasandín, A.R.; Pérez, I. Mechanical properties of hot-mix asphalt made with recycled concrete aggregates coated with bitumen emulsion. *Constr. Build. Mater.* **2014**, *55*, 350–358. [CrossRef]
129. Hou, Y.; Ji, X.; Su, X.; Zhang, W.; Liu, L. Laboratory investigations of activated recycled concrete aggregate for asphalt treated base. *Constr. Build. Mater.* **2014**, *65*, 535–542. [CrossRef]
130. Pasandín, A.R.; Pérez, I.P. Laboratory evaluation of hot-mix asphalt containing construction and demolition waste. *Constr. Build. Mater.* **2013**, *43*, 497–505. [CrossRef]
131. Grabiec, A.; Starzyk, J.; Zawal, D.; Krupa, D. Modification of recycled concrete aggregate by calcium carbonate biodeposition. *Constr. Build. Mater.* **2012**, *34*, 145–150. [CrossRef]

132. Zhu, J.; Wu, S.; Zhong, J.; Wang, D. Investigation of asphalt mixture containing demolition waste obtained from earthquake-damaged buildings. *Constr. Build. Mater.* **2012**, *29*, 466–475. [CrossRef]
133. Lee, C.H.; Du, J.C.; Shen, D.H. Evaluation of pre-coated recycled concrete aggregate for hot mix asphalt. *Constr. Build. Mater.* **2012**, *28*, 66–71. [CrossRef]
134. Wong, Y.; Sun, D.; Lai, D. Value-added utilisation of recycled concrete in hot-mix asphalt. *Waste Manag.* **2007**, *27*, 294–301. [CrossRef] [PubMed]
135. Katz, A. Treatments for the Improvement of Recycled Aggregate. *J. Mater. Civ. Eng.* **2004**, *16*, 597–603. [CrossRef]
136. Radević, A.; Despotović, I.; Zakić, D.; Orešković, M.; Jevtić, D. Influence of acid treatment and carbonation on the properties of recycled concrete aggregate. *Chem. Ind. Chem. Eng. Q.* **2018**, *24*, 23–30. [CrossRef]
137. Taboada, G.L.; Seruca, I.; Sousa, C.; Pereira, Á. Exploratory Data Analysis and Data Envelopment Analysis of Construction and Demolition Waste Management in the European Economic Area. *Sustainability* **2020**, *12*, 4995. [CrossRef]
138. European Commission. *EU Construction and Demolition Waste Management Protocol (in Portuguese)*; European Commission: Brussels, Belgium, 2016.
139. Marcellus-Zamora, K.A.; Gallagher, P.M.; Spatari, S. Can Public Construction and Demolition Data Describe Trends in Building Material Recycling? Observations From Philadelphia. *Front. Built Environ.* **2020**, *6*, 131.0968-41070968-4107. [CrossRef]
140. Hu, W.; Dong, J.; Xu, N. Multi-period planning of integrated underground logistics system network for automated construction-demolition-municipal waste collection and parcel delivery: A case study. *J. Clean. Prod.* **2022**, *330*, 129760. [CrossRef]
141. Neto, G.A.D.S.; de Oliveira, J.P.V.; Salles, P.V.; Barros, R.T.V.; Paulino, M.T.; Dos Santos, W.J. Influence of heterogeneity, typology, and contaminants of recycled aggregates on the properties of concrete. *Open Construct. Build. Technol. J.* **2020**, *14*, 382–399. [CrossRef]
142. Rangel, C.S.; Toledo Filho, R.D.; Amario, M.; Pepe, M.; de Castro Polisseni, G.; Puente de Andrade, G. Generalized quality control parameter for heterogenous recycled concrete aggregates: A pilot scale case study. *J. Clean. Prod.* **2019**, *208*, 589–601. [CrossRef]
143. Sampaio, C.H.; Ambrós, W.M.; Cazacliu, B.G.; Moncunill, J.O.; Veras, M.M.; Miltzarek, G.L.; Silva, L.F.O.; Kuerten, A.S.; Liendo, M.A. Construction and demolition waste recycling through conventional jig, air jig, and sensor-based sorting: A comparison. *Minerals* **2021**, *11*, 904. [CrossRef]
144. APA. Guia de Boas Práticas Para Uma Adequada Gestão de Resíduos de Construção e Demolição. Available online: https://apambiente.pt/sites/default/files/_Residuos/FluxosEspecificosResiduos/RCD/BP_v2.pdf (accessed on 23 November 2021).
145. EPA. *Best Practice Guidelines for the Preparation of Resource & Waste Management Plans for Construction & Demolition Projects*; 978-1-80009-007-1; Environmental Protection Agency: Wexford, Ireland, 2021.
146. Villoria Sáez, P.; Del Río Merino, M.; Porras-Amores, C.; Santa Cruz Astorqui, J.; González Pericot, N. Analysis of Best Practices to Prevent and Manage the Waste Generated in Building Rehabilitation Works. *Sustainability* **2019**, *11*, 2796. [CrossRef]
147. Sun, J.; Chen, J.; Liao, X.; Tian, A.; Hao, J.; Wang, Y.; Tang, Q. The Workability and Crack Resistance of Natural and Recycled Aggregate Mortar Based on Expansion Agent through an Environmental Study. *Sustainability* **2021**, *13*, 491. [CrossRef]
148. Vega, A.D.L.; Santos, J.; Martinez-Arguelles, G. Life cycle assessment of hot mix asphalt with recycled concrete aggregates for road pavements construction. *Int. J. Pavement Eng.* **2022**, *23*, 923–936. [CrossRef]
149. Nwakaire, C.M.; Yap, S.P.; Onn, C.C.; Yuen, C.W.; Ibrahim, H.A. Utilisation of recycled concrete aggregates for sustainable highway pavement applications; a review. *Constr. Build. Mater.* **2020**, *235*, 117444. [CrossRef]
150. IRCOW. Final Report Summary-IRCOW (Innovative Strategies for High-Grade Material Recovery from Construction and Demolition Waste). Available online: https://cordis.europa.eu/project/id/265212/reporting (accessed on 14 September 2021).
151. Abadías, A.I.; Guzmán, B. Recycling of bituminous materials: Conclusions of the DIRECT-MAT project (in Spanish, Técnicas de demolición y reciclado de materiales para carretera). *Asf. Y Paviment.* **2013**, *III*, 11–25.
152. Sánchez-Cotte, E.H.; Fuentes, L.; Martinez-Arguelles, G.; Quintana, H.A.R.; Walubita, L.F.; Cantero-Durango, J.M. Influence of recycled concrete aggregates from different sources in hot mix asphalt design. *Constr. Build. Mater.* **2020**, *259*, 120427. [CrossRef]
153. Gopalam, J.; Giri, J.P.; Panda, M. Influence of binder type on performance of dense bituminous mixture prepared with coarse recycled concrete aggregate. *Case Stud. Constr. Mater.* **2020**, *13*, e00413. [CrossRef]
154. Kanoungo, S.; Sharma, U.; Singh, S. Assessment of treatment methods of recycled aggregates for utilization in flexible pavements. *Mater. Today Proc.* **2020**, *43*, 1320–1324. [CrossRef]
155. Radević, A.; Isailović, I.; Wistuba, M.P.; Zakić, D.; Orešković, M.; Mladenović, G. The Impact of Recycled Concrete Aggregate on the Stiffness, Fatigue, and Low-Temperature Performance of Asphalt Mixtures for Road Construction. *Sustainability* **2020**, *12*, 3949. [CrossRef]
156. Polo-Mendoza, R.; Peñabaena-Niebles, R.; Giustozzi, F.; Martinez-Arguelles, G. Eco-friendly design of Warm mix asphalt (WMA) with recycled concrete aggregate (RCA): A case study from a developing country. *Constr. Build. Mater.* **2022**, *326*, 126890. [CrossRef]
157. Abass, B.J.; Albayati, A.H. Influence of recycled concrete aggregate treatment methods on performance of sustainable warm mix asphalt. *Cogent Eng.* **2020**, *7*, 1718822. [CrossRef]
158. Neves, J.; Lameirão, J.; Freire, A.C. Physical and mechanical properties of warm mix asphalt composed of recycled concrete aggregates. In Proceedings of the 7th International Conference on Bituminous Mixtures and Pavements, Thessaloniki, Greece, 12–14 June 2019; pp. 194–200.

159. Monu, K.; Gondaimei Ransinchung, R.N.; Pandey, G.S.; Singh, S. Performance Evaluation of Recycled-Concrete Aggregates and Reclaimed-Asphalt Pavements for Foam-Mix Asphalt Mixes. *J. Mater. Civ. Eng.* **2020**, *32*, 04020295. [CrossRef]
160. Daquan, S.; Yang, T.; Guoqiang, S.; Qi, P.; Fan, Y.; Xingyi, Z. Performance evaluation of asphalt mixtures containing recycled concrete aggregates. *Int. J. Pavement Eng.* **2018**, *19*, 422–428. [CrossRef]
161. Galan, J.J.; Silva, L.M.; Pasandín, A.R.; Pérez, I. Evaluation of the Resilient Modulus of Hot-Mix Asphalt Made with Recycled Concrete Aggregates from Construction and Demolition Waste. *Sustainability* **2020**, *12*, 8551. [CrossRef]
162. Zhang, Z.; Wang, K.; Liu, H.; Deng, Z. Key performance properties of asphalt mixtures with recycled concrete aggregate from low strength concrete. *Constr. Build. Mater.* **2016**, *126*, 711–719. [CrossRef]
163. Takamura, K.; Lok, K.P.; Wittlinger, R.; Aktiengesellschaft, B. Microsurfacing for preventive maintenance: Eco-efficient strategy. In Proceedings of the International Slurry Seal Association Annual Meeting, Maui, HI, USA, 4–8 March 2001; pp. 63–72.
164. Paranavithana, S.; Mohajerani, A. Effects of recycled concrete aggregates on properties of asphalt concrete mixture. *J. Mater. Civ. Eng.* **2006**, *17*, 400–406.
165. Mills-Beale, J.; You, Z. The mechanical properties of asphalt mixtures with Recycled Concrete Aggregates. *Constr. Build. Mater.* **2010**, *24*, 230–235. [CrossRef]
166. Zulkati, A.; Wong, Y.; Sun, D. Effects of Fillers on Properties of Asphalt-Concrete Mixture. *J. Transp. Eng.* **2012**, *138*, 902–910. [CrossRef]
167. Rafi, M.M.; Qadir, A.; Siddiqui, S. Experimental testing of hot mix asphalt mixture made of recycled aggregates. *Waste Manag. Res. J. Int. Solid Wastes Public Clean. Assoc. ISWA* **2011**, *29*, 1316–1326. [CrossRef]
168. Zhang, H.; Mao, X.; Li, W.; Gou, J. Effect of gradation and fineness on performance of rubber asphalt mixture at low temperature. In Proceedings of the 6th International Conference on Machinery, Materials, Environment, Biotechnology and Computer, Tianjin, China, 11–12 June 2016; pp. 1506–1509.
169. Dhir, R.; Brito, J.; Silva, R.V.; Lye, C.Q. Use of Recycled Aggregates in Road Pavement Applications. In *Sustainable Construction Materials*; Dhir, R., Brito, J., Silva, R.V., Lye, C.Q., Eds.; Woodhead Publishing: Sawston, UK, 2019; pp. 451–494. [CrossRef]
170. Qasrawi, H.; Asi, I. Effect of bitumen grade on hot asphalt mixes properties prepared using recycled coarse concrete aggregate. *Constr. Build. Mater.* **2016**, *121*, 18–24. [CrossRef]
171. Kowalski, K.J.; Król, J.; Radziszewski, P.; Casado, R.; Blanco, V.; Pérez, D.; Viñas, V.M.; Brijsse, Y.; Frosch, M.; Le, D.M.; et al. Eco-friendly Materials for a New Concept of Asphalt Pavement. *Transp. Res. Procedia* **2016**, *14*, 3582–3591. [CrossRef]
172. Abedalqader, A.; Shatarat, N.; Ashteyat, A.; Katkhuda, H. Influence of temperature on mechanical properties of recycled asphalt pavement aggregate and recycled coarse aggregate concrete. *Constr. Build. Mater.* **2021**, *269*, 121285. [CrossRef]
173. Fournier, J.M.; Acosta Álvarez, D.; Aenlle, A.A.; Tenza-Abril, A.J.; Ivorra, S. Combining reclaimed asphalt pavement (RAP) and recycled concrete aggregate (RCA) from Cuba to obtain a coarse aggregate fraction. *Sustainability* **2020**, *12*, 5356. [CrossRef]
174. Coban, H.S.; Cetin, B.; Ceylan, H.; Edil, T.B.; Likos, W.J. Evaluation of Engineering Properties of Recycled Aggregates and Preliminary Performance of Recycled Aggregate Base Layers. *J. Mater. Civ. Eng.* **2022**, *34*, 04022053. [CrossRef]
175. Pasandín, A.R.; Pérez, I. Fatigue performance of bituminous mixtures made with recycled concrete aggregates and waste tire rubber. *Constr. Build. Mater.* **2017**, *157*, 26–33. [CrossRef]
176. Pérez, I.; Pasandín, A.R. Moisture damage resistance of hot-mix asphalt made with recycled concrete aggregates and crumb rubber. *J. Clean. Prod.* **2017**, *165*, 405–414. [CrossRef]
177. Giri, J.P.; Panda, M.; Sahoo, U.C. Use of waste polyethylene for modification of bituminous paving mixes containing recycled concrete aggregates. *Road Mater. Pavement Des.* **2020**, *21*, 289–309. [CrossRef]
178. Chen, M.J.; Wong, Y.D. Porous asphalt mixture with 100% recycled concrete aggregate. *Road Mater. Pavement Des.* **2013**, *14*, 921–932. [CrossRef]
179. Arabani, M.; Azarhoosh, A.R. The effect of recycled concrete aggregate and steel slag on the dynamic properties of asphalt mixtures. *Constr. Build. Mater.* **2012**, *35*, 1–7. [CrossRef]
180. Roque, A.J.; da Silva, P.F.; Rodrigues, G.; Almeida, R. Recycling of CDW and Steel Slag in Drainage Layers of Transport Infrastructures. *Procedia Eng.* **2016**, *143*, 196–203. [CrossRef]

sustainability

MDPI

Article

Effect of Palm Oil Clinker (POC) Aggregate on the Mechanical Properties of Stone Mastic Asphalt (SMA) Mixtures

Ali Mohammed Babalghaith [1], Suhana Koting [1,*], Nor Hafizah Ramli Sulong [1,2], Mohamed Rehan Karim [1], Syakirah Afiza Mohammed [1] and Mohd Rasdan Ibrahim [1]

1. Center for Transportation Research, Department of Civil Engineering, Faculty of Engineering, University of Malaya, Kuala Lumpur 50603, Malaysia; bablgeath@hotmail.com (A.M.B.); hafizah.ramlisulong@qut.edu.au (N.H.R.S.); mrehan57@gmail.com (M.R.K.); syakirahafiza@unimap.edu.my (S.A.M.); rasdan@um.edu.my (M.R.I.)
2. School of Civil and Environmental Engineering, Science and Engineering Faculty, Queensland University of Technology, 2 George Street, Brisbane, QLD 4000, Australia

* Correspondence: suhana_koting@um.edu.my

check for updates

Received: 29 February 2020; Accepted: 26 March 2020; Published: 30 March 2020

Abstract: Aggregate composition has a pivotal role in ensuring the quality of pavement materials. The use of waste materials to replace the aggregate composition of asphalt pavement leads to green, sustainable, and environmentally friendly construction, which ultimately preserves nature by reducing the need to harvest materials from natural sources. Using the Marshall mix design, the main objective of this paper is to investigate the effects of waste palm oil clinker (POC) as fine aggregates replacement on the properties of stone mastic asphalt (SMA) mixture. Six groups of asphalt mixtures were prepared using different percentages of palm oil clinker content (0%, 20%, 40%, 60%, 80%, and 100%). To determine the Marshall properties and select the optimum binder content, asphalt mixture samples with different percentages of asphalt binder content (5.0%, 5.5%, 6.0%, 6.5%, and 7.0%) were prepared for each group. The results showed that the palm oil clinker was appropriate for use as a fine aggregate replacement up to 100% in SMA mixture and could satisfy the mix design requirements in terms of Marshall stability, flow, quotient, and volumetric properties. However, the percentage of palm oil clinker replacement as fine aggregate has merely influenced the optimum binder content. Furthermore, there were improvements in the drain down, resilient modulus and indirect tensile fatigue performances of the SMA mixture. In conclusion, the use of POC as fine aggregates replacement in SMA mixture indicates a good potential to be commercialized in flexible pavement construction.

Keywords: palm oil clinker; hot mix asphalt; Marshall properties; stone mastic asphalt; waste materials; sustainable pavement

---

1. Introduction

In recent times, natural resources are being quickly depleted due to the enormous amount of raw materials being consumed worldwide. With the growth in the population, there are concomitant increases in waste generated by the increasing demand of new agricultural, manufacturing and construction industries, which results in a large amount of waste materials being deposited in landfills every year. Hence, researchers are trying to find new solutions to reduce the negative impact of these wastes on the environment. For example, waste materials are used to replace the original constructions materials such as binders, aggregates, and fibers. This leads to green, sustainable, and environmentally friendly construction. The use of waste materials in asphalt pavement minimizes the construction costs and preserves nature by reducing the need to harvest aggregates from natural sources [1]. Using waste materials in the asphalt pavement not only reduces the environmental problems but also improves some of the properties of the asphalt pavement [2]. With respect to asphalt pavement, many researchers have studied the use of waste materials as aggregate in asphalt mixtures, such as crushed glass [3], steel slag [4], plastics [5], recycled pavement [6], recycled concrete demolition [7–9], coal bottom ash [10], and mining waste [11].

Palm oil trees, as shown in Figure 1, are planted in many countries for food applications such as cooking oil and margarines, and for non-food applications such as animal feed, biodiesel and energy generation. Malaysia, Indonesia, and Thailand produced almost 91% of the total world's palm oil in 2015. Hence, Malaysia, being one of the largest producers and manufacturers of palm oil products contributes 32% of the world supply of palm oil [12]. Malaysia alone has produced 19.86 million tons of palm oil and exported 18.47 million tons of palm oil in 2019 [13]. Palm oil processing activities produce only 10% oils from fresh fruit and kernel while the remaining 90% remain in the form of waste is still not used for the industry [14]. The processing of palm oil generates various types and forms of waste materials and by-products, such as empty fruit bunches, palm oil mill effluent, sterilizer condensate, fibers and kernel shells [15]. Malaysia produced about 50 million tonnes of palm oil biomass every year and it is expected to be 100 million tonnes in 2020 [16]. The oil palm biomass, such as palm oil shell and fibers, was used as boiler fuel for power generation purposes [17,18]. This process creates large amounts of a new waste material called a palm oil clinker—an aggregate like a porous stone, grey in color, irregular, and flaky [19].

Figure 1. Palm oil tree.

Many studies have been conducted on the use of palm oil clinker (POC) as a replacement to concrete construction materials, such as aggregate [20–23], and cement [24–27]. In highway pavement, a number of researchers have studied the use of palm oil waste as a filler [28–32]. Ahmad et al. [31] investigated the practical use of Palm oil fuel ash (POFA) at different percentages as a filler in asphalt pavement. They found that the addition of 3% POFA by weight of total aggregate could increase the stability and resilient modulus of asphalt pavement. Moreover, Kamaluddin [29] studied the use of palm oil fuel ash in stone mastic asphalt (SMA) mixture as a filler material, and the results indicated that the optimum value of POFA by weight of filler content is 50%.

To date, there are very limited studies that have investigated the use of palm oil clinker as aggregate replacement in highway pavement. This research presents an experimental research on the use of palm oil clinker waste with different percentages in SMA as fine aggregate replacement. The main objective was to study the effect of aggregate replacement on the properties of SMA, and to compare with the control mixture containing normal granite aggregate. The Marshall stability, Marshall flow, and volumetric tests were conducted on five different asphalt mixtures containing different contents of palm oil clinker (20%, 40%, 60%, 80%, and 100%) and one asphalt mixture containing original aggregate (control mix). Finally, based on optimum binder content results, different sets of samples were prepared and tested for drain down test, resilient modulus test, and indirect tensile fatigue test.

2. Materials

2.1. Asphalt Binder

In this study, 80/100 penetration grade bitumen was used as the binder. To maintain the quality of the binder during binder testing preparations, the 80/100 binder was transferred from the drum to one-liter containers. The properties of the asphalt binder are presented in Table 1.

Table 1. Properties of 80/100 asphalt binder.

Property	Reference	Value	Requirement
Penetration @ 25 °C (0.1mm)	ASTM D0005-13	87	80–100
Softening Point (°C)	ASTM D0036-95	46	45–52
Rotational Viscosity @ 135 °C, (cps)	ASTM D4402-02	312	<3000
Rotational Viscosity @ 165 °C, (cps)	ASTM D4402-02	100	<3000
Specific Gravity @ 25 °C	ASTM D0070-09	1.020	-
G*/sinδ @58 °C (kpa)	ASTM D7175-15	1.576	>1.0

2.2. Aggregates

The crushed granite aggregates used in this study were collected from the same aggregate supplier to maintain the quality and results reliability. The dried materials were sieved, and the materials retained on each sieve were transferred to a container and labeled with the aggregate size. The aggregate gradation used in this study was stone mastic asphalt (SMA) 20. The gradation of SMA20 was according to the specification of the Malaysian Public Works-Road Department (Jabatan Kerja Raya-JKR) [33]; as shown in Table 2 and Figure 2.

Table 2. Stone mastic asphalt (SMA)20 aggregate gradation.

Sieve Size (mm)	% Passing			% Retained	Weight (g)
	Min.	Max.	Mid.		
19	100	100	100.0	0.0	0.0
12.5	85	95	90.0	10.0	110
9.5	65	75	70.0	20.0	220
4.75	20	28	24.0	46.0	506
2.36	16	24	20.0	4.0	44
0.600	12	16	14.0	6.0	66
0.300	12	15	13.5	2.0	5.5
0.075	8	10	9.0	3.0	49.5
Pan				9.0	99
				100	**1100**

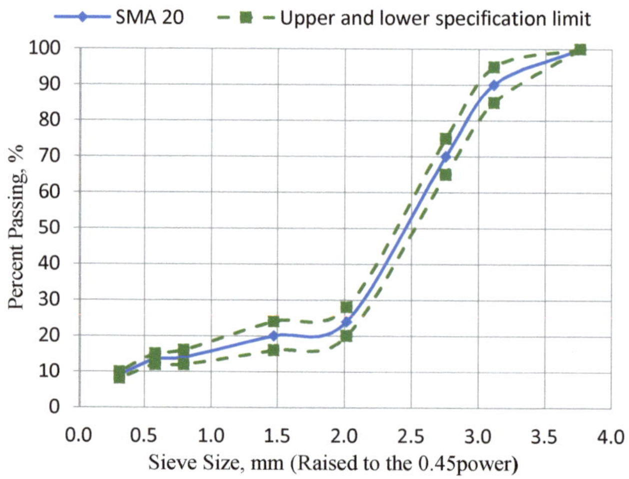

Figure 2. SMA20 aggregate gradation.

Meanwhile, the physical properties tests for SMA20 aggregate were also carried out. The physical properties values for the aggregate are tabulated in Table 3.

Table 3. Physical properties of SMA20 aggregate.

Property	Test Method	Value (%)	Requirement
Los Angeles abrasion	ASTM: C131-14	19.8	<30%
Flakiness index	BS 182: Part3	7.3	<20%
Elongation index	BS 182: Part3	13.4	<20%
Impact value	BS 812: Part3	10.3	<15%

2.3. Palm Oil Clinker and Fuel Ash

The palm oil fuel ash and clinker were obtained from a local factory that processes palm oil. The palm oil clinker was collected as large chunk clinker, as shown in Figure 3a. The large chunk clinker was then crushed to become a fine aggregate, as shown in Figure 3b. The crushed materials were sieved and each sieve size was transferred to a separate container. Palm oil fuel ash was collected in powder form and sieved through a 0.075 mm (#200) sieve.

(a) (b)

Figure 3. Palm oil clinker; (**a**) large chunk clinker, (**b**) crushed clinker.

3. Experimental Design

The mix design was used to obtain the optimum binder content for the mixtures to achieve high resistance to deformation and cracking. The method consisted of three basic steps: Aggregate selection, asphalt binder selection, and optimum binder content determination. In this study, the Marshall mix design method was used, as it is considered to be one of the most common methods, well proven, and requires relatively light, portable, and inexpensive equipment. The general steps of the experimental work are illustrated in Figure 4.

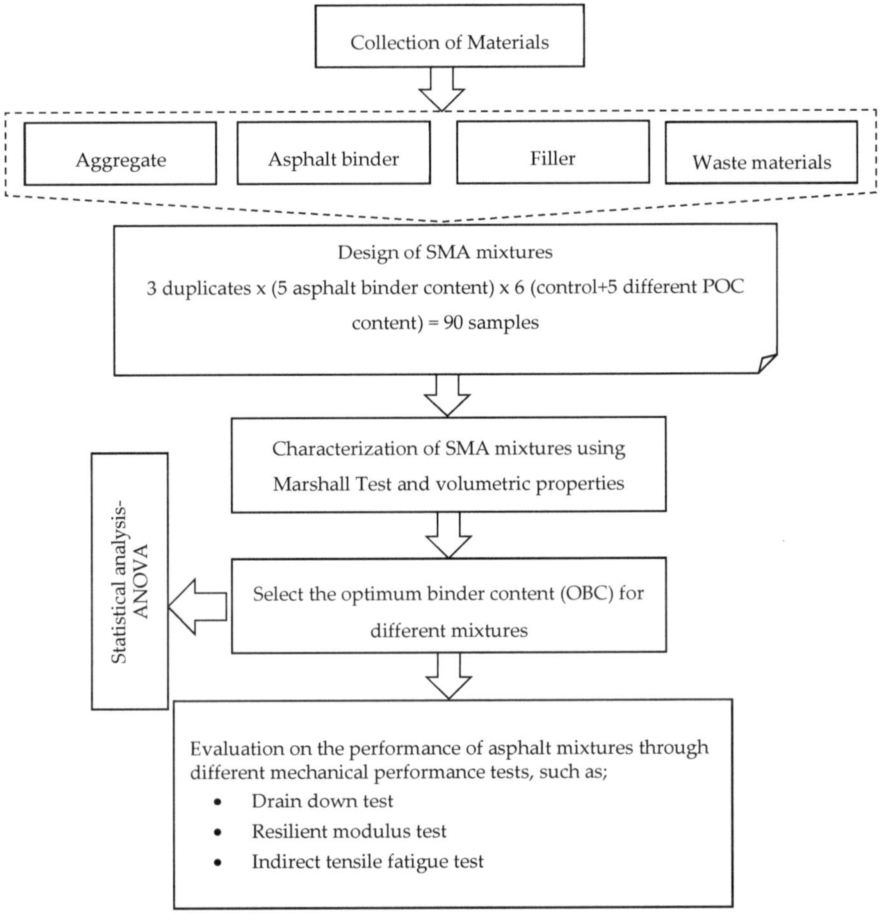

Figure 4. Experimental work.

In this study, six groups of asphalt mixtures were prepared using different contents of palm oil clinker (0%, 20%, 40%, 60%, 80%, and 100%) for the granite fine aggregate (passing sieve size 4.75 mm) replacement. The group with 0.0% waste materials was used as the control mixture. For each group, three duplicate samples with a constant content of palm oil clinker and different asphalt binder content (5.0%, 5.5%, 6.0%, 6.5%, and 7.0%) were prepared to determine the Marshall properties and select the optimum binder content. Then, the statistical analysis, namely, the analysis of variance (ANOVA) test was carried out using SPSS software. Finally, to evaluate the mechanical performance of optimum

SMA mixtures, different sets for each group were prepared. The SMA mixtures samples were tested for drain down, resilient modulus, and indirect tensile fatigue performance.

3.1. Preparation of Samples

Samples of approximately 1100 grams were weighed according to the combination of aggregates, then heated to a temperature of 160–170 °C for 3 hours. At the same time, the asphalt binder was heated in another oven at a mixing temperature of 150–155 °C for 1 hour. Using the dry process method, all ingredients were mixed together, then the required asphalt binder content (5.0%, 5.5%, 6.0%, 6.5%, and 7.0% by weight of mix) was added and mixed thoroughly to form homogenous mixtures. The blended mixture was put into a Marshall mold and moved to the Marshall compactor. Then, each sample was subjected to 50 blows on each side at a compaction temperature of 140 °C–145 °C according to the specifications of the Malaysian Public Works Department for road works [33]. Finally, they were left to cool for 24 hours at a normal temperature then removed from the mold using a jack and kept in the laboratory until further testing.

3.2. Marshall Test

3.2.1. Stability, Flow and Quotient

The Marshall stability of asphalt mixture is defined as maximum load value during the test before the compacted sample failure at a temperature of 60 °C, and a loading rate of 5.08 cm per minute [34]. It was determined according to ASTM D6927 [35]. The plastic flow at the failure point of the mix is called the Marshall flow value, which is expressed in units of 0.25 mm and is measured by the sample vertical deformation in the direction of the applied load. The Marshall flow indicates the plasticity and flexibility properties of mixtures. According to the JKR specifications for SMA, the acceptable range is from 2 to 4 mm [33]. The minimum value controls the brittleness and strength, while the maximum value dictates the plasticity and maximum asphalt binder content of the mixture [36]. The Marshall stability value divided by the Marshall flow value is called the Marshall quotient, which can be used as an indicator of resistance to rutting or permanent deformation of the asphalt mixture [11].

3.2.2. Volumetric Properties

The volume of asphalt binder and aggregates is affected by the volumetric properties of the required asphalt mixture. The volumetric properties of the asphalt mixture are among the important factors that affect the pavement performance and durability [37]. In this study, the volumetric properties, such as density, air voids, and voids filled with asphalt binder, were calculated. The density was determined according to ASTM D2726 [38], and the air voids and voids filled with asphalt binder were determined according to ASTM D3203 [39].

3.2.3. Optimum Asphalt Content Selection

To optimize the asphalt mixtures, the Marshall procedure (ASTM D6926 [40]) was applied. Five percentages of asphalt binder (5.0%, 5.5%, 6.0%, 6.5%, and 7.0% by weight of mix) were used, and, for each percentage, three duplicate samples were prepared using the Marshall compactor. In general, the selection of the optimum binder content (OBC) depends on the stability, flow, specific gravity, and voids in the mixture. However, in the SMA mixtures, the air voids were considered to be the main factor to select the optimum binder content [41–44]. Moreover, the asphalt binder percentage that produces approximately 4% air voids will provide less asphalt binder bleeding and better rutting resistance [43]. In this study, the optimum asphalt contents for all SMA mixtures were selected to produce 4% air voids and also checked for the mix design criteria specified by JKR [33]. Statistical analysis was done using SPSS software at the level of 5% significance.

3.3. Mixture Performance Tests

To evaluate the performance and mechanical properties of POC as fine aggregates replacement in SMA20 mixture, drain down, resilient modulus, and indirect tensile fatigue tests were performed. Each test is described in the following sections.

3.3.1. Drain Down Test

Drain down situation is the portion of the mixture, namely, fines and asphalt binder, which separates and flows downward through the mixture, and is more significant for SMA mixtures than conventional (dense-graded) mixtures. This test was carried out in accordance with ASTM D6390 [45]. This test is to simulate the conditions that the mixture is likely to encounter as it is produced, stored, transported, and placed. The test is carried out on loose mixtures at the optimum binder content. This is to ensure that the engineering properties of the binder drain down of the SMA mixture is within the require or acceptable level. The main procedure of the test is to place loose SMA mixture in a wire basket, which has been fabricated using standard 6mm sieve cloth. The basket and the loose SMA mixture are then placed in an oven for one hour at 170 °C. The mass of portion from fines and asphalt binder draining from the mixture was measured. This mass was recorded as a percentage of the initial sample weight. The maximum drain down of SMA mixture is 0.3% by weight of the mixture [33].

3.3.2. Resilient Modulus Test

The resilient modulus test is the most appropriate and common test to measure the stiffness modulus of asphalt mixtures. It was carried out using a universal materials testing apparatus (UMATTA) machine, in accordance with ASTM D4123 [46]. An average of 101.7 mm diameter and 65 ± 1 mm thickness sample was prepared for resilient modulus test. The test was carried out at four temperatures, namely 5 °C, 25 °C, 35 °C, and 40 °C. The samples were left in the chamber at the desired temperature for a minimum of 3 hours before the test. The stiffness modulus of the samples was calculated automatically after the test, which is based on the following equation:

$$\text{Resilient Modulus (MPa)} = \frac{P(v + 0.25)}{HxT} \tag{1}$$

where, P is the peak load (N); v is the Poisson's ratio; T is the average thickness of the sample (mm) and H is total recoverable deformation on the horizontal axis (mm). In this test, all samples were tested at three different points of vertical loading. Finally, the average result from the three samples was calculated and recorded.

3.3.3. Indirect Tensile Fatigue Test

Asphalt mixture fatigue performance is linked to the service life of the roadway. The better the fatigue performance of the asphalt mixture, the longer the service life of the pavement. The fatigue life of asphalt mixture depends on the mixture properties such as grade and amount of asphalt binder, aggregates gradation, and mixture air voids [47]. The fatigue performance of the asphalt mixture can be characterized by four-point bending fatigue test or indirect tensile fatigue test [48]. Indirect tensile fatigue test is an effective method that has been used in many studies to investigate the fatigue performance [3,47,49,50]. This test can be carried out in two modes, either in controlled stress or controlled strain.

In the controlled stress mode, the applied stress is kept constant, and the strain values are changeable; while in controlled strain mode, the strain value is kept constant and the stress values are changeable. In this study, indirect tensile fatigue test was performed by a UMATTA machine in controlled stress mode according to EN 12697-24 [51]. The loading shape used was a haversine signal loading force, i.e., 2600 N. The loading time was 0.1s, follows by a rest period of 0.4s. The test was carried out at 25 °C. The vertical deformation of the sample was recorded during the test. The fatigue life was defined as the number of load cycles reached when the sample cracked, or when the permanent vertical deformation reached the maximum value of 9 mm [47].

4. Results and Discussion

4.1. Marshall Stability

The average Marshall stability for the three duplicate samples from each asphalt mixture are shown in Table 4 and presented in Figure 5. The results indicate that the values of stability for all asphalt binder contents vary in tandem with the palm oil clinker content. The stability values increased with higher percentage of POC replacement (with slight reduction at 80% replacement). This improvement might be due to the fiber content in the palm oil clinker, where the fiber increased the strength and ductility of the mixtures. Moreover, the addition of asphalt binder into the mixture resulted in a decrease in the stability value because of reduced contact point between the aggregates within the asphalt mixture [34].

Table 4. Marshall stability with different ratios of binder content and palm oil clinker (POC).

Binder Content (%)	Marshall Stability (kN)					
	POC Content (%)					
	0	20	40	60	80	100
5.0	8.69	9.34	9.70	9.94	9.26	11.21
5.5	8.27	9.52	8.81	10.22	8.61	10.75
6.0	8.34	8.16	7.62	8.10	9.11	9.31
6.5	8.61	8.33	7.76	8.91	8.67	10.13
7.0	8.33	8.11	7.06	8.30	8.24	9.22

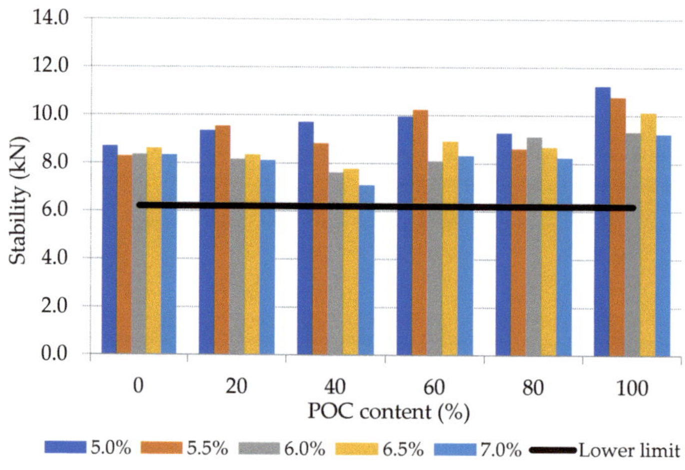

Figure 5. Marshall stability for different clinker and binder contents.

Table 4 shows that the control mixture and all aggregate replacement mixtures satisfied the recommended minimum stability value of 6.2 kN according to the JKR specifications for SMA20 [33]. Therefore, palm oil clinker has proven to be an alternative waste material for fine aggregate replacement in SMA20.

4.2. Marshall Flow and Quotient

The Marshall flow values versus palm oil clinker contents for each asphalt binder content are summarized in Table 5 and illustrated in Figure 6. There was no clear trend on the influence of POC contents on the flow values. However, all of the asphalt mixtures Marshall flow values were within the acceptable range of values (2–4 mm) recommended by JKR for SMA20 [33].

Table 5. Marshall flow with different ratios of binder content and POC.

Binder Content (%)	Marshall Flow (mm)					
	POC Content (%)					
	0	**20**	**40**	**60**	**80**	**100**
5.0	3.22	3.04	4.18	2.81	3.05	2.61
5.5	3.65	3.33	3.82	3.55	3.29	3.53
6.0	3.70	3.59	3.00	4.22	2.84	3.73
6.5	3.67	3.69	3.19	3.10	3.74	4.20
7.0	2.61	3.40	2.98	3.48	3.12	3.96

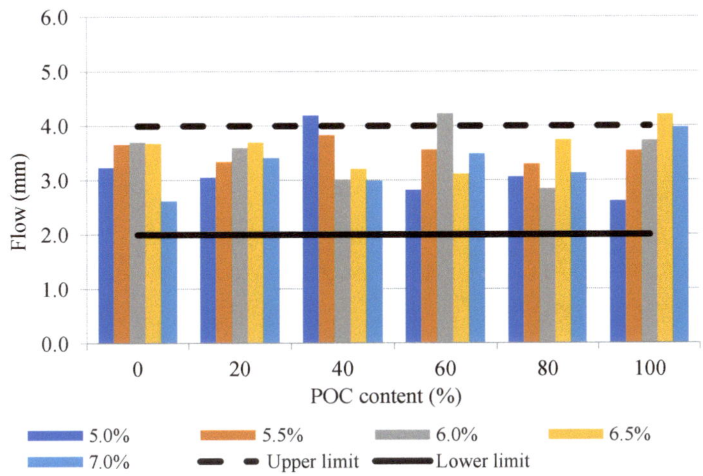

Figure 6. Marshall flow for the different POC and binder contents.

Figure 7 shows the Marshall quotient values versus palm oil clinker for different asphalt binder contents. Most of the asphalt mixtures quotient values were within the acceptable range of 2–4 kN/mm, except for the mix with 60% clinker and 6.0% asphalt binder for which the value was slightly lower than the prescribed lower limit. Therefore, it can be concluded that the asphalt mixtures with POC as aggregate replacement have the required strength and stiffness to resist the permanent deformation under traffic loads.

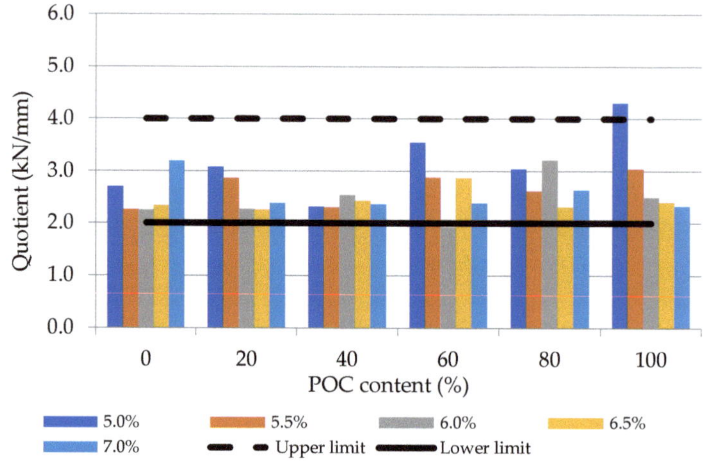

Figure 7. Marshall quotient for different POC and binder contents.

4.3. Mixture Density

The density values of all the mixtures versus the palm oil clinker for different asphalt binder contents are graphically presented in Figure 8.

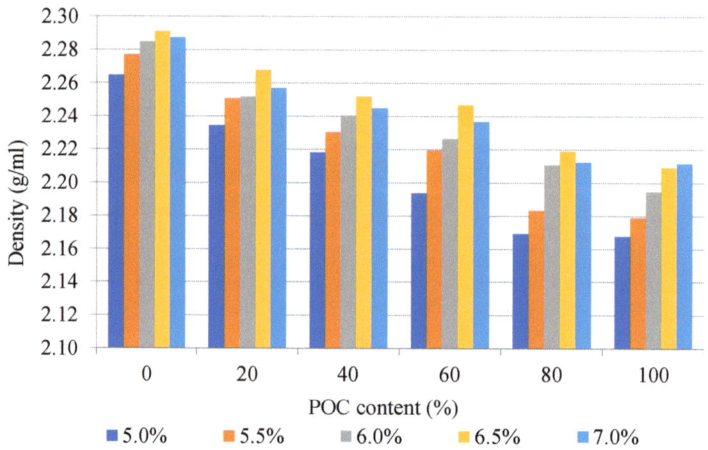

Figure 8. Mixture density for different POC and binder contents.

The results show that the palm oil clinker aggregate affected the density of the mixture and had a similar trend, and that, for any asphalt binder content, the density value of the compacted sample decreased with an increase in the clinker aggregate replacement. Moreover, all the mixtures replaced with the clinker aggregate had lower density values compared to the control mix, which is because of the lower specific gravity of palm oil clinker compared to the normal aggregates.

Moreover, for each particular POC content, the trend of density variation with different alsphalt binder contents followed almost a similar feature. The density kept increasing until 6.5% asphalt binder content, then reduced afterwards. The main reason for the increment in density is due to the asphalt binder filling the air voids between the aggregates.

4.4. Air Voids and the Void Filled with Binder

A void in the mix is one of the important parameters for asphalt mix designs as it is used to determine the optimum asphalt binder for the SMA mixes [41–43]. The performance of the asphalt mixture is dependent on the air voids in the mix. Too many air voids in the mixture lead to asphalt cracking due to the low asphalt binder content, which fails to coat the aggregates in the mixture, while too few air voids in the mixture may result in more deformation and asphalt binder bleeding [52].

The mixtures air void values versus palm oil clinker for different asphalt binder contents are tabulated in Table 6 and graphically presented in Figure 9.

Table 6. Air voids in mixtures with different ratios of binder content and POC.

Binder Content (%)	Mixture Air Void (%)					
	POC Content (%)					
	0	20	40	60	80	100
5.0	6.62	6.46	6.49	7.01	7.46	7.00
5.5	5.44	5.10	5.32	5.27	6.23	5.89
6.0	4.43	4.40	4.35	4.32	4.40	4.57
6.5	3.48	3.04	3.07	2.76	3.40	3.29
7.0	2.96	2.84	2.72	2.56	3.04	2.56

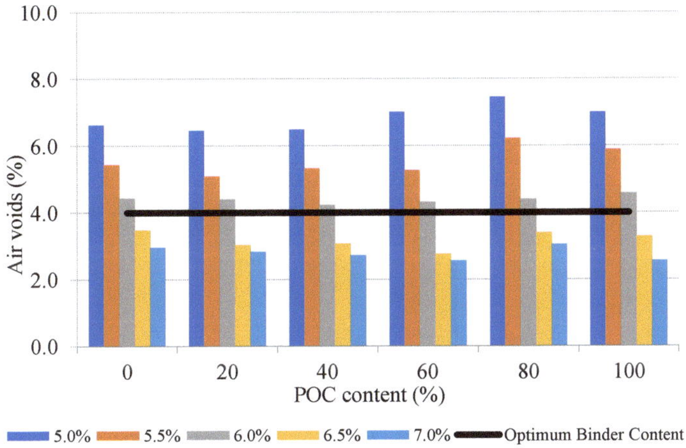

Figure 9. Air voids for different POC and binder contents.

From Figure 9, it can be seen that with an increase in the palm oil clinker aggregate replacement, the air void values of the mixtures had a similar trend and were approximately stable, which means that replacing the original aggregate with palm oil clinker aggregate has no effect on the mixtures air void values. However, there was a significant effect on the mixtures' air voids for different asphalt binder content, where the values decreased as the asphalt binder contents increased. The same behavior was observed by several previous studies [5,9,36,41,42,52]. The reason behind this decrement is due to the asphalt binder filling the air voids. For all POC replacement mixtures, 6.5% and 7.0% binders were not acceptable as the air voids' percentages were less than the desired value of 4.0%, which may have resulted in more deformation and asphalt binder bleeding. On the other hand, too many air voids, as shown in 5.0% and 5.5% binders, is also not preferable as low asphalt binder content may cause cracking in asphalt pavement.

The results for voids filled with asphalt (VFA) versus palm oil clinker for different asphalt binder contents are tabulated in Table 7 and presented in Figure 10. It shows that, for all asphalt binder contents, an increase in palm oil clinker content caused the VFA to approximately level off, which means that a change in the clinker content did not affect the VFA values. However, there was a significant effect on the VFA values versus the asphalt binder content. This shows that the VFA values increased when the asphalt binder contents increased. This behavior was confirmed in other researches [5,36,41].

Table 7. Mixtures voids filled with asphalt (VFA) with different ratios of binder content and POC.

Binder Content (%)	Mixture VFA (%) POC Content (%)					
	0	**20**	**40**	**60**	**80**	**100**
5.0	62.64	62.93	62.64	60.55	58.79	60.32
5.5	69.31	70.49	69.32	69.45	65.42	66.65
6.0	75.22	75.07	75.84	75.22	74.74	73.95
6.5	80.77	82.64	82.36	83.83	80.63	81.07
7.0	84.18	84.53	84.98	85.73	83.32	85.63

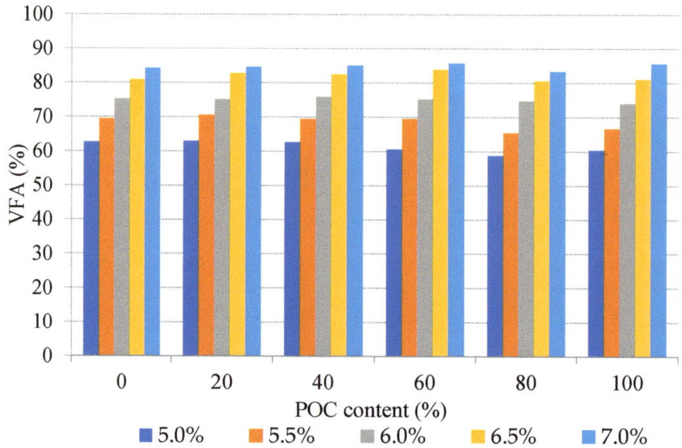

Figure 10. VFA for different POC and binder contents.

4.5. Asphalt Content Optimization

For each asphalt mixture group, five percentages of asphalt binder (5.0%, 5.5%, 6.0%, 6.5%, and 7.0% by weight of mix) were used to determine the optimum asphalt content. The optimum asphalt contents for all SMA mixtures that produce 4% air voids were selected and are shown in Figure 11.

Statistical analysis was done using analysis of variance (ANOVA) to evaluate the effect of POC on the optimum asphalt content, and the hypothesis test was utilized to test whether the differences in the optimum asphalt content between all the mixtures were statistically significant. In this hypothesis test, the null hypothesis H0 is $\mu1=\mu2=\mu3=\mu4=\mu5=\mu6$ and the alternative hypothesis H1 is $\mu1\neq\mu2\neq\mu3\neq\mu4\neq\mu5\neq\mu6$, where $\mu1$, $\mu2$, $\mu3$, $\mu4$, $\mu5$, and $\mu6$ represent the mean value of the optimum asphalt content for the different mixtures. The descriptive statistics results are shown in Table 8, and the ANOVA results are shown in Table 9. It can be seen that the differences in the optimum asphalt content were not statistically significant since the p-value (0.578) obtained from the statistical analysis was higher than the alpha value, 0.05, used in the statistical analysis. Also, the F-ratio (0.787) value was less than 1, which immediately indicates that the difference was not significant.

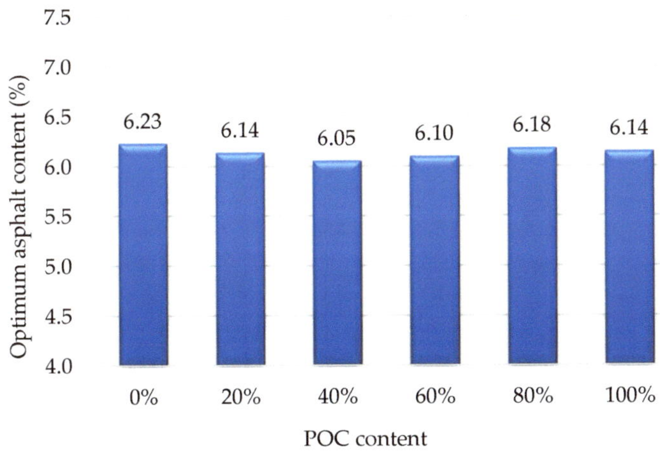

Figure 11. Optimum binder content for different asphalt mixtures.

Table 8. Descriptive statistics results for optimum binder content.

Mixtures	N	Mean	Std. Deviation	Std. Error	95% Confidence Interval for Mean		Minimum	Maximum
					Lower Bound	Upper Bound		
0%	3	6.23	0.038	0.022	6.13	6.32	6.18	6.25
20%	3	6.14	0.070	0.041	5.96	6.31	6.07	6.21
40%	3	6.05	0.194	0.112	5.57	6.53	5.91	6.27
60%	3	6.10	0.046	0.026	5.98	6.21	6.06	6.15
80%	3	6.18	0.094	0.055	5.94	6.41	6.07	6.26
100%	3	6.14	0.175	0.101	5.71	6.58	5.98	6.33
Total	18	6.14	0.116	0.027	6.08	6.20	5.91	6.33

Table 9. ANOVA results for optimum binder content.

	Sum of Squares	DF	Mean Square	F Ratio	*p*-Value
Between Groups	0.056	5	0.011	0.787	0.578
Within Groups	0.172	12	0.014		
Total	0.228	17			

4.6. Drain Down Test Results

Throughout this test, the susceptibility of the SMA mixtures with different POC replacement to drain down was investigated and evaluated. The results of drain down performance for control mix and SMA mixtures with POC replacement are presented in Figure 12.

It can be observed that all SMA mixtures with different POC replacement exhibited lower drain down values as compared to the maximum allowable limit, i.e., 0.3% by weight of the mixture. The percentage of drain down decreased significantly with the increment of POC content. The SMA mixture with 100% POC replacement showed the lowest result of drain down performance, i.e., 0.038% by weight of mixture. For all POC replacement mixtures, the result was lower compared to the control mixture (0% POC replacement). This indicates that POC act or perform as a stabilizing additive in SMA mixtures. The lower values of drain down for SMA mixtures with POC replacement are due to the existence of fibers from the palm oil waste. The fibers will reinforce the SMA mixtures and help to generate a three-dimensional network, thus improving the mix adhesion and reduce the asphalt binder drain down [53]. This finding is also consistent with that of Panda, Suchismita and Giri [42], who found that the use of the coconut fiber in SMA mixture prevented the usual draining of asphalt binder.

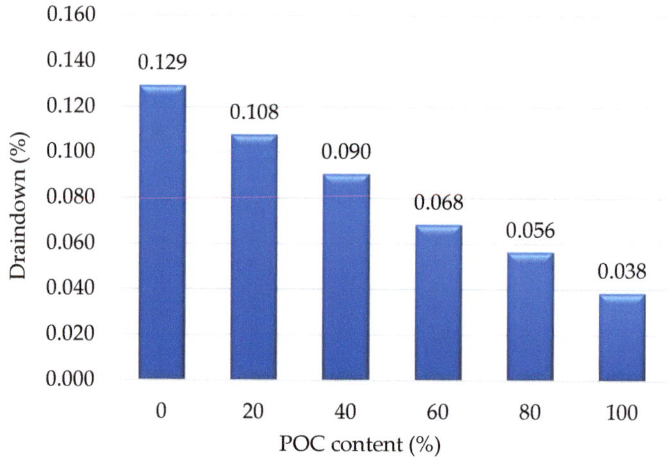

Figure 12. Drain down performance for different proportion of SMA mixtures.

4.7. Resilient Modulus Test Results

The average value of resilient modulus result for each SMA mixture with POC replacement and also from the control SMA mixture is presented in Figure 13.

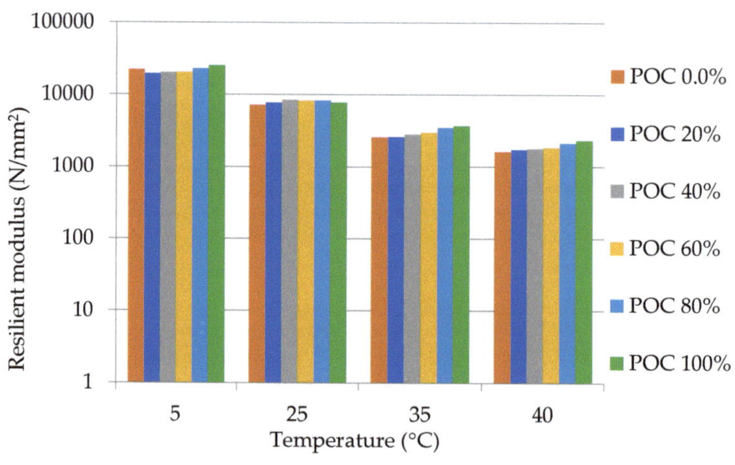

Figure 13. Resilient modulus for different SMA mixtures at different temperatures.

It is exhibited that resilient modulus values increased by POC replacement increment. This trend was apparently similar for different temperature values. It is desired to have higher stiffness values at high temperature to improve rutting resistance, while low stiffness at low temperature for thermal cracking resistance. The increment of stiffness values at high temperature for SMA mixtures with POC replacement indicates a good potential of using POC as fine aggregate in the asphalt mixtures. The increment in the resilient modulus performance is due to the existence of fibrous material from crushed POC. Fibers help the mixture to resist tensile strength, which avoids the expansion of cracks [53,54], hence increasing the stiffness of the samples. In addition, the effect of temperature susceptibility on the resilient modulus performance exhibited a similar trend. As the temperature increased, the resilient modulus value of the control mixture and SMA mixtures with POC replacement decreased. This is due to the decrement in the asphalt binder viscosity, as a result from the increment in temperatures which causes particles slippage in asphalt mixtures. The same trend was observed at a temperature of 25 °C by previous studies conducted in SMA mixture using recycled concrete as fine aggregate replacement [7], and using polyester and rockwool fibers as additive [54].

4.8. Indirect Tensile Fatigue Test Results

During the fatigue test, the sample deformation was recorded and the curve was plotted corresponding to the number of load cycles (fatigue life). For each SMA mixture proportion, three samples were tested and the results were calculated and averaged. The numbers of cycles versus deformation of SMA mixtures with various POC contents are shown in Figure 14.

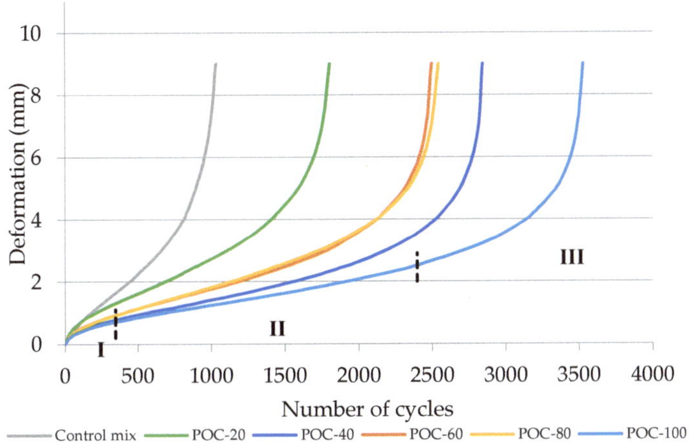

Figure 14. Deformation vs fatigue life for different SMA mixtures.

Obviously, all SMA mixtures reached the third phase of the displacement curve. For the first phase of the displacement curve, the rate of deformation increment was relatively high. This is due to the compression of current air voids in the asphalt mixtures. The second phase was the elastic zone, which showed low increment of displacement and the fatigue curve was in a linear trend. In addition, the horizontal length of the second stage (elastic zone) for the SMA mixtures with POC replacement was significantly longer as compared to the control mixture. Therefore, it can be noted that the use of POC as fine aggregates replacement enhanced the elastic property and was more likely to have plastic fracture in comparison with the control mixture, which was more likely to have brittle fracture. Meanwhile, the third phase is called the plastic zone, which is categorized as the unstable phase where the displacement increased rapidly due to crack growth in the SMA mixtures.

According to EN 12697-24 [51], the fracture life is equal to the total number of cycles that led to complete splitting of the sample. Figure 15 shows the fracture life for all the SMA mixtures with and without POC replacement.

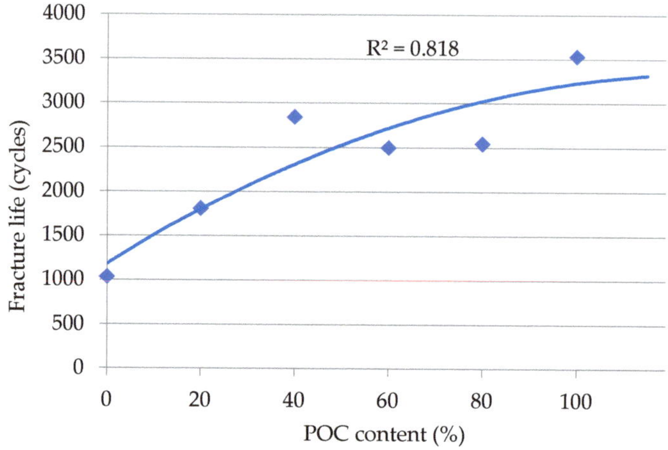

Figure 15. Fracture life for different SMA mixtures.

When compared with the control mixture, the fracture life of SMA mixtures with POC replacement increased. For example, the fracture life for of SMA mixtures with POC replacement were 1.7 times (20% POC), 2.7 times (40% POC), 2.4 times (60% POC), 2.4 times (80% POC), and 3.4 times (100% POC) compared to the fracture life of the control mixture. Therefore, it can be indicated that the fatigue performance of the SMA mixtures can be enhanced by replacing the fine aggregates with POC. This is due to the improvement in elastic properties, dispersion, and absorption of concentrated stress, which is produced by fatigue loading. This will delay the progress of micro-cracks, which in turn postpones the asphalt sample failure.

5. Conclusions

In this study, the feasibility of using palm oil clinker as fine aggregate in asphalt mixture is examined. Based on the results and analysis, the following conclusions are offered:

1. From the Marshall test, it can be concluded that using palm oil clinker as fine aggregate has less significant effect on the Marshall stability, flow, or optimum binder content, but it could affect the density, as the density value of the compacted sample decreased with an increase in the clinker aggregate replacement. This is because of the lower specific gravity of the palm oil clinker as compared to the normal aggregate.
2. The quotient values of all the asphalt mixtures, except for the mix with 60% clinker and 6.0% asphalt binder, were within the acceptable range of 2–4 kN/mm. Therefore, the asphalt mixtures have the required strength and stiffness to resist permanent deformation under traffic loads.
3. Overall, the palm oil clinker was found to be appropriate to be used as a fine aggregate replacement up to 100% in the stone mastic asphalt mixture and could satisfy the mix design requirements. These findings can reduce the negative impact of wastes on the environment and lead to a pavement design that is green, sustainable, and environmentally friendly.
4. For drain down performance, the drain down values decreased significantly with the increment of POC content. This indicates that POC is performing its function as a stabilizing additive. Furthermore, the drain down values for all the SMA mixtures with POC replacement are below the maximum required value, i.e., 0.3% by weight of the mixture.
5. For resilient modulus performance, the values increased by the increment of fine aggregates replacement especially at high temperatures, which indicated that the SMA mixtures with POC replacement will result in a high rutting resistance.

6. From indirect tensile fatigue test, the length of the elastic stage for the SMA mixtures with POC replacement is significantly longer as compared to the control mixture. Hence, adding POC to SMA mixtures enhanced the elastic properties and makes them more inclined toward plastic fracture. The fracture life of asphalt mixtures is also increased by increasing the POC replacement.

Author Contributions: Conceptualization, A.M.B. and S.K.; methodology, A.M.B., S.K., N.H.R.S., and M.R.K.; formal analysis, A.M.B. and S.K.; investigation, A.M.B. and S.A.M.; resources, S.K. and M.R.I.; writing—original draft preparation, A.M.B.; writing—review and editing, S.K., N.H.R.S., and M.R.K.; visualization, A.M.B.; supervision, S.K., N.H.R.S. and M.R.K.; project administration, S.K.; funding acquisition, S.K. All authors have read and agreed to the published version of the manuscript.

Funding: This research was funded by University of Malaya, Kuala Lumpur, through the Research University-Faculty Grant, grant number GPF066A-2018.

Acknowledgments: The authors would like to acknowledge the University of Malaya, Kuala Lumpur, for providing support on laboratory works. The first author would also like to express his gratitude to the Hadhramout Foundation, Yemen, for their support in tuition fees.

Conflicts of Interest: The authors declare no conflicts of interest.

References

1. Bolden, J.; Abu-Lebdeh, T.; Fini, E. Utilization of recycled and waste materials in various construction applications. *Am. J. Environ.l Sci.* **2013**, *9*, 14–24. [CrossRef]
2. Hasan, M.R.M.; Chew, J.-W.; Jamshidi, A.; Yang, X.; Hamzah, M.O. Review of sustainability, pretreatment, and engineering considerations of asphalt modifiers from the industrial solid wastes. *J. Traffic Transport. Eng.* **2019**. [CrossRef]
3. Arabani, M.; Pedram, M. Laboratory investigation of rutting and fatigue in glassphalt containing waste plastic bottles. *Constr. Build. Mater.* **2016**, *116*, 378–383. [CrossRef]
4. Hainin, M.R.; Yusoff, N.I.M.; Mohammad Sabri, M.F.; Abdul Aziz, M.A.; Sahul Hameed, M.A.; Farooq Reshi, W. Steel slag as an aggregate replacement in Malaysian hot mix asphalt. *ISRN Civil Eng.* **2012**, *2012*. [CrossRef]
5. Abdul-Rahman, M.N.; Abdul-Wahab, A.F. Green pavement using recycled polyethylene terephthalate (PET) as partial fine aggregate replacement in modified asphalt. *Procedia Eng.* **2013**, *53*, 124–128. [CrossRef]
6. Tapsoba, N.; Baaj, H.; Sauzéat, C.; Di Benedetto, H.; Ech, M. 3D analysis and modelling of thermal stress restrained specimen test (TSRST) on asphalt mixes with RAP and roofing shingles. *Constr. Build. Mater.* **2016**, *120*, 393–402. [CrossRef]
7. Pourtahmasb, M.S.; Karim, M.R. Performance evaluation of stone mastic asphalt and hot mix asphalt mixtures containing recycled concrete aggregate. *Adv. Mater. Sci. Eng.* **2014**, *2014*. [CrossRef]
8. Galan, J.J.; Silva, L.M.; Pérez, I.; Pasandín, A.R. Mechanical behavior of hot-mix asphalt made with recycled concrete aggregates from construction and demolition waste: A design of experiments approach. *Sustainability* **2019**, *11*, 3730. [CrossRef]
9. Acosta Álvarez, D.; Alonso Aenlle, A.; Tenza-Abril, A.J.; Ivorra, S. Influence of Partial Coarse Fraction Substitution of Natural Aggregate by Recycled Concrete Aggregate in Hot Asphalt Mixtures. *Sustainability* **2020**, *12*, 250. [CrossRef]
10. Pandey, Y.; Tare, V. Utilization of Coal Mixed Waste Aggregates Available at Thermal Power Plants for GSB and Asphalt Mixtures. *Procedia Eng.* **2016**, *143*, 170–177. [CrossRef]
11. Gautam, P.K.; Kalla, P.; Nagar, R.; Agrawal, R.; Jethoo, A.S. Laboratory investigations on hot mix asphalt containing mining waste as aggregates. *Constr. Build. Mater.* **2018**, *168*, 143–152. [CrossRef]
12. Lam, M.K.; Jamalluddin, N.A.; Lee, K.T. Chapter 23-Production of biodiesel using palm oil. In *Biofuels: Alternative Feedstocks and Conversion Processes for the Production of Liquid and Gaseous Biofuels*, 2nd ed.; Pandey, A., Larroche, C., Dussap, C.-G., Gnansounou, E., Khanal, S.K., Ricke, S., Eds.; Academic Press: Cambridge, MA, USA, 2019; pp. 539–574.
13. Malaysian Palm Oil Board. Monthly Production of Oil Palm Products Summary for the Month of December. Available online: http://bepi.mpob.gov.my/index.php/en/production/production-2019/production-of-oil-palm-products-2019.html (accessed on 10 January 2020).

14. Dungani, R.; Aditiawati, P.; Aprilia, S.; Yuniarti, K.; Karliati, T.; Suwandhi, I.; Sumardi, I. Biomaterial from Oil Palm Waste: Properties, Characterization and Applications. In *Palm Oil*, 1st ed.; Waisundara, V., Ed.; IntechOpen: London, UK, 2018; pp. 31–52.
15. Yusoff, S. Renewable energy from palm oil–innovation on effective utilization of waste. *J. Clean. Prod.* **2006**, *14*, 87–93. [CrossRef]
16. Aghamohammadi, N.; Reginald, S.S.; Shamiri, A.; Zinatizadeh, A.A.; Wong, L.P.; Sulaiman, N.; Binti, N.M. An investigation of sustainable power generation from oil palm biomass: A case study in Sarawak. *Sustainability* **2016**, *8*, 416. [CrossRef]
17. Hamzah, N.; Tokimatsu, K.; Yoshikawa, K. Solid fuel from oil palm biomass residues and municipal solid waste by hydrothermal treatment for electrical power generation in Malaysia: A review. *Sustainability* **2019**, *11*, 1060. [CrossRef]
18. Yusof, M.; Hanim, S.J.; Roslan, A.M.; Ibrahim, K.N.; Abdullah, S.; Soplah, S.; Zakaria, M.R.; Hassan, M.A.; Shirai, Y. Life Cycle Assessment for Bioethanol Production from Oil Palm Frond Juice in an Oil Palm Based Biorefinery. *Sustainability* **2019**, *11*, 6928. [CrossRef]
19. Ahmad, H.; Hilton, M.; Mohd Noor, N. Physical properties of local palm oil clinker and fly ash. In Proceedings of the 1st Engineering Conference on Energy and Environment, Kuching, Sarawak, 27–28 December 2007.
20. Nayaka, R.R.; Alengaram, U.J.; Jumaat, M.Z.; Yusoff, S.B.; Ganasan, R. Performance evaluation of masonry grout containing high volume of palm oil industry by-products. *J. Clean. Prod.* **2019**, *220*, 1202–1214. [CrossRef]
21. Kanadasan, J.; Fauzi, A.F.A.; Razak, H.A.; Selliah, P.; Subramaniam, V.; Yusoff, S. Feasibility studies of palm oil mill waste aggregates for the construction industry. *Materials* **2015**, *8*, 6508–6530. [CrossRef]
22. Ibrahim, H.A.; Razak, H.A. Effect of palm oil clinker incorporation on properties of pervious concrete. *Constr. Build. Mater.* **2016**, *115*, 70–77. [CrossRef]
23. Ibrahim, H.A.; Razak, H.A.; Abutaha, F. Strength and abrasion resistance of palm oil clinker pervious concrete under different curing method. *Constr. Build. Mater.* **2017**, *147*, 576–587. [CrossRef]
24. Karim, M.R.; Hashim, H.; Razak, H.A.; Yusoff, S. Characterization of palm oil clinker powder for utilization in cement-based applications. *Constr. Build. Mater.* **2017**, *135*, 21–29. [CrossRef]
25. Nayaka, R.R.; Alengaram, U.J.; Jumaat, M.Z.; Yusoff, S.B.; Alnahhal, M.F. High volume cement replacement by environmental friendly industrial by-product palm oil clinker powder in cement–lime masonry mortar. *J. Clean. Prod.* **2018**, *190*, 272–284. [CrossRef]
26. Kanadasan, J.; Abdul Razak, H. Utilization of palm oil clinker as cement replacement material. *Materials* **2015**, *8*, 8817–8838. [CrossRef] [PubMed]
27. Karim, M.R.; Hashim, H.; Razak, H.A. Thermal activation effect on palm oil clinker properties and their influence on strength development in cement mortar. *Constr. Build. Mater.* **2016**, *125*, 670–678. [CrossRef]
28. Rusbintardjo, G.; Hainin, M.R.; Yusoff, N.I.M. Fundamental and rheological properties of oil palm fruit ash modified bitumen. *Constr. Build. Mater.* **2013**, *49*, 702–711. [CrossRef]
29. Kamaluddin, N.A. Evaluation of Stone Mastic Asphalt Using Palm Oil Fuel Ash as Filler Material. Master's Thesis, Universiti Teknologi Malaysia, Johor Bahru, Malaysia, 2008.
30. Borhan, M.N.; Ismail, A.; Rahmat, R.A. Evaluation of palm oil fuel ash (POFA) on asphalt mixtures. *Aust. J. Basic Appl. Sci.* **2010**, *4*, 5456–5463.
31. Ahmad, J.; Yunus, M.; Nizam, K.; Kamaruddin, M.; Hidayah, N.; Zainorabidin, A. The practical use of palm oil fuel ash as a filler in asphalt pavement. In Proceedings of the International Conference on Civil and Environmental Engineering Sustainability (IConCEES), Johor Bahru, Malaysia, 3–5 April 2012.
32. Maleka, A.M.; Hamad, A.W.; PutraJaya, R. Effect of Palm Oil Fuel Ash (POFA) on the durability of asphaltic concrete. *Appl. Mech. Mater.* **2014**. [CrossRef]
33. Jabatan Kerja Raya Malaysia (JKR). Standard Specification for Road Work. In *Section 4: Flexible Pavement*; Malaysian Public Works Department: Kuala Lumpur, Malaysia, 2008.
34. Mashaan, N.S.; Ali, A.H.; Koting, S.; Karim, M.R. Performance evaluation of crumb rubber modified stone mastic asphalt pavement in Malaysia. *Adv. Mater. Sci. Eng.* **2013**, *2013*. [CrossRef]
35. ASTM. Standard test method for Marshall stability and flow of asphalt mixtures. In *ASTM D6927-15*; ASTM International: West Conshohocken, PA, USA, 2015.
36. Akbulut, H.; Gürer, C. Use of aggregates produced from marble quarry waste in asphalt pavements. *Build. Environ.* **2007**, *42*, 1921–1930. [CrossRef]

37. Jenks, C.; Jencks, C.; Harrigan, E.; Adcock, M.; Delaney, E.; Freer, H. *NCHRP Report 673: A manual for Design of Hot Mix Asphalt with Commentary*; Advanced Asphalt Technologies: Washington, DC, USA, 2011.
38. ASTM. Standard test method for bulk specific gravity and density of non-absorptive compacted bituminous mixtures. In *ASTM D2726*; ASTM International: West Conshohocken, PA, USA, 2009.
39. ASTM. Standard test method for percent air voids in compacted dense and open bituminous paving mixtures. In *ASTM D3203*; ASTM International: West Conshohocken, PA, USA, 2011.
40. ASTM. Standard test method for preparation of bituminous specimens using Marshall apparatus. In *ASTM D6926*; ASTM International: West Conshohocken, PA, USA, 2010.
41. Asi, I.M. Laboratory comparison study for the use of stone matrix asphalt in hot weather climates. *Constr. Build. Mater.* **2006**, *20*, 982–989. [CrossRef]
42. Panda, M.; Suchismita, A.; Giri, J. Utilization of ripe coconut fiber in stone matrix asphalt mixes. *Int. J. Transp. Sci. Technol.* **2013**, *2*, 289–302. [CrossRef]
43. National Asphalt Pavement Association. *Designing and Constructing SMA Mixtures: State of the Practice*; National Asphalt Pavement Association (NAPA): Lanham, MA, USA, 2002.
44. Choudhary, R.; Kumar, A.; Murkute, K. Properties of waste polyethylene terephthalate (PET) modified asphalt mixes: Dependence on PET size, PET content, and mixing process. *Periodica Polytechnica Civ. Eng.* **2018**, *62*, 685–693. [CrossRef]
45. ASTM. Standard test method for determination of draindown characteristics in uncompacted asphalt mixtures. In *ASTM D6390*; ASTM International: West Conshohocken, PA, USA, 2017.
46. ASTM. Standard test method for indirect tension test for resilient modulus of bituminous mixtures. In *ASTM D4123*; ASTM International: West Conshohocken, PA, USA, 1995.
47. Moghaddam, T.B.; Karim, M.R.; Syammaun, T. Dynamic properties of stone mastic asphalt mixtures containing waste plastic bottles. *Constr. Build. Mater.* **2012**, *34*, 236–242. [CrossRef]
48. Moghaddam, T.B.; Karim, M.R.; Abdelaziz, M. A review on fatigue and rutting performance of asphalt mixes. *Sci. Res. Essays* **2011**, *6*, 670–682. [CrossRef]
49. Modarres, A.; Rahmanzadeh, M.; Ayar, P. Effect of coal waste powder in hot mix asphalt compared to conventional fillers: Mix mechanical properties and environmental impacts. *J. Clean. Prod.* **2015**, *91*, 262–268. [CrossRef]
50. Mashaan, N.S.; Karim, M.R.; Abdel Aziz, M.; Ibrahim, M.R.; Katman, H.Y.; Koting, S. Evaluation of fatigue life of CRM-reinforced SMA and its relationship to dynamic stiffness. *Sci. World J.* **2014**, *2014*. [CrossRef] [PubMed]
51. British Standards Institution. Bituminous mixtures –Test methods–Part 24: Resistance to fatigue. In *EN 12697-24*; The British Standards Institution: London, UK, 2018.
52. Ahmadinia, E.; Zargar, M.; Karim, M.R.; Abdelaziz, M.; Shafigh, P. Using waste plastic bottles as additive for stone mastic asphalt. *Mater. Design.* **2011**, *32*, 4844–4849. [CrossRef]
53. Slebi-Acevedo, C.J.; Lastra-González, P.; Pascual-Muñoz, P.; Castro-Fresno, D. Mechanical performance of fibers in hot mix asphalt: A review. *Constr. Build. Mater.* **2019**, *200*, 756–769. [CrossRef]
54. Lavasani, M.; Namin, M.L.; Fartash, H. Experimental investigation on mineral and organic fibers effect on resilient modulus and dynamic creep of stone matrix asphalt and continuous graded mixtures in three temperature levels. *Constr. Build. Mater.* **2015**, *95*, 232–242. [CrossRef]

sustainability

Article

Evaluation of Resilience Parameters of Soybean Oil-Modified and Unmodified Warm-Mix Asphalts—A Way Forward towards Sustainable Pavements

Muhammad Akhtar Tarar [1,2], Ammad Hassan Khan [2,*], Zia ur Rehman [2], Wasim Abbass [3], Ali Ahmed [3], Elimam Ali [4,5], Mohamed Mahmoud Sayed [6] and Mubashir Aziz [7,8]

1 Department of Civil Engineering, The University of Lahore, Lahore 54000, Pakistan; muhammad.akhtar@ce.uol.edu.pk
2 Department of Transportation Engineering and Management, University of Engineering and Technology Lahore, Lahore 54890, Pakistan; gzia718@uet.edu.pk
3 Department of Civil Engineering, University of Engineering and Technology Lahore, Lahore 54890, Pakistan; wabbass@uet.edu.pk (W.A.); ali@uet.edu.pk (A.A.)
4 Department of Civil Engineering, Prince Sattam Bin Abdul Aziz University, Al-Kharj 16273, Saudi Arabia; i.ali@psau.edu.sa
5 Department of Civil Engineering, College of Engineering, Mansoura University, Mansoura 35516, Egypt
6 Department of Architecture, Faculty of Engineering and Technology, Future University, Cairo 11835, Egypt; mohamed.mahmoud@fue.edu.eg
7 Department of Civil and Environmental Engineering, King Fahd University of Petroleum and Minerals, Dhahran 31261, Saudi Arabia; mubashir.aziz@kfupm.edu.sa
8 Interdisciplinary Research Center for Construction and Building Materials, King Fahd University of Petroleum and Minerals, Dhahran 31261, Saudi Arabia
* Correspondence: chair-tem@uet.edu.pk

Citation: Tarar, M.A.; Khan, A.H.; Rehman, Z.u.; Abbass, W.; Ahmed, A.; Ali, E.; Sayed, M.M.; Aziz, M. Evaluation of Resilience Parameters of Soybean Oil-Modified and Unmodified Warm-Mix Asphalts—A Way Forward towards Sustainable Pavements. *Sustainability* **2022**, *14*, 8832. https://doi.org/10.3390/su14148832

Academic Editors: Joel R.M. Oliveira, Hugo Silva, R. Christopher Williams and Zejiao Dong

Received: 5 May 2022
Accepted: 19 June 2022
Published: 20 July 2022

Abstract: The sustainable design and construction of highways is indispensable for the economic growth and progress of any region. Highway pavements are one of the core transportation infrastructures that require energy efficient materials with durability and an optimized lifecycle. Recent research has proven that warm-mix asphalt pavements prepared with renewable bio-binders are less susceptible to distresses. This study aims to investigate the resilience characteristics (load time, deformation time) of soybean oil modified and unmodified warm-mix asphalts. Aggregates, asphalt binders and asphalt mixes were characterized in accordance with the Superpave Mix Design Criteria. The resilient modulus tests were performed as per ASTM D7369. The test results indicated that the soybean-modified warm asphalt mix samples showed a 20% to 32% reduction in load-carrying capacity than unmodified warm asphalt mixes. The values of the horizontal and vertical recoverable deformations observed in the soybean-modified mixes were found to be 3% to 7% more than in the unmodified mixes. A slight variability (up to 7%) was also observed in the time-response spectra, i.e., peak load, unload and rest periods, in the soybean-modified mixes compared with the unmodified mixes. The Pearson correlation coefficient showed a significant trend between the resilient modulus test parameters for the soybean-modified warm asphalt mixes, i.e., load deformation, load time and deformation time. Soybean oil showed sustainable behavior as a bio-binder, particularly in the deformation-time response for the warm asphalt mixes. However, the effect of soybean in terms of the reduction of the load-carrying capacity from a sustainability perspective needs to be investigated.

Keywords: transportation infrastructures; sustainable pavements; durable pavements; warm-mix asphalts; bio-binders; lifecycle

1. Introduction

Highway infrastructures have a significant influence on the socioeconomic development of countries [1]; therefore, the investment in these infrastructures provides opportunities for the economic growth [2,3] and development of a region [4,5]. The lack of transport infrastructure in developing countries is one of the major hinderances to accessing international markets [6], which highlights the global significance of transport infrastructure [7]. Non-conventional and environmentally friendly materials are beneficial for sustainable construction in the highway industry [8]. The use of these materials enhances the quality of environmental control measures and the development of durable transport infrastructures [9,10]. The sustainable construction of highways is indispensable for the transportation of people and goods [11]. Researchers have taken the motivation of using renewable resource-derived materials and utilized it in the modification of asphalt binders. The asphalt mixes produced using these modified binders exhibit merits over unmodified binders, such as emerging cost, environmental issues and the short supply of materials based on nonrenewable resources [12,13]. The properties of asphalt binders have a considerable effect on the performance of asphalt mixes [14]; therefore, to cope with the evolving issues related to pavement distresses, the modification of asphalt binders is indispensable. The utilization of bio-oils in asphalt binders reduces the stiffness of asphalt mixes, and thereby lessens the cracks that develop in the pavements [15–17]. Soybean-derived oil-based asphalt modification improves the mechanical properties of the asphalt binders [18–22].

The asphalt mixes (wearing and base) used in pavement surfacing primarily comprise asphalt binder and aggregates [23–28]. These asphalt mixes, termed as warm asphalt mixes, are generally prepared between temperatures of 140 °C and 160 °C [18]. The key objective of the warm asphalt mix design is to obtain the optimum combination of different constituents of the mix [29]. The asphalt mixes exhibit viscoelastic, viscoplastic, and time- and stress-dependent behavior when subjected to repeated loadings [23,30–34]. Therefore, pavement surface courses face different distresses during their service life, such as rutting, fatigue and thermal cracking. For the assessment of the viscoelastic behavior of asphalt mixes, the resilient modulus test can be performed [35,36].

This study aims to evaluate the effects of soybean as a bio-binder on the resilient modulus of warm asphalt mixes. The objectives of this research were: (1) to determine the effects of soybean oil on the load time and deformation time behavior of warm-mix asphalt during resilient modulus tests, (2) to compare the resilient modulus of soybean-modified and unmodified warm asphalt mixes, and (3) to assess the correlation dependency trends of different parameters (on each other) obtained in soybean-modified warm-mix asphalts' resilient modulus tests and compare these with unmodified warm-mix asphalt trends.

2. Materials and Methods

Commercially available soybean was processed to extract the soybean oil used in this study. The unmodified and soybean oil-based asphalt binders were selected in accordance with the details reported by Tarar et al. [37]. Two unmodified asphalt binders, PG 64-16 and PG 64-22, were labeled as A and B, whereas the two soybean oil (5% by weight of binder)-modified asphalt binders, PG 52-22 and PG 52-28, were categorized as A_o and B_o, respectively. The binders' characteristics such as high and low temperatures, performance grade, viscosity, mass change, penetration, softening point, ductility, flash and fire point were evaluated in laboratory based on respective American Association of State Highway and Transportation Officials (AASHTO)/American Society for Testing and Materials (ASTM) standards.

Two crushed aggregate sources, i.e., Sargodha (S) and Margalla (M), were used. The properties of the aggregates such as soundness, water absorption, Los Angeles abrasion (C131), elongation and flakiness index, fractured faces, uncompacted voids and sand equivalent were determined in the laboratory as per prevailing ASTM standards.

The Superpave (Sup-1 and Sup-2) and National Highway Authority (NH-A and NH-B) gradations were used. The wearing and base course mixes were designated as W1 to W24 and B1 to B8, respectively. The test matrix of the mixes is summarized in Table 1.

Table 1. Summary of the test matrix of warm asphalt mixes.

Mix ID	Asphalt Binders		Aggregates		Gradations
	A	Ao	S	M	
W1	√	-	√	-	SUP-1
W2	√	-	-	√	
W3	-	√	√	-	
W4	-	√	-	√	
W5	√	-	√	-	NH-A
W6	√	-	-	√	
W7	-	√	√	-	
W8	-	√	-	√	
	B	**Bo**	**S**	**M**	
W9	√	-	√	-	SUP-1
W10	√	-	-	√	
W11	-	√	√	-	
W12	-	√	-	√	
W13	√	-	√	-	NH-A
W14	√	-	-	√	
W15	-	√	√	-	
W16	-	√	-	√	
	B	**Bo**	**S**	**M**	
B1	√	-	-	√	SUP-2
B2	-	√	-	√	
B3	√	-	-	√	NH-A
B4	-	√	-	-	

Note: W = wearing course, B = base course, S = Sargodha aggregate, M = Margalla aggregate, Sup = Superpave, NH = National Highway Authority.

To determine the optimum binder contents (OBC), the mixes were tested according to the Marshall Mix test (ASTM D6926). The mixing and compaction temperatures of the binders were determined using a rotational viscometer (RV) test at 135 °C to 165 °C before the mix preparation. The binders were mixed with aggregate in a controlled mechanical mixer at 145 °C. The Superpave gyratory compactor (SGC) was used to compact the samples while keeping the air voids at 7 ± 0.5%. The indirect tensile strength and modulus of resilience test specimens were fabricated at 101.6 mm (4 inches) in diameter and 63.5 mm (2.5 inches) in thickness.

Modulus of resilience (M_R) describes the mechanical properties of asphalt mixes subjected to dynamic (traffic) loading. The asphalt mixes were tested according to ASTM D6931 for indirect tensile strength determination before M_R testing. The M_R tests were performed according to ASTM D7369 using an environmentally controlled universal testing machine: Cooper Research Technology HYD25 II.

The test temperature was set at 25 °C. The load was applied in the form of a haversine shape, i.e., $(1 - \cos\ \theta)/2$, as shown in Figure 1.

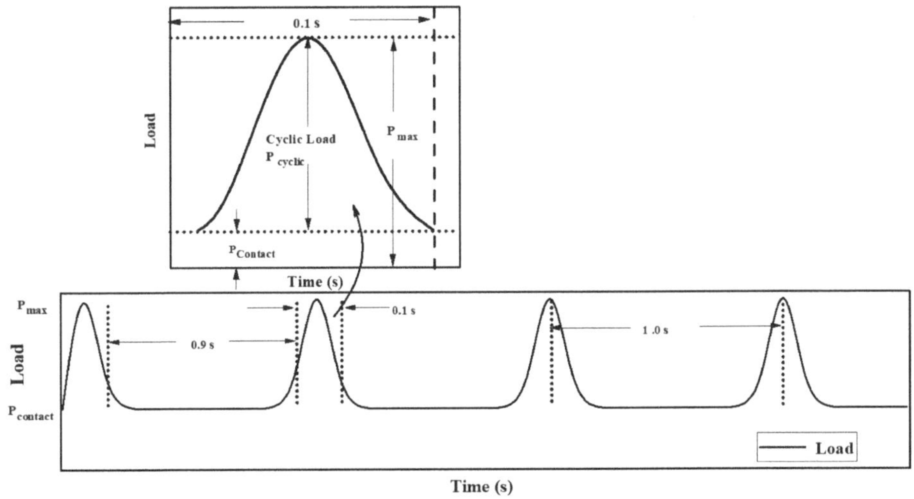

Figure 1. Typical load-time cycles with rest periods during M_R tests.

The instantaneous deformation, total deformation, Poisson's ratio and M_R were calculated according to the equations below.

$$Y = a + bx, \tag{1}$$

where Y is deformation value, x is time and a and b are regression constants.

$$Y = a + \frac{b}{x} \tag{2}$$

where Y is deformation value, x is time and a and b are regression constants.

$$\mu = \frac{I_4 - I_1 \times \left(\frac{\delta_v}{\delta_h}\right)}{I_3 - I_2 \times \left(\frac{\delta_v}{\delta_h}\right)}, \tag{3}$$

where μ is Poisson's ratio, I_1, I_2 I_3 and I_4 are constants and δ_v and δ_h are vertical and horizontal recoverable deformations, respectively.

$$M_R = \frac{P_{Cyclic}}{\delta_h t}(I_1 - I_2\delta), \tag{4}$$

where M_R is resilient modulus, P_{Cyclic} is the cyclic load applied to the specimen and t is the thickness of the specimen.

3. Results and Discussion

The physical properties of the soybean oil are summarized in Table 2a. The properties of the asphalt binders are shown in Table 2b.

Table 2. (**a**) Soybean oil physical properties [22]. (**b**) Summary of unmodified and soybean oil-modified asphalt binders' properties [22].

(a)

Description	**Soybean Oil**
Flash point (°C), ASTM D93	320
Fire point (°C), ASTM D93	354
Carbon residue (%), ASTM D189	0.37
Dynamic viscosity @ 25 °C(Pa.S), AASHTO T-316	0.062
Cloud point (°C), ASTM D5551	−9
Melting point (°C), ASTM D5440	0.5

(b)

Test Description	**Type of Asphalt Binder**			
	A	**Ao**	**B**	**Bo**
Original asphalt binder (high temperature °C)AASHTO T315	68.9	54.1	65.3	53.6
BBR (low temperature), AASHTO T313	−17	−24	−23	−29
Performance grades (PG), AASHTO M320	64–16	52–22	64–22	52–28
Viscosity (Pa.s) at 135 °C, AASHTO T316	0.462	0.250	0.445	0.242
Viscosity (Pa.s) at 165 °C, AASHTO T316	0.116	0.125	0.110	0.115
VTS	−3.557	−1.890	−3.381	−2.053
Mass change (%), AASHTO T240	0.078	0.083	0.056	0.068
Penetration (1/10th mm), ASTM D5	43	49	65	68
Softening point (°C), ASTM D36	54	47.1	48	45.6
Ductility (cm), ASTM D113	100+	100+	100+	100+
Flash and fire point (°C), ASTM D113	300	317	307	315

By the addition of soybean oil in binders A and B, few properties showed a decrease, i.e., high and low temperatures, viscosity at 135 °C and softening point, while others showed an increase, i.e., viscosity at 165 °C, mass change, penetration, flash and fire point, viscosity temperature susceptibility (VTS). The performance grade after the addition of soybean oil altered from 64–16 to 52–22 in one sample and 64–22 to 52–28 in another. However, overall, the penetration grade of the asphalt binder remained unchanged with the addition of the soybean oil to the asphalt binders. Soybean oil blended into the asphalt binder proved to have significant potential as a bio-binder.

The physical properties of the aggregates are summarized in Table 3.

Table 3. Summary of aggregate physical properties [11].

Description	**Type of Aggregate**		**Standards**
	S	**M**	
Water absorption (%)	0.95	0.93	ASTM C 127
Soundness (fine) (%)	3.8	4.5	ASTM C 88
Soundness (coarse) (%)	4.65	6.98	ASTM C 88
Los Angeles aberration (%)	23	24.5	ASTM C 131
Elongation index (%)	7	3	ASTM D 4791
Flakiness index (%)	9	5	ASTM D 4791
Fractured faces (%)	100	100	ASTM D 5821
Uncompacted voids (fine) (%)	45	44	ASTM C 1252
Sand equivalent (%)	71	74	ASTM D 2419

Note: S = Sargodha aggregate, M = Margalla aggregate.

The aggregates S and M have water absorption values of 0.95% and 0.93% respectively. The water content in the aggregates affects the performance of the asphalt mixes [38–41]. The optimum binder content is affected by the higher water absorption of the aggregates [42]. The soundness values of S and M are 3.8 and 4.5, respectively. The soundness value signifies the resistance of the aggregates against weathering. The Los Angeles abrasion values of S and M are 23 and 24.5, respectively, which specifies that the M aggregate source has higher abrasive resistance than S. The long-term performance of the pavement exposed to traffic loadings depends upon the abrasion resistance of the aggregates [43–45]. The elongation indices values of S and M are 7 and 3. The flakiness index values of S and M are 9 and 5. Researchers have reported that higher values of elongated and flaky particles reduce the strength of asphalt mixes [42,46,47]. The morphological properties of the aggregates affect the performance of asphalt mixes [48–51]. The aggregate gradation can also affect the modulus of resilience [52]. The uncompacted voids of S and M are 45 and 44, and the sand equivalents are 71 and 74, respectively. The engineering properties of both M and S aggregates qualify the acceptable limits for possible use in warm asphalt mixes. The consistency in the engineering properties of the aggregates is desirable, as it influences the resilient modulus of sustainable pavement structures [53,54]. The resilient modulus value affects the service life of the material and its resistance against pavement damage [55,56]. The energy absorption of soybean-modified mixes can be calculated based on the hysteresis loop response of samples under repeated loads [57]. This can be used as a potential advantage of soybean-modified mixes by researchers in the future.

The results in Table 4 show a summary of the different parameters obtained in the resilient modulus test, as illustrated in Figures 1 and 2.

Table 4. Summary of load, deformation and their corresponding time parameters obtained during the resilient modulus tests.

Mix ID	Load (kN)	T_m (s)	T_1 (s)	T_2 (s)	T_c (s)	T_{55} (s)	T_D (s)	T_e (s)	T_f (s)	δ_h (mm)	δ_v (mm)	δ_{total} (mm)
W1	1198.29	0.11000	0.09900	0.09240	0.06600	0.05280	0.02310	0.02750	0.01870	0.00140	0.08550	0.08690
W2	1165.23	0.11100	0.09980	0.09310	0.06650	0.05320	0.02330	0.02770	0.01880	0.00141	0.08620	0.08760
W3	939.72	0.11900	0.10700	0.09970	0.07120	0.05700	0.02490	0.02970	0.02020	0.00145	0.09230	0.09380
W4	907.75	0.11900	0.10700	0.10000	0.07160	0.05730	0.02510	0.02980	0.02030	0.00146	0.09280	0.09430
W5	1222.13	0.11100	0.09980	0.09310	0.06650	0.05320	0.02330	0.02770	0.01880	0.00139	0.08620	0.08760
W6	1210.35	0.11100	0.09990	0.09320	0.06660	0.05330	0.02330	0.02770	0.01890	0.00140	0.08630	0.08770
W7	954.65	0.11200	0.10100	0.09420	0.06730	0.05390	0.02360	0.02810	0.01910	0.00144	0.08720	0.08860
W8	923.15	0.11600	0.10400	0.09700	0.06930	0.05540	0.02430	0.02890	0.01960	0.00145	0.08980	0.09120
W9	1096.39	0.11600	0.10400	0.09710	0.06940	0.05550	0.02430	0.02890	0.01970	0.00143	0.08990	0.09130
W10	1067.33	0.11600	0.10500	0.09790	0.06990	0.05590	0.02450	0.02910	0.01980	0.00144	0.09060	0.09200
W11	913.52	0.11900	0.10700	0.09980	0.07130	0.05700	0.02490	0.02970	0.02020	0.00147	0.09240	0.09390
W12	889.45	0.11900	0.10700	0.10000	0.07140	0.05710	0.02500	0.02980	0.02020	0.00148	0.09250	0.09400
W13	1132.23	0.11100	0.09990	0.09320	0.06660	0.05330	0.02330	0.02770	0.01890	0.00140	0.08630	0.08770
W14	1109.35	0.11700	0.10500	0.09790	0.07000	0.05600	0.02450	0.02920	0.01980	0.00141	0.09070	0.09210
W15	852.65	0.11600	0.10400	0.09700	0.06930	0.05540	0.02430	0.02890	0.01960	0.00144	0.08980	0.09120
W16	823.15	0.11700	0.10500	0.09790	0.07000	0.05600	0.02450	0.02920	0.01980	0.00145	0.09070	0.09210
B1	584.45	0.11600	0.10400	0.09700	0.06930	0.05540	0.02430	0.02890	0.01960	0.00146	0.09060	0.09200
B2	494.87	0.11900	0.10700	0.09980	0.07130	0.05700	0.02490	0.02970	0.02020	0.00149	0.09250	0.09400
B3	623.25	0.11700	0.10500	0.09790	0.07000	0.05600	0.02450	0.02920	0.01980	0.00147	0.08720	0.08860
B4	514.34	0.11700	0.10500	0.09790	0.07000	0.05600	0.02450	0.02920	0.01980	0.00151	0.08610	0.08760

Note: peak load time (T_m), straight portion of unloading path between points T_1 and T_2, 40% rest period (T_c), 55% rest period (T_{55}), 90% rest period (T_d), time for 85% rest period (T_e), time for 95% rest period (T_f) in measurement units of second (s).

It is evident that the soybean-modified mixes took lesser loads (20% to 32%) than unmodified mixes in both wearing and base-course samples. In addition, the peak load time (T_m) was observed to be higher (2% to 7%) in the soybean-modified mixes than in the unmodified mixes. The straight portion of the unloading path T_1 and T_2 values were lower (2% to 7%) in the unmodified samples than in the soybean-modified samples. The time spectra of the rest periods (T_c, T_{55}, T_d, T_e and T_f) were also noted to be higher (2% to 7%) in the soybean-modified samples than in the unmodified mixes. The soybean-modified mixes exhibited improved horizontal (3% to 6%) and vertical (6% to 7%) recoverable deformations in comparison to the unmodified mixes.

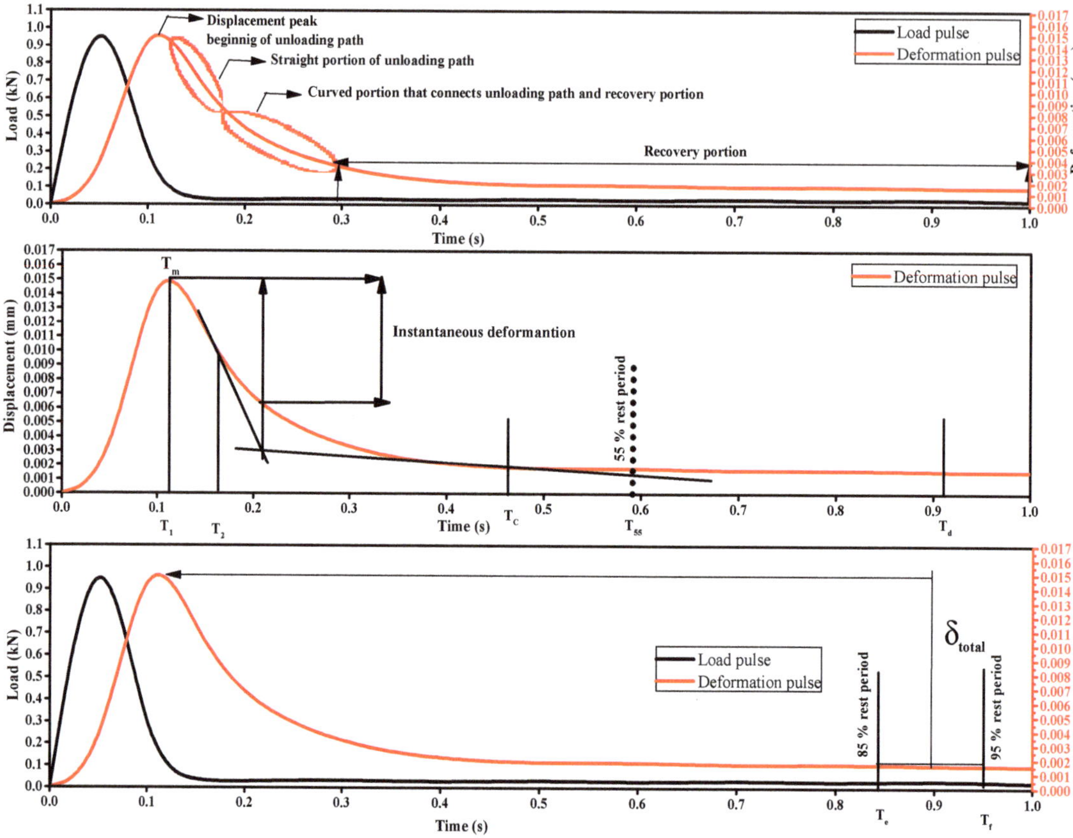

Figure 2. Typical load-time and deformation-time plots for a single cycle with time parameter explanation during M_R tests, as per ASTM D7369.

The M_R values for all wearing and base-course mixes were determined using Equation (4), as shown in Figures 3–5.

The M_R value of S for the Superpave and NH gradations was higher than for M. Figure 3 shows that the M_R values of S and M for the Superpave gradation and asphalt binder A were 7049 MPa and 6802 MPa, respectively, while the soybean oil-modified asphalt binders with Superpave gradations showed M_R values of 5063 MPa and 4751 MPa, respectively. The NH gradation exhibited an M_R value for the asphalt binder A and S and M of 7350–7224 MPa. On the other hand, the M_R values for the A_o and NH gradation were 5086–4850 MPa.

Figure 4 indicates that the M_R values of S and M for the Superpave gradation and B asphalt binder were 6344 and 6129 MPa, respectively, while the B_o asphalt binders with Superpave gradations showed M_R values of 4823 and 4665 MPa, respectively. The NH gradation exhibited M_R values for the B asphalt binder and S and M of 6708 and 6512 MPa, respectively. On the other hand, the M_R values for the B_o asphalt binders and NH gradation were 4538 and 4349 MPa, respectively.

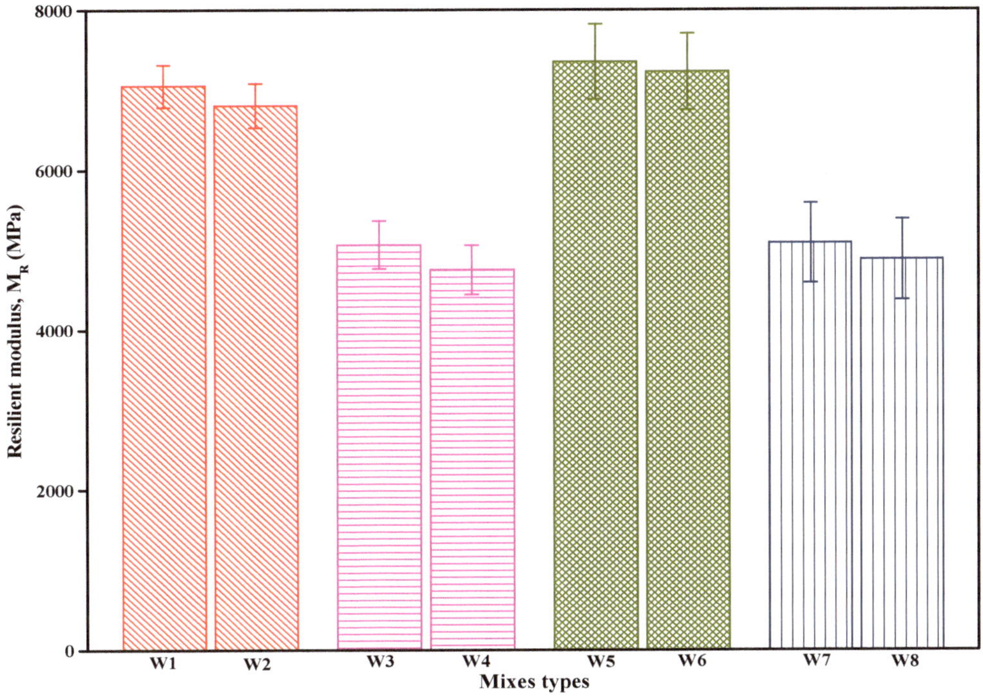

Figure 3. Resilient modulus of modified and unmodified asphalt mixes (W1–W8).

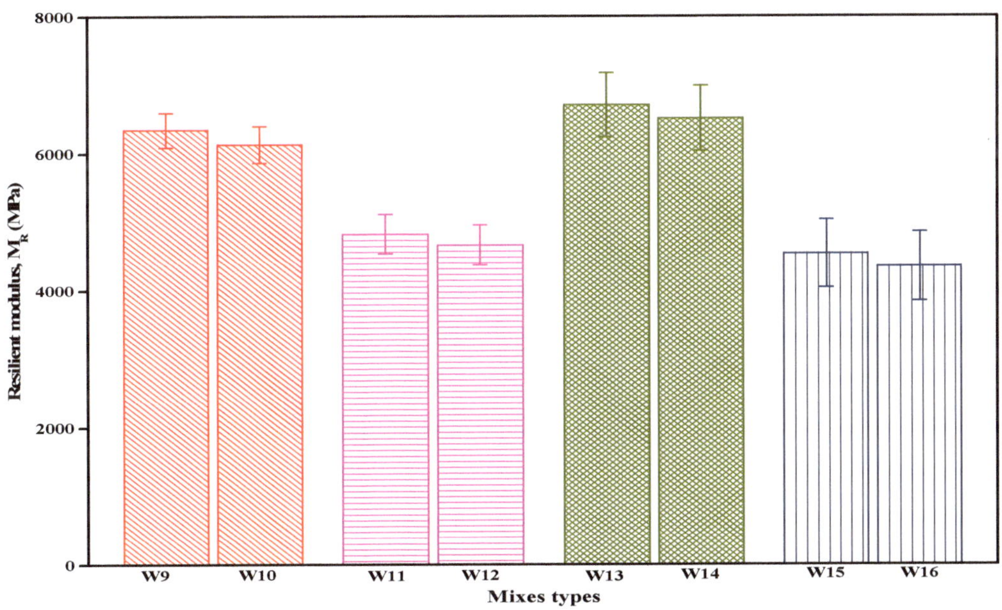

Figure 4. Resilient modulus of modified and unmodified asphalt mixes (W9–W16).

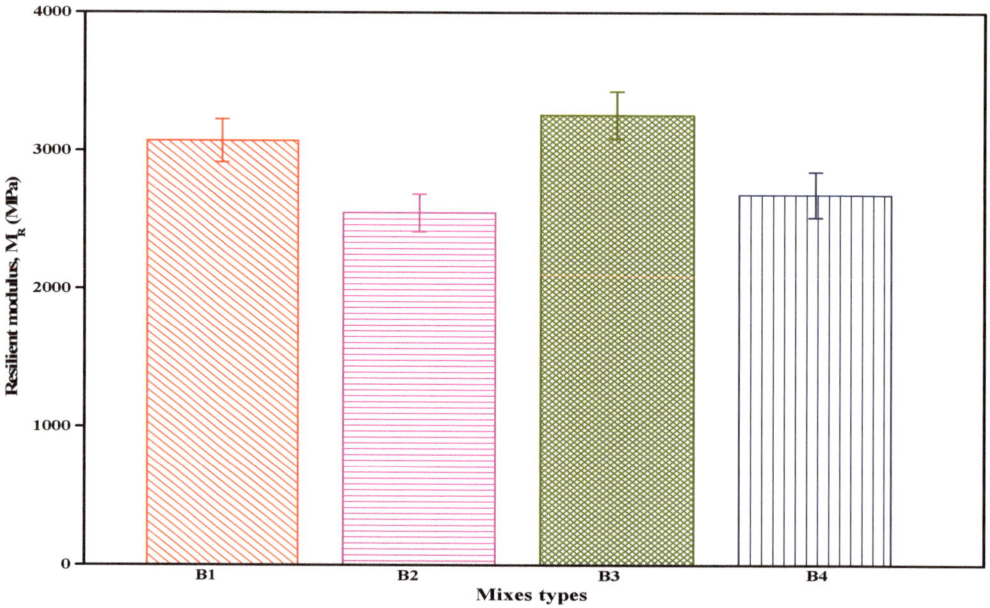

Figure 5. Resilient modulus of modified and unmodified asphalt mixes (B1–B4).

Figure 5 shows that the M_R values of M for the Superpave gradation and B asphalt binder were in the range of 3068 MPa, while the soybean oil-modified asphalt binders with Superpave gradations exhibited M_R values in the range of 2545 MPa. The NH gradation exhibited an M_R value for the asphalt binder B and M of 2911 MPa. On the other hand, the M_R value of the B_o asphalt binder with NH gradation was shown to be 2619 MPa.

Figures 3–5 show that the addition of soybean oil decreased the M_R values of both the wearing and base-course asphalt mixes.

Table 5 shows a summary of the statistical analysis carried out using the Origin software from OriginLab®. The different parameters (load, T_m, T_1, T_2, ...) obtained in the M_R tests were correlated with each other to assess the trend and possible dependency. The Pearson correlation and the respective significance values are summarized in Table 5. The values of the Pearson correlation indicate the strength of the relationship (linear) between the different variables. A positive Pearson correlation value indicates that two parameters have a direct relationship—if one parameter increases, then the other increases, and vice versa, while a negative Pearson correlation value indicates that both of the parameters have an inverse relationship—if one parameter increases, then other decreases, and vice versa. It can be seen from Table 5 that the load deformation, load time and deformation time showed reasonable significance (shaded regions) for both the modified and unmodified mixes, in line with typical trends, as shown in Figures 1 and 2.

Soybean oil showed sustainable behavior as bio-binder, particularly in the deformation-time response for warm asphalt mixes. However, the effect of soybean in the reduction of the load-carrying capacity from a sustainability perspective needs to be investigated. The minimal requirement of M_R for asphalt mixes was reported in ASTM 7369. An M_R obtained with the 5% addition of soybean as an asphalt binder falls well within the optimal acceptable stiffness range, especially for pavements subjected to light to medium traffic loading.

Table 5. Summary of the statistical analysis on M_R test parameters using Origin software from OriginLab®.

		Load (kN)	T_m (s)	T_1 (s)	T_2 (s)	T_c (s)	T_{55} (s)	T_D (s)	T_e (s)	T_f (s)	δ_h (mm)	δ_v (mm)	δ_{tota} (mm)
Load(kN)	Pearson Corr.	1	−0.37007	−0.35666	−0.35665	−0.35684	−0.35421	−0.36084	−0.36444	−0.34458	−0.88334	−0.34865	−0.35739
	Sig.	–	0.02631	0.03274	0.03275	0.03265	0.03405	0.03062	0.02887	0.03959	9.88321×10^{-13}	0.03716	0.03236
T_m (s)	Pearson Corr.	−0.37007	1	0.99595	0.99612	0.99626	0.9957	0.99507	0.99679	0.99267	0.55234	0.9961	0.99579
	Sig.	0.02631	–	0	0	0	0	0	0	0	4.78096×10^{-4}	0	0
T_1 (s)	Pearson Corr.	−0.35666	0.99595	1	0.9994	0.99893	0.99854	0.99748	0.99876	0.99733	0.54857	0.99885	0.99881
	Sig.	0.03274	0	–	0	0	0	0	0	0	5.32052×10^{-4}	0	0
T_2 (s)	Pearson Corr.	−0.35665	0.99612	0.9994	1	0.9997	0.99944	0.99876	0.9992	0.9982	0.55396	0.99967	0.99966
	Sig.	0.03275	0	0	–	0	0	0	0	0	4.5638×10^{-4}	0	0
T_c (s)	Pearson Corr.	−0.35684	0.99626	0.99893	0.9997	1	0.99982	0.99922	0.99932	0.99856	0.55177	0.99987	0.99987
	Sig.	0.03265	0	0	0	–	0	0	0	0	4.85904×10^{-4}	0	0
T_{55} (s)	Pearson Corr.	−0.35421	0.9957	0.99854	0.99944	0.99982	1	0.99907	0.99919	0.99895	0.54864	0.99978	0.99978
	Sig.	0.03405	0	0	0	0	–	0	0	0	5.3103×10^{-4}	0	0
T_D (s)	Pearson Corr.	−0.36084	0.99507	0.99748	0.99876	0.99922	0.99907	1	0.99882	0.99728	0.55504	0.99916	0.99893
	Sig.	0.03062	0	0	0	0	0	–	0	0	4.42388×10^{-4}	0	0
T_e (s)	Pearson Corr.	−0.36444	0.99679	0.99876	0.9992	0.99932	0.99919	0.99882	1	0.99725	0.55563	0.99908	0.99898
	Sig.	0.02887	0	0	0	0	0	0	–	0	4.34876×10^{-4}	0	0
T_f (s)	Pearson Corr.	−0.34458	0.99267	0.99733	0.9982	0.99856	0.99895	0.99728	0.99725	1	0.54036	0.99856	0.9986
	Sig.	0.03959	0	0	0	0	0	0	0	–	6.69006×10^{-4}	0	0
δ_h (mm)	Pearson Corr.	−0.88334	0.55234	0.54857	0.55396	0.55177	0.54864	0.55504	0.55563	0.54036	1	0.54437	0.55288
	Sig.	9.88321×10^{-13}	4.78096×10^{-4}	5.32052×10^{-4}	4.5638×10^{-4}	4.85904×10^{-4}	5.3103×10^{-4}	4.42388×10^{-4}	4.34876×10^{-4}	6.69006×10^{-4}	–	5.98648×10^{-4}	4.70652×10^{-4}
δ_v (mm)	Pearson Corr.	−0.34865	0.9961	0.99885	0.99967	0.99987	0.99978	0.99916	0.99908	0.99856	0.54437	1	0.99982
	Sig.	0.03716	0	0	0	0	0	0	0	0	5.98648×10^{-4}	–	0

4. Conclusions

In this study, the effect of soybean oil on the resilient modulus of asphalt mixes was evaluated using the ASTM D7369 procedure. The statistical analysis was performed to check the correlations between the different parameters obtained in the M_R tests. The following conclusions can be drawn from the above findings:

1. The soybean-modified warm asphalt mixes showed a 20% to 32% reduction in load-carrying capacity, i.e., for the resilient modulus than the unmodified warm asphalt mixes.
2. The values of the horizontal and vertical recoverable deformations remained comparable (3% to 7%) in both the soybean-modified and unmodified warm asphalt mixes.
3. A slight variability (2% to 7%) was observed in the time-response spectra, i.e., peak, unload, rest periods of loads and deformations during the resilient modulus tests performed on the soybean-modified and unmodified warm asphalt mixes.
4. Each parameter obtained in the soybean-modified warm-mix asphalt resilient modulus test showed a reasonable correlation trend with the others, as depicted by the Pearson coefficient. Hence, the trends of the soybean-modified and unmodified warm-mix asphalt in resilient modulus tests are comparable.
5. Soybean oil showed sustainable behavior as bio-binder, particularly in the deformation-time response for warm asphalt mixes. However, the effect of soybean in the reduction of the load-carrying capacity from a sustainability perspective needs to be investigated.

Author Contributions: Conceptualization, A.H.K.; Data curation, M.A.T. and A.H.K.; Formal analysis, M.A.T., Z.u.R., W.A., A.A., E.A. and M.M.S.; Funding acquisition, A.H.K.; Investigation, M.A.T.; Methodology, M.A.T.; Project administration, A.H.K. and Z.u.R.; Resources, A.H.K., W.A., A.A., E.A., M.M.S. and M.A.; Supervision, A.H.K. and Z.u.R.; Writing—original draft, M.A.T.; Writing—review & editing, A.H.K., Z.u.R., W.A., A.A., E.A., M.M.S. and M.A. All authors have read and agreed to the published version of the manuscript.

Funding: This research was funded by the Higher Education Commission of Pakistan, grant number NRPU 9639.

Institutional Review Board Statement: Not applicable.

Informed Consent Statement: Not applicable.

Data Availability Statement: Not applicable.

Conflicts of Interest: The authors declare no conflict of interest.

References

1. Rahman, I.; Sharma, B.P.; Fetuu, E.; Yousaf, M. Do Roads Enhance Regional Trade? Evidence Based on China's Provincial Data. *J. Asian Financ. Econ. Bus.* **2020**, *7*, 657–664. [CrossRef]
2. Javid, M. Public and Private Infrastructure Investment and Economic Growth in Pakistan: An Aggregate and Disaggregate Analysis. *Sustainability* **2019**, *11*, 3359. [CrossRef]
3. Holl, A. Transport Infrastructure, Agglomeration Economies, and Firm Birth: Empirical Evidence from Portugal. *J. Reg. Sci.* **2004**, *44*, 693–712. [CrossRef]
4. Gibbons, S.; Lyytikäinen, T.; Overman, H.; Sanchis-Guarner, R. New road infrastructure: The effects on Arms. *J. Urban Econ.* **2019**, *110*, 35–50. [CrossRef]
5. Lopez, M.A.G.; Holl, A.; Marsal, E.V. Suburbanization and highways in Spain when the Romans and the Bourbons still shape its cities. *J. Urban Econ.* **2015**, *85*, 52–67. [CrossRef]
6. Coşar, A.K.; Demir, B. Domestic road infrastructure and international trade: Evidence from Turkey. *J. Dev. Econ.* **2016**, *118*, 232–244. [CrossRef]
7. Ghani, E.; Goswami, A.G.; Kerr, W.R. Highway to Success: The Impact of the Golden Quadrilateral Project for the Location and Performance of Indian Manufacturing. *Econ. J.* **2015**, *126*, 317–357. [CrossRef]
8. Thom, N.; Dawson, A. Sustainable Road Design: Promoting Recycling and Non-Conventional Materials. *Sustainability* **2019**, *11*, 6106. [CrossRef]
9. Zhao, Y.; Goulias, D.; Peterson, D. Recycled Asphalt Pavement Materials in Transport Pavement Infrastructure: Sustainability Analysis & Metrics. *Sustainability* **2021**, *13*, 8071. [CrossRef]
10. Lee, J.; Edil, T.B.; Benson, C.H.; Tinjum, J. Building Environmentally and Economically Sustainable Transportation Infrastructure: Green Highway Rating System. *J. Constr. Eng. Manag.* **2013**, *139*, A4013006. [CrossRef]
11. Ibrahim, A.H.; Shaker, M.A. Sustainability index for highway construction projects. *Alex. Eng. J.* **2019**, *58*, 1399–1411. [CrossRef]
12. Yang, X.; You, Z.; Dia, Q.; Beale, J.M. Mechanical performance of asphalt mixtures modified by bio-oils derived from waste wood resources. *Constr. Build. Mater.* **2014**, *51*, 424–431. [CrossRef]
13. Williams, R.C.; Peralta, J.; Puga, K.L.N. *Development of Non-Petroleum-Based Binders for Use in Flexible Pavements—Phase II (Report No. IHRB Project TR-650)*; Iowa Department of Transportation, Iowa State University: Ames, IA, USA, 2015.
14. Saravanan, U. On the use of linear viscoelastic constitutive relations to model asphalt. *Int. J. Pavement Eng.* **2012**, *13*, 360–373. [CrossRef]
15. Elseifi, M.A.; Mohammad, L.N.; Cooper, S.B., III. Laboratory evaluation of asphalt mixtures containing sustainable technologies. *J. Assoc. Asph. Paving Technol.* **2011**, *80*, 227–254.
16. Hajj, E.; Souliman, M.; Alavi, M.; Loría Salazar, L. Influence of hydro green bio asphalt on viscoelastic properties of reclaimed asphalt mixtures. *Transp. Res. Rec. J. Transp. Res. Board* **2013**, *2371*, 13–22. [CrossRef]
17. Zaumanis, M.; Mallick, R.B.; Frank, R. Evaluation of Rejuvenator's Effectiveness with Conventional Mix Testing for 100% Reclaimed Asphalt Pavement Mixtures. *Transp. Res. Rec. J. Transp. Res. Board* **2013**, *2370*, 17–25. [CrossRef]
18. Podolsky, J.H.; Buss, A.; Williams, R.C.; Cochran, E.W. Effect of bio-derived/chemical additives on warm mix asphalt compaction and mix performance at low temperature. *Cold Reg. Sci. Technol.* **2017**, *136*, 52–61. [CrossRef]
19. Elkashef, M.; Podolsky, J.; Williams, R.C.; Cochran, E.W. Introducing a soybean oil-derived material as a potential rejuvenator of asphalt through rheology, mix characterization and Fourier Transform Infrared analysis. *Road Mater. Pavement Des.* **2017**, *19*, 1–21. [CrossRef]
20. Podolsky, J.H.; Williams, R.C.; Cochran, E. Effect of corn and soybean oil derived additives on polymer-modified HMA and WMA master curve construction and dynamic modulus performance. *Int. J. Pavement Res. Technol.* **2018**, *11*, 541–552. [CrossRef]
21. Podolsky, J.H.; Chen, C.; Buss, A.F.; Williams, R.C.; Cochran, E.W. Effect of bio-derived/chemical additives on HMA and WMA compaction and dynamic modulus performance. *Int. J. Pavement Eng.* **2019**, *22*, 613–624. [CrossRef]

22. Tarar, M.A.; Khan, A.H.; Rehman, Z.U. Evaluation of effects of soybean derived oil and aggregate petrology on the performance of asphalt mixes. *Road Mater. Pavement Des.* **2020**, *23*, 308–334. [CrossRef]
23. Swamy, A.K.; Daniel, J.S. Effect of Mode of Loading on Viscoelastic and Damage Properties of Asphalt Concrete. *Transp. Res. Rec. J. Transp. Res. Board* **2012**, *2296*, 144–152. [CrossRef]
24. Nejad, F.M.; Azarhoosh, A.R.; Hamedi, G.H. Laboratory Evaluation of Using Recycled Marble Aggregates on the Mechanical Properties of Hot Mix Asphalt. *J. Mater. Civ. Eng.* **2013**, *25*, 741–746. [CrossRef]
25. Wen, H.; Bhusal, S.; Wen, B. Laboratory evaluation of waste cooking oil-based bio asphalt as an alternative binder for hot mix asphalt. *J. Mater. Civ. Eng.* **2013**, *25*, 1432–1437. [CrossRef]
26. Feng, H.; Pettinari, M.; Hofko, B.; Stang, H. Study of the internal mechanical response of an asphalt mixture by 3-D discrete element modeling. *Constr. Build. Mater.* **2015**, *77*, 187–196. [CrossRef]
27. Pan, P.; Kuang, Y.; Hu, X.; Zhang, X. A Comprehensive Evaluation of Rejuvenator on Mechanical Properties, Durability, and Dynamic Characteristics of Artificially Aged Asphalt Mixture. *Materials* **2018**, *11*, 1554. [CrossRef]
28. Islam, R.; Kalevela, S.A.; Mendel, G. How the Mix Factors Affect the Dynamic Modulus of Hot-Mix Asphalt. *J. Compos. Sci.* **2019**, *3*, 72. [CrossRef]
29. Kim, Y.; Lee, J.; Baek, C.; Yang, S.; Kwon, S.; Suh, Y. Performance Evaluation of Warm- and Hot-Mix Asphalt Mixtures Based on Laboratory and Accelerated Pavement Tests. *Adv. Mater. Sci. Eng.* **2012**, *2012*, 1–9. [CrossRef]
30. Ouf, M.S.; Abdolsamed, A.A. Controlling Rutting Performance of Hot Mix Asphalt. *Int. J. Sci. Eng. Res.* **2016**, *6*, 2229–5518.
31. Al-Qadi, I.L.; Yoo, P.J.; Elseifi, M.A.; Nelson, S. Creep Behavior of Hot-Mix Asphalt due to Heavy Vehicular Tire Loading. *J. Eng. Mech.* **2009**, *135*, 1265–1273. [CrossRef]
32. Ahmad, J.; Rahman, M.Y.A.; Hainin, M.R. Rutting Evaluation of Dense Graded Hot Mix Asphalt Mixture. *Int. J. Eng. Technol.* **2011**, *11*, 48–52.
33. Huang, Y.; Wang, X.; Liu, Z.; Li, S. Dynamic modulus test and master curve analysis of asphalt mix with trapezoid beam method. *Road Mater. Pavement Des.* **2017**, *18*, 1–11. [CrossRef]
34. Rahman, A.A.S.M.; Islam, M.R.; Tarefder, R.A. Assessment and modification of nationally-calibrated dynamic modulus predictive model for the implementation of Mechanistic-Empirical design. *Int. J. Pavement Res. Technol.* **2018**, *11*, 502–508. [CrossRef]
35. Khedr, S.A.; Breakah, T.M. Rutting parameters for asphalt concrete for different aggregate structures. *Int. J. Pavement Eng.* **2011**, *12*, 13–23. [CrossRef]
36. Ezzat, H.; El-Badawy, S.; Gabr, A.; Zaki, S.; Breakah, T. Predicted performance of hot mix asphalt modified with nano-montmorillonite and nano-silicon dioxide based on Egyptian conditions. *Int. J. Pavement Eng.* **2018**, *21*, 642–652. [CrossRef]
37. Tarar, M.A.; Khan, A.H.; Rehman, Z.; Inam, A. Changes in the rheological characteristics of asphalt binders modified with soybean-derived materials. *Int. J. Pavement Eng.* **2019**, *22*, 233–248. [CrossRef]
38. Airey, G.D.; Choi, Y. State of the Art Report on Moisture Sensitivity Test Methods for Bituminous Pavement Materials. *Road Mater. Pavement Des.* **2002**, *3*, 355–372. [CrossRef]
39. Apeagyei, A.K.; Grenfell, J.R.A.; Airey, G.D. Moisture-induced strength degradation of aggregate–asphalt mastic bonds. *Road Mater. Pavement Des.* **2014**, *15*, 239–262. [CrossRef]
40. Apeagyei, A.K.; Grenfell, J.R.A.; Airey, G.D. Influence of aggregate absorption and diffusion properties on moisture damage in asphalt mixtures. *Road Mater. Pavement Des.* **2015**, *16*, 404–422. [CrossRef]
41. Goel, G.; Sachdeva, S.N. Stripping Phenomenon in Bituminous Mixes: An Overview. *Int. J. Math. Sci. Appl.* **2016**, *6*, 353–360.
42. El-Tahan, D.; Gabr, A.; El-Badawy, S.; Shetawy, M. Evaluation of recycled concrete aggregate in asphalt mixes. *Innov. Infrastruct. Solut.* **2018**, *3*, 20. [CrossRef]
43. Hamzah, M.O.; Hasan, M.R.M.; Ismail, M.R.; Shahadan, Z. Effects of Temperature on Abrasion Loss of Porous and Dense Asphalt Mixes. *Eur. J. Sci. Res.* **2010**, *40*, 589–597.
44. Mohajerani, A.; Nguyen, B.T.; Tanriverdi, Y.; Chandrawanka, K. A new practical method for determining the LA abrasion value for aggregates. *Soils Found.* **2017**, *57*, 840–848. [CrossRef]
45. Wua, J.; Hou, Y.; Wang, L.; Guo, M.; Meng, L.; Xiong, H. Haocheng Xiong a Analysis of coarse aggregate performance based on the modified Micro Deval abrasion test. *Int. J. Pavement Res. Technol.* **2018**, *11*, 185–194. [CrossRef]
46. Mahmud, M.Z.H.; Yaacobb, H.; Jayab, R.P.; Hassanb, N.A. Laboratory investigation on the effects of flaky aggregates on dynamic creep and resilient modulus of asphalt mixtures. *J. Teknol. (Sci. Eng.)* **2014**, *70*, 107–110. [CrossRef]
47. Tahmoorian, F.; Samali, B. Laboratory investigations on the utilization of RCA in asphalt mixtures. *Int. J. Pavement Res. Technol.* **2018**, *11*, 627–638. [CrossRef]
48. Al-Rousan, T.; Masad, E.; Tutumluer, E.; Pan, T. Evaluation of image analysis techniques for quantifying aggregate shape characteristics. *Constr. Build. Mater.* **2007**, *21*, 978–990. [CrossRef]
49. Arasan, S.; Hasiloglu, S.A.; Akbulut, S. Shape Properties of Natural and Crushed Aggregate using Image Analysis. *Int. J. Civ. Struct. Eng.* **2010**, *1*, 221–233.
50. Wang, H.; Bu, Y.; Wang, Y.; Yang, X.; You, Z. The Effect of Morphological Characteristic of Coarse Aggregates Measured with Fractal Dimension on Asphalt Mixture's High-Temperature Performance. *Adv. Mater. Sci. Eng.* **2016**, *2016*, 1–9. [CrossRef]
51. Galan, J.; Silva, L.; Pasandín, A.; Pérez, I. Evaluation of the Resilient Modulus of Hot-Mix Asphalt Made with Recycled Concrete Aggregates from Construction and Demolition Waste. *Sustainability* **2020**, *12*, 8551. [CrossRef]

52. Rizvi, M.A.; Khan, A.H.; Rehman, Z.U.; Inam, A.; Masoud, Z. Evaluation of Linear Deformation and Unloading Stiffness Characteristics of Asphalt Mixtures Incorporating Various Aggregate Gradations. *Sustainability* **2021**, *13*, 8865. [CrossRef]
53. Mackiewicz, P.; Szydło, A. Viscoelastic Parameters of Asphalt Mixtures Identified in Static and Dynamic Tests. *Materials* **2019**, *12*, 2084. [CrossRef] [PubMed]
54. White, G. A Synthesis on the Effects of Two Commercial Recycled Plastics on the Properties of Bitumen and Asphalt. *Sustainability* **2020**, *12*, 8594. [CrossRef]
55. Sun, Y.; Gu, B.; Gao, L.; Li, L.; Guo, R.; Yue, Q.; Wang, J. Viscoelastic Mechanical Responses of HMAP under Moving Load. *Materials* **2018**, *11*, 2490. [CrossRef] [PubMed]
56. Czech, K.R.; Gardziejczyk, W. Dynamic Stiffness of Bituminous Mixtures for the Wearing Course of the Road Pavement—A Proposed Method of Measurement. *Materials* **2020**, *13*, 1973. [CrossRef]
57. Arulrajah, A.; Naeini, M.; Mohammadinia, A.; Horpibulsuk, S.; Leong, M. Recovered plastic and demolition waste blends as railway capping materials. *Transp. Geotech.* **2020**, *22*, 100320. [CrossRef]

sustainability

MDPI

Article

Impact Analysis Using Life Cycle Assessment of Asphalt Production from Primary Data

Giuseppe Sollazzo, Sonia Longo, Maurizio Cellura and Clara Celauro *

Department of Engineering, University of Palermo, Viale delle Scienze, 90128 Palermo, Italy; giuseppe.sollazzo@unipa.it (G.S.); sonia.longo@unipa.it (S.L.); maurizio.cellura@unipa.it (M.C.)
* Correspondence: clara.celauro@unipa.it; Tel.: +39-09123899716

Received: 16 November 2020; Accepted: 4 December 2020; Published: 9 December 2020

Abstract: Road construction and maintenance have a great impact on the environment, owing to the huge volumes of resources involved. Consequently, current production procedures and technologies must be properly investigated, for identifying and quantifying the life cycle environmental impacts produced. In this paper, primary data, i.e., site-specific data directly collected or measured on a reference plant, are analyzed for calculating the impact of the production of a hot mix asphalt. The analysis is performed in a from "cradle to gate" approach to estimate the environmental burdens of the production process in an average plant, representative of the existing technology in Italy and Southern Europe. The research outcomes are useful to increase reliability in quantification of asphalt production impacts and the contribution of each component. The results represent a reference basis for producers, designers, and contractors in the decisional phases, identifying the most critical aspects in the current practice and the possible improvements for reducing impacts of road industries. In this regard, efficient energy technologies for reducing the production temperature (such as warm mix asphalt) and burned fuels are proven to assure relevant improvements in the environmental performance.

Keywords: asphalt production; foreground data; LCA; emissions; energy consumption; environmental impacts

1. Introduction

The infrastructure construction industry, continuously growing in recent years all over the world for both construction and maintenance activities, requires relevant consumption of precious materials, with huge consequences for the natural ecosystem [1]. These environmental problems are also due to the huge volumes involved in pavement constructions and to the required technology for material extraction and production, traditionally characterized by strong impacts [2]. It may be considered that, in the transportation sector, road construction determines a contribution up to 10% of total greenhouse gas (GHG) emissions, representing a useful indicator for sustainability assessment. For example, according to the Federal Highway Administration estimates, pavement construction, maintenance, and rehabilitation in the United States roughly produce 75 million tons CO_{2eq} per year [3]. In the last decades, relevant attention to these issues and environment preservation has remarkably risen. In this regard, numerous researchers have investigated and analyzed the possible sustainability issues caused by traditional approaches and technologies in pavement construction and maintenance phases, and proposed modern and more efficient sustainable solutions that may effectively reduce consumption of energy and resources, as well as the production of waste [4–6].

From this perspective, in order to assure comprehensive and exhaustive analysis of product and technology impacts, life cycle assessment (LCA) represents an appropriate and accurate methodology for any kind of material and process, regarding the whole life cycle of the products. The International

Organization for Standardization (ISO) provided accurate and general instructions and regulations for proper performing of LCA in its 14040 series publications [7,8]. Concerning the pavement industry, LCA can actually be adopted for estimating impacts of different asphalt mixtures and technologies, for both production and construction. The assessment can assure an overview of the entire life cycle of the material by taking into consideration production of raw materials and mixtures, construction, effective exercise and final disposal, listing all the involved resources and evaluating, with relevant accuracy, and the produced impacts on the environment in "cradle to grave" approaches.

With specific regard to existing literature for the road construction sector, many scientific researchers investigated material production and pavement construction, maintenance and rehabilitation processes, and technologies, using LCA to understand the main drawbacks of the current practice and evaluating possible improvements in methodologies and operative solutions [9–18]. These studies provided a helpful and various reference for a practical application of the methodology and, further, they have sometimes evidenced critical aspects and positive strategic solutions for industries, road agencies, and government subjects aiming to improve the entire process. However, in these studies, several research gaps can be identified [19] and two main issues emerged: first, several assessments rely on secondary data for production processes and technologies; secondly, most of the literature focused on specific models and case studies that may be relevant only for the evaluated scenario or the reference country [20]. Indeed, as any data-based analysis, reference information reliability remains remarkable for homogeneous markets/contexts only.

Owing to these complications and in order to directly define and quantify impacts of road pavement construction and maintenance, usually researchers focused mainly on these stages of the product life cycle only, in a from "gate to grave" approach. If production steps are kept in consideration, LCA applications commonly rely on numerical values and quantities extracted from available datasets, with possible inconsistencies due to the origin or the quality of the different data [21]. Indeed, the lack of reliable and solid primary data on road materials and technology components, provided with significant accuracy by manufacturing companies, generally represents a remarkable issue [16,18,22]. Though some attempts were performed in foreign contexts for accurate primary data collection [23–27], in few cases LCAs related to the Italian or Southern European contexts have been based (sometimes partially or with strong hypothesis) on foreground data directly collected on site [15,16,22,28,29]. Consequently, as the definition of the eco-profiles of the considered products and processes generally relies on secondary data derived from different (foreign) contexts, the accuracy and reliability of the derived results should be questioned. Instead, a complete and exhaustive analysis of all the involved inputs and outputs assures good accuracy in the determination of the from "cradle to gate" impact on several environmental indicators. Obviously, as accurate calculations of pavement impacts require the examination of every phase and process, each of them should be properly analyzed, considering the specific process conditions and scenarios, for assuring exhaustive and reliable results.

The production stage of the asphalt mixture naturally determines a relevant contribution to the overall impacts of the product [18,30]. Further, it can strongly influence the quality of the product and determines the consumption of natural resources (both raw components and fuels) during the production cycles. Then, the production phases of the asphalt mixture have to be properly investigated and analyzed for assuring adequate quality and reliability of the whole life cycle impact interpretation and also for identifying strategies aimed at increasing the eco-efficiency of the asphalt mixture production process and the resource productivity [31].

Consequently, the main goal of this research is to carry out a complete environmental LCA of the production of an asphalt mixture in an Italian plant, fully representative of the average technology common in the Italian and Southern European area, following a from "cradle to gate" approach based on foreground data collected in the field, determining a useful benchmark for the Italian context. This goal is of particular interest, considering that over 26 million of tons of asphalt mixtures have been produced only in 2018 in Italy for maintenance and construction needs, with an increase of more than 10% with respect to 2017 [32], while in 2017 over 250 millions of tons were produced in Europe, according to the European Asphalt Pavement Association [33]. Based on a preliminary and accurate phase of primary data collection and on the definition of the eco-profile of the selected product, the LCA analysis is performed to evidence the most impactful elements in the process. All the relevant data related to input/output for the asphalt production were measured and quantified, including raw materials and energy consumption, emissions, and solid wastes. In summary, indeed, the study mainly aims to define a reliable context-related impact analysis of hot mix asphalt (HMA) production phase based on foreground data directly collected on a representative plant for the analyzed geographical area and the related market. More details regarding novelty and motivation of the present research are provided in "Goal and scope definition" section. Moreover, for improving result quality, a preliminary sensitivity analysis was also performed, to better evaluate the effect of the different components and, further, to control the influence of eventual slight inaccuracies in the provided data. Moreover, a comparative analysis evidences possible benefits and improvements assured by different configurations of the plant, especially from an energetic perspective. In detail, different alternative energy sources and technological solutions are compared for estimating impact variations, relying on the primary data acquired at the plant. After a brief discussion on the LCA framework, the following sections present, first, goal and scope definition, and inventory analysis, then, life cycle impact evaluation and result analysis are provided and discussed.

2. Methodological Approach

2.1. Life Cycle Assessment

LCA is a powerful methodological framework to determine in a reliable way impacts on the environment of specific products, processes, and technologies. The general framework for a correct application of the methodology is provided in the ISO 14040 and ISO 14044 standards [7,8].

In general, LCAs are cradle-to-grave analyses of the products or services and include four basic steps:

1. Goal definition and scoping, required for identifying the goal of the study, the system boundaries, the functional or declared unit (FU or DU), the target audience, etc.;
2. Inventory analysis, for the quantification of resource consumption, waste flows, and emissions for the reference unit attributable to the different processes in the life cycle system;
3. Impact assessment, providing useful environmental characterizations, for a better and deeper understanding of the inventory data;
4. Interpretation, stating specific conclusions and strategic recommendations for improving the analysis or evidencing relevant aspects and critical issues for the decision makers.

2.2. Technology Benchmarking

According to the aim of the paper, the first step for performing the LCA of HMA production is the selection of the reference technology and production processes. Therefore, at the beginning of the research activity, a preliminary analysis of the existing technology in Italy and Europe (significantly different from the American context, for which several studies are available) was performed in order to select a representative plant as reference. As known, there are several configurations and technologies for asphalt mixture production, characterized by the mixing process and technology. Generally, the batch-mix plant is the most common solution in Europe [34,35] and, in particular, in Italy [36], especially due to its adaptability with regards to design mix and productivity. Considering the Italian scenario, in 2010, almost 90% of the asphalt plants (606 of 693 active plants) relied on the batch-mix technology [36] and this scenario can be considered constant for the last 10 years. Furthermore, since it is reasonable to think that relevant practical technology will not affect the existing plant configuration in the next decade (especially for traditional HMA production), the technological characteristics as well as the operating conditions of batch-mix plants refer to an existing large-scale technology in Italy, and in Southern Europe too. Further, the strong crisis of the bitumen/asphalt market in the last decade [37] and its recent evolution towards minor maintenance and rehabilitation projects have discouraged possible modifications in the current technology. In terms of adopted heating technologies in the production process, according to U.S. Environmental Protection Agency [38], in batch mix HMA plants, the rotary dryer for heating the aggregates are typically oil- or gas-fired.

Thus, according to the above-mentioned information, the average technology in Italy and in Southern Europe is represented by a batch-mix asphalt, including oil and gas burners for aggregate and bitumen heating. For the purpose of this study, in order to assure technological representativeness, an Italian batch-mix plant was selected as reference, representing a reliable "average plant". Further details on the selected plant are provided in Section 3.2.3.

3. Case study: Production of Asphalt Mixture in a Southern Italy Batch-Mix Plant

3.1. Goal and Scope Definition

3.1.1. Goal of the Study

The main goal of this study is to assess the environmental impacts of the production of a hot mix asphalt (HMA) mixture in an Italian plant, useful as a benchmark for the Italian, mainly, and Southern European context and for providing accurate results to be included in from "cradle to grave" LCAs of asphalt pavements. For performing this task, an accurate preliminary data collection campaign was performed in the considered plant. Further objectives of the investigation were: (1) to assess possible variations in the final outcomes due to eventual small inaccuracies of the provided data; (2) to quantify the environmental advantages of operating the plant through electricity produced by renewable sources, instead of electricity produced by private generating-set or obtained from the grid; and (3) to estimate possible benefits of more sustainable technologies, such as warm mix asphalt (WMA).

The results of this research may provide a reference basis for improving LCA analysis of asphalt pavements, considering with more accuracy and reliability actual plant information. Further, as the considered plant represents the most common technology in the reference context, the results of the study may also constitute a benchmark for other plants, providing useful data on their sustainability to their owner, for possible technological modifications and improvements.

The study is performed in compliance with the LCA regulations provided by the international standards of series ISO 14040 [7,8] and by the Environmental Product Declaration (EPD) product category rules [39]. According to the reference methodology framework, input and output flows of each productive step are considered for the different processes in which they occur, considering an attributional approach [40].

3.1.2. Declared Unit and System Boundaries

The investigated product is an asphalt mixture for road pavement industry. The reference unit adopted in this study, according to the standards, is represented by the declared unit (DU), consisting of 1 metric ton of HMA mixture, suitable for binder courses in compliance with Italian regulations. The included raw materials are limestone aggregates and bitumen.

For the definition of the system boundaries, first it is fundamental to consider the entire life cycle of asphalt pavements, from production to disposal. In general, it is possible to identify five different stages in the pavement life cycle, as shown in Figure 1.

In this research, the analysis focuses on the production stage and, thus, it is stopped at the exit factory gate (from "cradle to gate"). Therefore, the considered processes are: bitumen supply (including oil extraction and its transformation); aggregate supply (including extraction and transformation); aggregate and bitumen transportation to the production plant; and plant production processes.

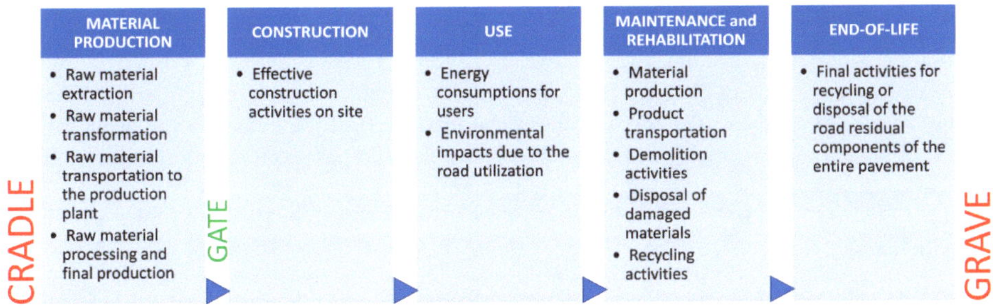

Figure 1. Stages of road pavement life cycle.

3.1.3. Impact Assessment Methodology and Impact Categories

Impact calculations were based on impact categories and characterization factors of the EPD 2018 method [41]. The following seven impact categories were calculated according to the above method in order to describe performance and effects of the considered product system:

- AC: Acidification (fate not incl.) (kg SO_{2eq});
- EU: Eutrophication (kg $PO_4{}^{3-}{}_{eq}$);
- GW: Global warming (100 year) (kg CO_{2eq});
- PO: Photochemical oxidation (kg NMVOC);
- OD: Ozone layer depletion (kg CFC_{11eq});
- AD and AD*: Abiotic depletion (total (kg SB_{eq}) and fossil fuels only (MJ), respectively).

3.2. *Life Cycle Inventory Analysis*

3.2.1. Data Collection and Quality

A crucial aspect of LCA is the nature of the considered data, that greatly affects the reliability and accuracy of the results. LCA applications should always define source and collection approaches for each type of data involved in the analysis. In the following, specific information regarding product and production plant, primary data collection, and eventual assumptions for the various stages are provided. Secondary data concerning raw material extraction and transformation (background processes) were directly derived from the Ecoinvent database [42]. Secondary data have been used for modeling energy generation from low Sulphur fuel (LSF), liquid petroleum gas (LPG), and diesel included the gas pollutants emitted during the fuel combustion [42]. The eco-profile of limestone was used for modeling crushed stones, sand, and filler, since they have the same geological nature that is the parameter that, above all, affects the energy for quarrying and mining operations, while the energy differences for further crushing and sieving between fractions may be neglected. Furthermore, the different aggregate fractions are always provided by the same quarry, thus it was reasonable to apply the "average" eco-profile for the same mineral deposit. For more clarity, the following list reports reference to the issues considered for the unitary impact data extraction from the Ecoinvent database: bitumen, at refinery; limestone, crushed, washed, production; heat, central or small-scale, other than natural gas, heat production, light fuel oil, at boiler 100 kW, non-modulating; heat, central or small-scale, natural gas, propane extraction, from liquefied petroleum gas; diesel, burned in diesel-electric generating set, market for; transport, freight, lorry 16–32 metric ton, EURO4.

Transportation effects have been properly computed, based on actual supplier locations and effective travelled distances (see Section 3.2.4). Other quantities and flows (regarding material components, fuels, etc.) consist of primary data, directly collected in the selected production plant.

3.2.2. The Examined Product

As anticipated, the DU of the study consisted of 1 ton of HMA, composed of different gradations of limestone aggregates and bitumen (50/70 pen grade), suitable for construction and maintenance activities of binder courses. The mix-design of the mixture (i.e., the actual recipe of the product), which fulfills typical technical specifications for road flexible pavements, requires three different classes of crushed stone (considering min ÷ max diameter of aggregates: 30% in class 15 ÷ 20 mm, 12% in 10 ÷ 15 mm, 10% in 5 ÷ 10 mm), sand (0 ÷ 6 mm—45%), filler (3%), and bitumen (5.5% on aggregate mass).

3.2.3. The Production Plant

Based on the context/market analysis in Italy and Southern Europe, a reference plant (located close to the city of Palermo, in Italy) was selected. The asphalt plant has an average hourly production of 65 ÷ 75 tons. A single production cycle is about 60 s long, while the relative mixture production is about 1400 kg, with a production temperature of 165 °C, on average.

According to the study aim, all the different production processes performed in the plant (material movements and proportioning, drying and heating, mixing, etc.) were specifically analyzed and included in the assessment by acquiring data and all the available information for deriving inputs and outputs of each step and of the entire production process.

Concerning energy needs for production, different fuels and energy sources were considered for the plant equipment and parts. Indeed, the largest energy contribution is assured by power energy, used for general plant operation and for internal material movements (conveyor belts, bucket elevators, etc.). Since there is no direct connection to the grid, a diesel engine produces the required energy in the analyzed plant. For aggregate drying and heating, a powerful LSF oil burner is adopted, while the bitumen is heated by means of an LPG burner. Finally, a wheel loader allows operators to load the various aggregate components in the cold feeder and, thus, this requires other diesel consumption. A representative scheme of the processes and elements involved in the production activity is shown in Figure 2: in detail, the main plant elements are depicted in blue, while transport processes are in gray; activities for production of components and fuels are in green (related information was extracted from secondary data).

The first element to be considered is the need for virgin raw materials required for production of the DU. According to the mix-design presented in Section 3.2.2, in Table 1 the related mass of the various components for 1 ton of asphalt mixture is provided. Practically, there is no significant raw material scrap, since in the continuous process all the aggregates effectively mixed in a single batch enter in the final product.

Table 1. Mass of the various components of the selected mixture for the DU.

Component	Mass (kg)
Crushed stone 1.5 (fraction 15–20 mm)	284.4
Crushed stone 1.0 (fraction 10–15 mm)	113.7
Crushed stone 0.5 (fraction 5–10 mm)	94.8
Sand (fraction 0–6 mm)	426.5
Filler (recovered)	28.4
Bitumen	52.1

In terms of the energy consumption for production of the DU, the electric power globally required by the plant is about 100 kW. In order to assure such a value, the generating set burns on average 260 kg/day or 0.6 ton/cycle of diesel. It should be considered that the generating set has a low efficiency (around 30–35%), as is common for similar equipment, determining high quantities of fuel burned for assuring low effective energy to plant equipment working. According to the effective production rates of the plant in a single hour, the various consumption values have been related to a single cycle and, thus, to the DU. In particular, for the generating set, the diesel consumption is equal to 0.44 kg/DU. Aggregate movement and loading in the hoppers is performed by means of a diesel wheel loader, operating for all the four aggregate classes. The loader diesel consumption is around 0.19 kg/DU. The consumption of LSF oil for aggregate heating is around 5 ton/day, thus 8.42 kg/DU. Regarding LPG for bitumen heating, available measurements prove a consumption of around 375 kg/day, thus 0.63 kg/DU. Taking into consideration the calorific value of each fuel (average values here assumed are in line with the typical calorific values for selected petroleum products [43,44]), the energy values in MJ have been calculated for each of them: diesel for electric generator set 18.8 MJ, diesel for loader 8.0 MJ, LPG 25.9 MJ, LSF oil 387.9 MJ.

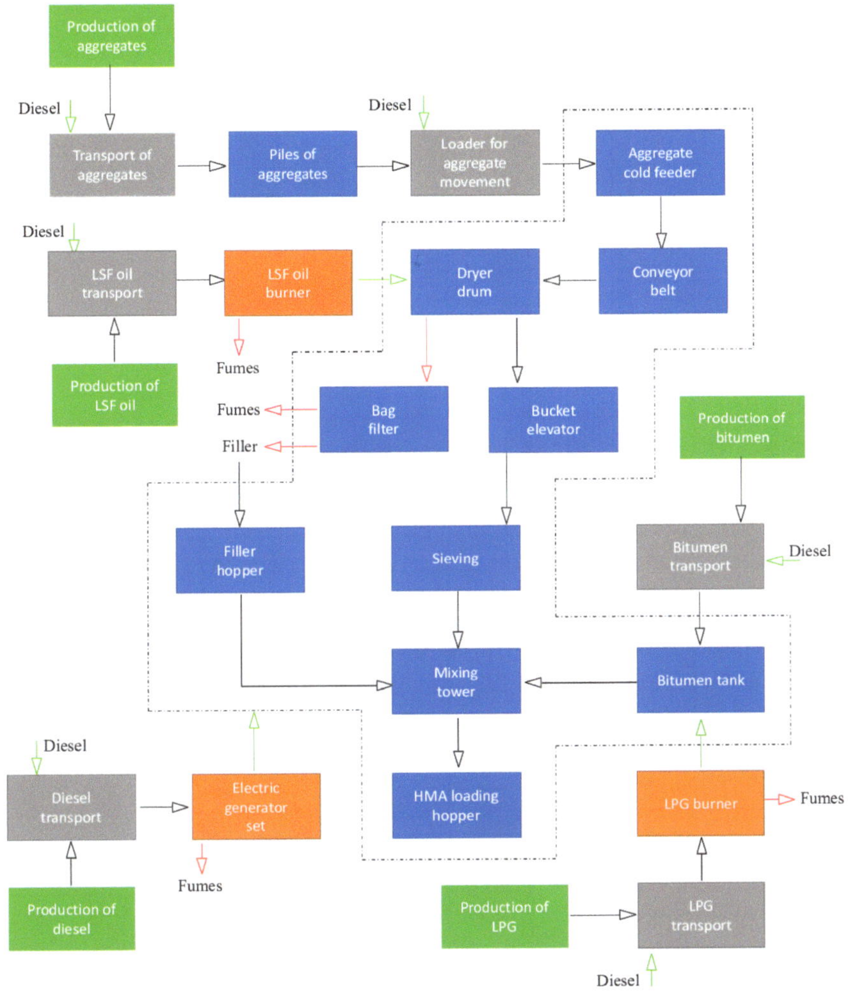

Figure 2. Production processes of a batch-mix asphalt production plant.

Furthermore, for a more accurate electrical analysis of the plant, the main elements were equipped with ammeters for continuous electricity measurements, during the production phases. The acquired measurements were used to calculate the required electricity for the DU, as shown in Figure 3. As mentioned, the overall plant involves a total electric power (including also secondary equipment and devices) of around 100 kW in normal operations, as evaluated by the plant direction. Consequently, the total required power energy for the DU is equal to 1.35 kWh, corresponding to 4.86 MJ/DU.

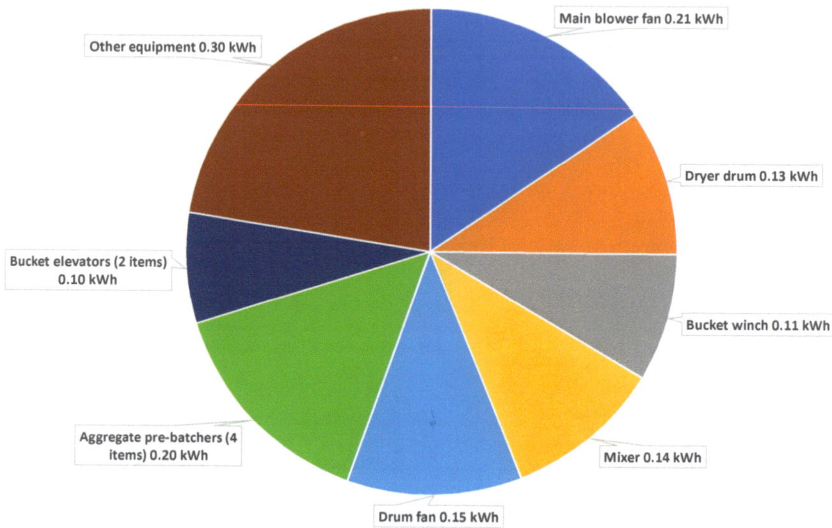

Figure 3. Electricity demand of the main plant components per declared unit (DU).

3.2.4. Raw material and Fuel Transportation

Regarding material and fuels transportation, all the components are provided to the plant by means of both Euro 4 and Euro 5 diesel trucks. Although all the material providers are located in Sicily, the different components are generally transported from different production sites (Figure 4), along paths of different lengths. Moreover, two different providers are involved both for bitumen and diesel, so calculations for both scenarios are presented. The reference unit for transportation calculations is represented by "ton-kilometer" (t·km), i.e., the mass of involved components transported for the relative distance for each DU. Data regarding transportations, acquired by the plant and analyzed in this study, are provided in Table 2.

It is straightforward to notice that transportation distance plays a fundamental role in this context: the exploitation of locally available materials may significantly reduce the truck travelled distances, producing relevant environmental advantages. According to this, both road engineers and industry managers should move towards more sustainable solutions, overpassing only direct economic considerations: raw materials and fuels supply from distant sources or plants should be avoided, their direct costs, in fact, do not take into account environmental drawbacks that, indeed, require significant attention and consideration. Therefore, pavement design and related technical specifications (including materials acceptance criteria) should also be carefully adapted to the local conditions and materials.

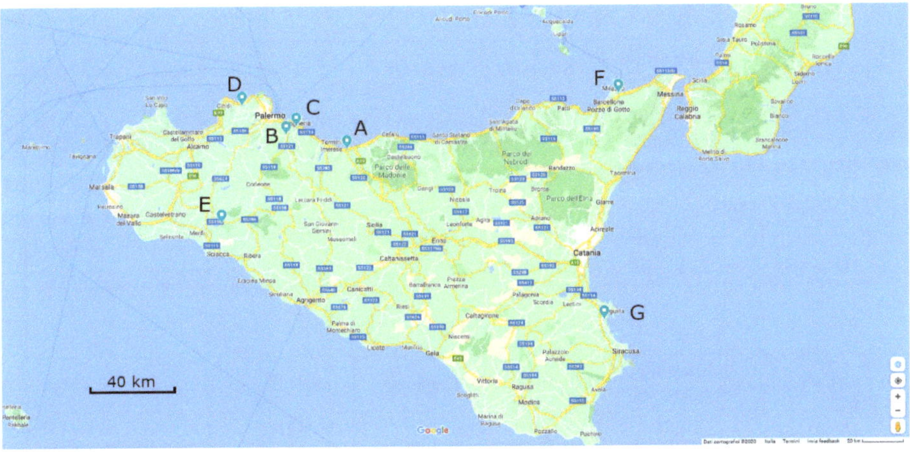

Figure 4. Location of the selected plant (A) and the various raw material providers: (B) diesel, (C) aggregates, (D) LPG, (E) diesel, (F) bitumen, LSF oil, (G) bitumen. Elaboration of map on [45].

Table 2. Transport information for each component and fuel involved in the production process (the letter in the distance column refers to plant location in Figure 4).

Component/Fuel	Distance from Asphalt Plant (km)	Mass for the DU (kg)	Ton-Kilometers for the DU (t·km)
Crushed stone 1.5	35.0 (C)	284.4	9.95
Crushed stone 1.0	35.0 (C)	113.7	3.98
Crushed stone 0.5	35.0 (C)	94.8	3.32
Sand	35.0 (C)	426.5	14.93
Filler (recovered)	35.0 (C)	28.4	0.99
Bitumen	152.0 (F)	52.1	7.92
	193.0 (G)	52.1	10.06
Diesel	123.0 (E)	0.6	0.08
	47.8 (B)	0.6	0.03
Low Sulphur fuel (LSF) oil	152.0 (F)	8.4	1.28
Liquid petroleum gas (LPG)	76.6 (D)	0.6	0.05

4. Life Cycle Impact Assessment and Discussion

4.1. Environmental Impacts

Results regarding impacts calculations are provided in the following figures and tables. In detail, first, the total values of the impact categories related to raw materials production and fuels production and use are listed in Table 3. Results related to transportation impacts are reported in Table 4, while Table 5 provides total impact values for each component (including raw materials production, fuels production and use, and transport). The obtained life cycle impact assessment results for AC and GW are of the same order of magnitude of those obtained by Mukherjee [24] for different asphalt mixtures (in [24] AC resulted equal to 2.95×10^{-1} $kgSO_{2eq}$ and GW to 6.46×10^{1} kg CO_{2eq}), but higher for OD (in [24] OD was 5.8×10^{-9} kg CFC_{11eq}). However, it is important to outline that, as stated in the "Introduction" section, Mukherjee examined asphalt mixtures produced in a different geographic context and characterized by different recipes than the DU of the present study and applied a different impact assessment method (TRACI 2.1).

Table 3. Impacts for raw materials and fuels for the DU.

Impact Category	Limestone Aggregates	Bitumen	Diesel	LSF oil	LPG	Total
AC (kg SO_{2eq})	4.47×10^{-2}	2.57×10^{-1}	3.08×10^{-2}	8.07×10^{-2}	5.37×10^{-3}	4.19×10^{-1}
EU (kg $PO_4{}^{3-}{}_{eq}$)	8.80×10^{-3}	4.19×10^{-2}	5.33×10^{-3}	1.06×10^{-2}	7.43×10^{-4}	6.74×10^{-2}
GW (kg CO_{2eq})	2.30×10^{0}	2.28×10^{1}	2.34×10^{0}	3.61×10^{1}	7.52×10^{-1}	6.43×10^{1}
PO (kg NMVOC)	5.30×10^{-2}	1.67×10^{-1}	4.25×10^{-2}	4.56×10^{-2}	3.28×10^{-3}	3.11×10^{-1}
OD (kg CFC_{11eq})	3.34×10^{-7}	2.30×10^{-5}	4.31×10^{-7}	6.65×10^{-6}	5.82×10^{-7}	3.10×10^{-5}
AD (kg SB_{eq})	4.29×10^{-6}	4.63×10^{-6}	4.05×10^{-7}	9.09×10^{-6}	3.43×10^{-7}	1.88×10^{-5}
AD* (MJ)	2.93×10^{1}	2.52×10^{3}	3.36×10^{1}	5.18×10^{2}	4.63×10^{1}	3.14×10^{3}

Table 4. Transport impacts for the DU.

Impact Category	Limestone Aggregates	Bitumen	Diesel	LSF oil	LPG	Total
AC (kg SO_{2eq})	2.46×10^{-2}	6.67×10^{-3}	3.98×10^{-5}	9.49×10^{-4}	3.59×10^{-5}	3.23×10^{-2}
EU (kg $PO_4{}^{3-}{}_{eq}$)	4.99×10^{-3}	1.35×10^{-3}	8.07×10^{-6}	1.93×10^{-4}	7.28×10^{-6}	6.56×10^{-3}
GW (kg CO_{2eq})	5.42×10^{0}	1.47×10^{0}	8.76×10^{-3}	2.09×10^{-1}	7.90×10^{-3}	7.11×10^{0}
PO (kg NMVOC)	2.93×10^{-2}	7.93×10^{-3}	4.73×10^{-5}	1.13×10^{-3}	4.27×10^{-5}	3.84×10^{-2}
OD (kg CFC_{11eq})	1.01×10^{-6}	2.73×10^{-7}	1.63×10^{-9}	3.89×10^{-8}	1.47×10^{-9}	1.32×10^{-6}
AD (kg SB_{eq})	1.64×10^{-5}	4.44×10^{-6}	2.65×10^{-8}	6.32×10^{-7}	2.39×10^{-8}	2.15×10^{-5}
AD* (MJ)	8.26×10^{1}	2.24×10^{1}	1.34×10^{-1}	3.19×10^{0}	1.21×10^{-1}	1.08×10^{2}

Table 5. Total impacts for the DU.

Impact Category	Limestone Aggregates	Bitumen	Diesel	LSF oil	LPG	Total
AC (kg SO_{2eq})	6.93×10^{-2}	2.64×10^{-1}	3.09×10^{-2}	8.16×10^{-2}	5.40×10^{-3}	4.51×10^{-1}
EU (kg $PO_4{}^{3-}{}_{eq}$)	1.38×10^{-2}	4.32×10^{-2}	5.34×10^{-3}	1.08×10^{-2}	7.50×10^{-4}	7.39×10^{-2}
GW (kg CO_{2eq})	7.72×10^{0}	2.43×10^{1}	2.35×10^{0}	3.63×10^{1}	7.59×10^{-1}	7.15×10^{1}
PO (kg NMVOC)	8.23×10^{-2}	1.75×10^{-1}	4.26×10^{-2}	4.67×10^{-2}	3.32×10^{-3}	3.50×10^{-1}
OD (kg CFC_{11eq})	1.34×10^{-6}	2.33×10^{-5}	4.33×10^{-7}	6.69×10^{-6}	5.83×10^{-7}	3.23×10^{-5}
AD (kg SB_{eq})	2.07×10^{-5}	9.07×10^{-6}	4.32×10^{-7}	9.72×10^{-6}	3.66×10^{-7}	4.02×10^{-5}
AD* (MJ)	1.12×10^{2}	2.54×10^{3}	3.38×10^{1}	5.21×10^{2}	4.64×10^{1}	3.25×10^{3}

For simplifying the impact analysis, Figures 5 and 6 show the contributions of the various elements to the impact categories, respectively for the raw materials and fuels only and for their transportation.

The analysis of the numerical outcomes evidences the most relevant components in the production of the DU of HMA mixture in a batch-mix plant. First, it appears that bitumen is the most relevant contributor to total impacts, almost for all categories, except for GW (Figure 6). Its incidence varies from 23% (AD) to 78% (AD*). LSF oil causes the main impact (51%) on GW while virgin aggregate consumption has the higher influence (51%) on AD. A negligible contribution (lower than 2%) comes from LPG for all the examined impact categories and from diesel for the impacts on global warming, ozone depletion, and abiotic depletion.

The above results strongly depend on the considered production technology, representing, as already underlined, the most common and adopted solution. In detail, the required quantities of bitumen in the mixture and of LSF oil for proper drying and heating of aggregates determine significant impacts. According to their nature, consumption of even small masses may cause huge consequences on environment. Obviously, based on current technologies, bitumen and LSF oil quantities used in the production cycle derive from specific mix-design and system efficiency evaluations. These quantities aim to assure proper mixing and workability to the mixture and, mostly, adequate mechanical performance to the final product. Then, in order to reduce relative and absolute impacts of asphalt mixture production in a similar plant, it is reasonable to boost innovative technological and technical solutions, requiring less volumes of raw materials (both bitumen and aggregates) and of fuels, with equivalent or comparable performance on field. Furthermore, replacing virgin aggregates with non-conventional aggregates (e.g., recycled aggregates) or using supplementary materials to reduce the binder content (obviously taking into account the specific recipe of the mixture) can represent other helpful strategies to promote sustainability [46]. In this regard, recent studies have proposed innovative non-petroleum-based binders for bitumen substitution for significantly improving sustainability of the product.

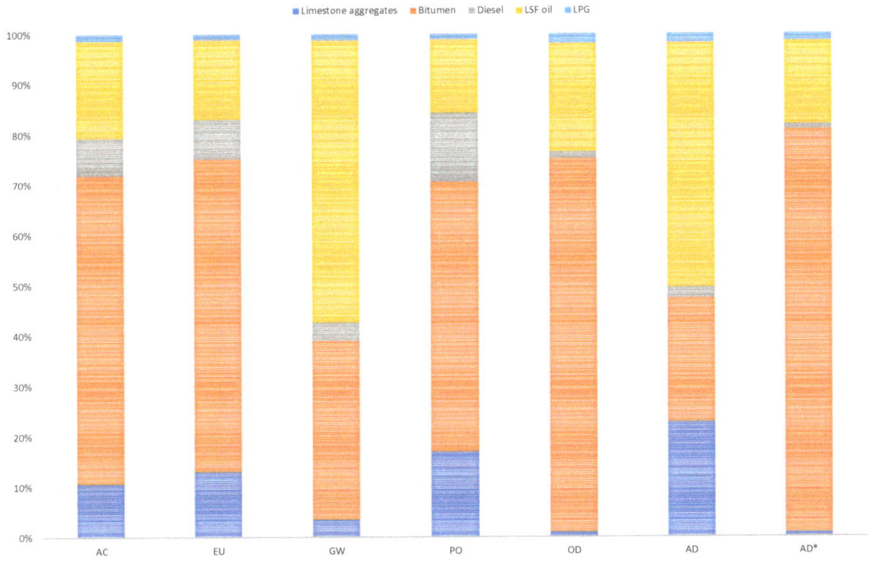

Figure 5. Raw materials and fuels impact for the DU.

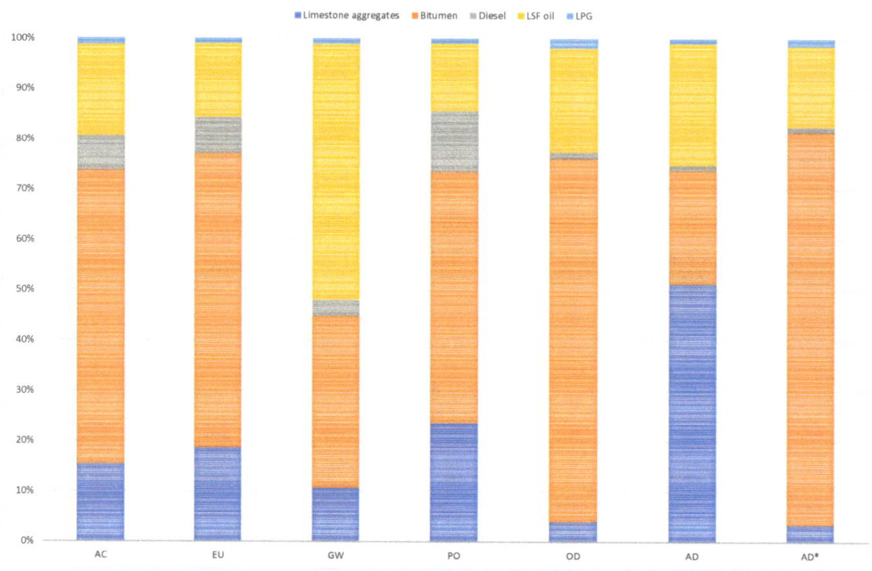

Figure 6. Total impacts of the components (including transport) for the DU.

Further, moving towards lower mixing temperatures—considering WMA production—will remarkably reduce the burned fuels for drying and heating aggregates, determining several benefits in terms of sustainability of the road industry sector [35,47–49], even though the effective consequences of additive adoption have to be properly taken into account [15,28]. Asphalt mixture with high percentages of reclaimed asphalt pavements (RAP) represents another possible positive sustainable solution, saving lot of natural virgin resources [6], even though Italian rates of RAP utilization are less than 25%, remarkably less than other countries [50].

Moreover, data also evidence how much transportation contributes to global impact values, generally less than 10%. Considering AD, the rate of incidence for transportation reaches 53% of the total impact. Although not only distance plays a role in transportation issues, efforts in reduction of transportation distances may increase product sustainability. This consciousness may effectively boost practical use and exploitation of locally available resources, even if slightly less performing than others, as long as they are proven to be adequate for the intended use. Sustainability evaluations, thus, should be effectively taken into consideration in mix-design procedures, in a continuous balance with mechanical performance. For instance, improved mixtures (including, for instance, specific additives) based on local products may assure equivalent service-life of traditional high-level solutions [51] but can increase the environmental efficiency of the asphalt mixture. Finally, recycling asphalt mixture from pavement dismission (both in plant and in place) may also appear as a strategic choice, as it may determine reduction of consumption of aggregate and binder and related transportation impacts in addition to the previously described benefits [29].

4.2. Sensitivity Analysis: Primary Data Variations

The numerical elaboration presented in this paper was mostly based on the analysis of primary data directly collected in the selected plant. However, due to the nature of the survey, provided and acquired data may be affected by some kinds of inaccuracies and errors [21,52]. In order to increase robustness of the numerical evidence provided in the previous paragraphs and, mostly, for assuring a clear overview on the considered scenario, the effects of eventual inaccuracies in the available primary input data were investigated. In particular, the influence on the impact values of five different variables has been investigated: (1) production rate per hour (t/h); (2) generating-set diesel consumption (kg); (3) loader diesel consumption (kg); (4) LPG consumption (kg); and (5) LSF oil consumption (kg). In detail, a 5% variation was applied to each of these variables, alternatively increasing and decreasing, as shown in Table 6, determining 10 different cases, called from A to E. Positive variations are indicated with symbol "+", while negative with "−": thus, "A+" and "A−" represent scenarios with production rate per hour respectively increased and decreased by 5%. Then, the values of the seven impact categories were recalculated for each case and compared to the reference ones (Tables 3–5), as shown in Figure 7.

Table 6. Reference and modified values for the five parameters. FU, functional unit.

Case	Parameter	Reference Value	+5%	−5%
A	Production rate per hour (t/h)	75.00	78.75	71.25
B	Generating-set diesel consumption (kg/FU)	0.44	0.46	0.42
C	Loader diesel consumption (kg/FU)	0.19	0.20	0.18
D	LPG consumption (kg/FU)	0.63	0.66	0.60
E	LSF oil consumption (kg/FU)	8.42	8.84	8.00

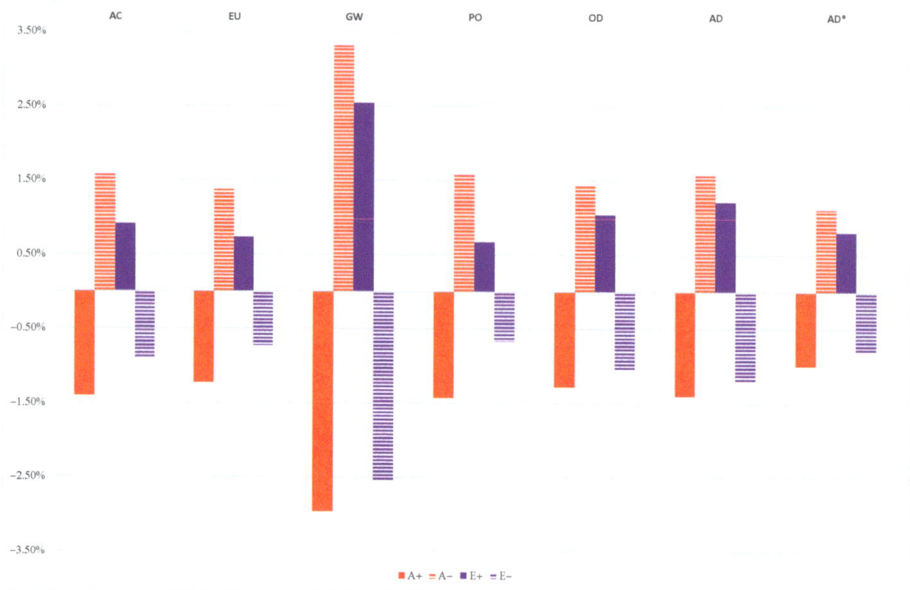

(**a**)

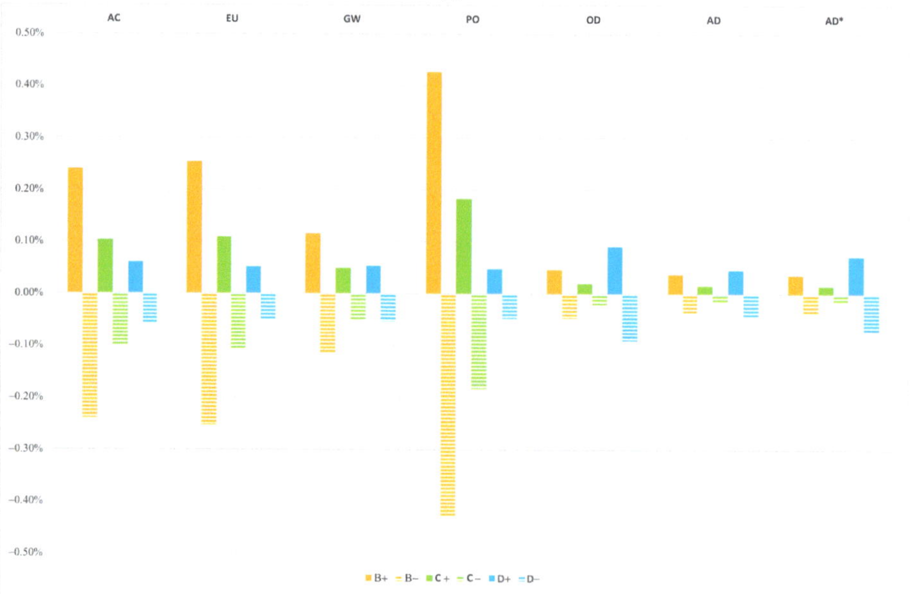

(**b**)

Figure 7. Impact variation for the examined scenarios: (**a**) scenarios A+, A−, E+, E−; (**b**) scenarios B+, B−, C+, C−, D+, D−.

The numerical results of Figure 7 show different influences of the various parameters. For more clarity, the results of the various scenarios are grouped according to the entity of variations: then, Figure 7a shows results related to scenarios A and E, while those of scenarios B, C, and D are shown in Figure 7b. Generally, if inaccuracies underestimate burned fuels (A−, B+, C+, D+, E+), the impact values increase, and vice versa. In detail, as expected, the most sensitive scenarios are A+, A−, E+, E−, i.e., those characterized by variations in plant productivity and in LSF oil consumption. For the former, larger (or smaller, in the opposite case) production rates are related to faster (or slower) production cycles, with consequent smaller (or higher) quantities of all the involved burned fuels. For scenarios E, as impacts related to LSF oil appeared as the most relevant among fuels, its variations have more effects on the results. GW is the most affected impact category, with a 3.31% increase in A− and 2.96% decrease in A+, thus for 5% variation in production rate. On average, 5% variations in production rate determine 1.70% variations in all impact categories. In cases E+ and E−, GW evidences 2.54% increase and decrease respectively, with an average 1.13% variation in all impact categories.

Considering the combined effects of all these possible variations (Figure 8a), results go towards two extreme scenarios: better (A+, B−, C−, D−, E−) or worse (A−, B+, C+, D+, E+) in terms of sustainability. For 5% change in production rate of the plant and of fuel consumptions, the impact categories show an average 2.8% decrease or 3.2% increase compared to the reference values, respectively in the better or worse scenario. GW undergoes the most relevant variations: it may reach up to 6.2% increase if all parameter values are actually more critical than measured, while it may be characterized by 5.5% decrease in the opposite scenario. These numerical outcomes are useful to prove that, despite the possible uncertainty due to the primary data collection in plant, the evidence provided in Section 4.1 still remains sound and reliable. Indeed, even taking into consideration significant variations in the original data, the derived impact values do not show particular alterations.

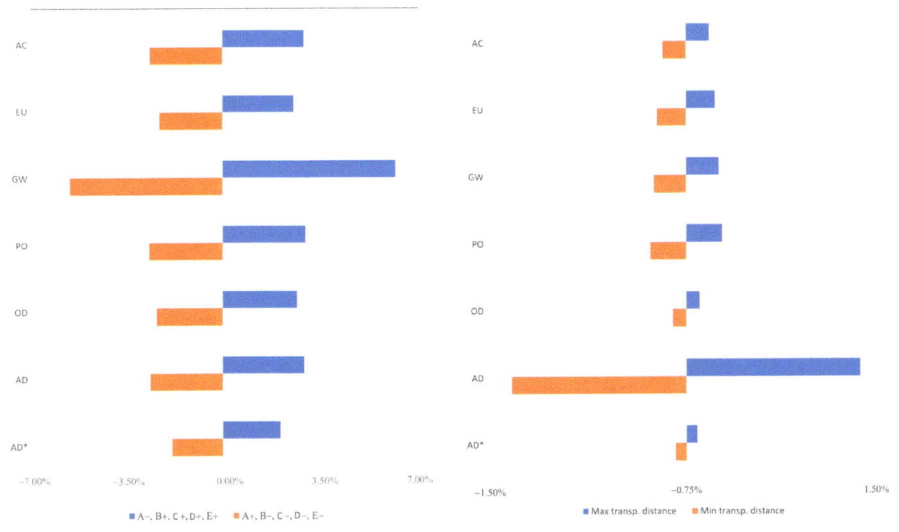

Figure 8. Impact variations: (**a**) combined effect of all parameter variations; (**b**) impact variations considering a single provider only for bitumen and diesel, respectively the closer and the farther.

Further, transportation distance was also numerically tested. In particular, for components with alternative providers, the effect of transportation distance maximization (and minimization) has been studied. As underlined in Section 3, bitumen and diesel are generally alternatively provided by two different production plants located in two different areas of Sicily (Figure 4, Table 2). In previous calculation, thus, the average distance was taken into consideration. In order to quantify possible effects of distance variations (even for these two components only), Figure 8b provides the variations for the various impact categories when both components are bought from the farther or the closer plant. AD results as the most influenced impact, with a positive or negative 1.35% change in the two cases. Although it seems that, compared to reference impacts of Figure 6, these variations are not relevant, it should be considered that passing from the farthest to the closest providers determines about 3% reduction in AD, confirming the advantages for limiting transportation distances.

4.3. Scenario Analysis: Effects of Changes in the Energy Production

In the evaluation of LCA of HMA mixture production in a plant, another crucial aspect is represented by the source of energy used for plant general operation. The considered plant is electrically powered using a diesel-based generating set. This choice derives from the specific configuration of the plant and some difficulties occurred in the grid connection to the factory. Considering the aim of the paper, in order to quantify possible improvements in impacts due to asphalt mixture production, the environmental impacts of the same plant powered by two different solutions were estimated: the electricity grid and a dedicated set of photovoltaic panels. Data for unit impact assessment were derived from the Ecoinvent database, as for other secondary data previously. For quantifying the required energy (in MJ), a 30% efficiency rate for the electric generator was considered. Consequently, based on data provided in Section 3.2.3, the effective required energy for plant operation is considered, prudently, equal to 5.63 MJ. Figure 9 shows impact variations for the two alternative options compared to the actual energy supply (raw materials and fuels only), while Figure 10 reports total impact variations, including transportation.

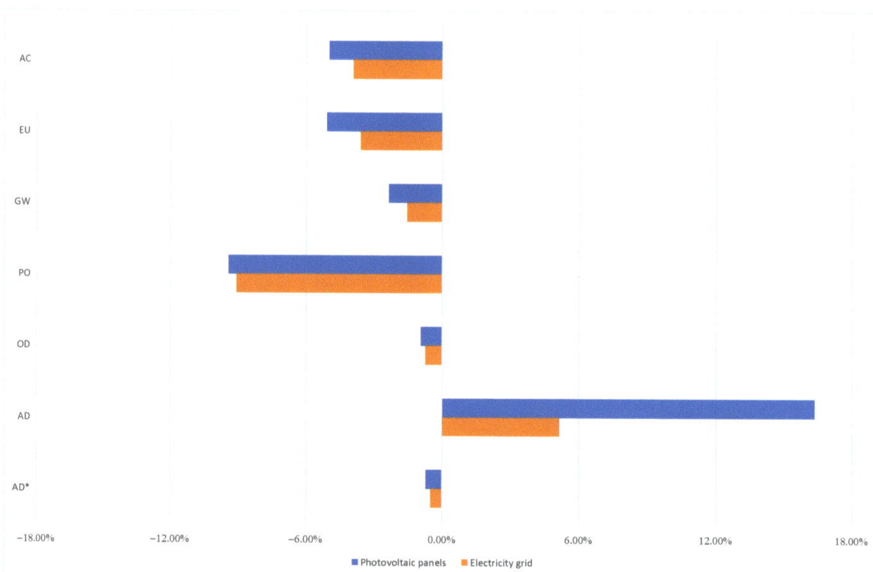

Figure 9. Variations of impact of raw materials and fuels for different energy supply technologies.

The provided results generally confirm a noticeable improvement in the sustainable performance of the selected production plant. From a global perspective, despite a not negligible increase of AD—especially for the photovoltaic panels—the two alternative ways of feeding energy assure important reductions for all the selected impact categories. PO is the most reduced category, with a peak around 8% for both solutions, followed by AC and EU (around 5% reduction for photovoltaic panels and 3% for the grid). However, in both cases, AD shows a remarkable increase, but this is substantially due to the specific reference technology. In particular, for the photovoltaic panels, AD is mainly related to the rare and precious minerals quantities required for panel production [53]. In summary, despite this aspect, the choice of a different source for electricity supply appears convenient from an environmental perspective, assuring improvements for the considered impact categories. Also taking into account the avoided impacts due to the diesel transportation (Figure 10), an increase in AD is observed as well as a higher decrease of the other impact categories. This result further confirms the improvement achievable by substituting electricity from the diesel-based generating set with electricity from the grid or from photovoltaic.

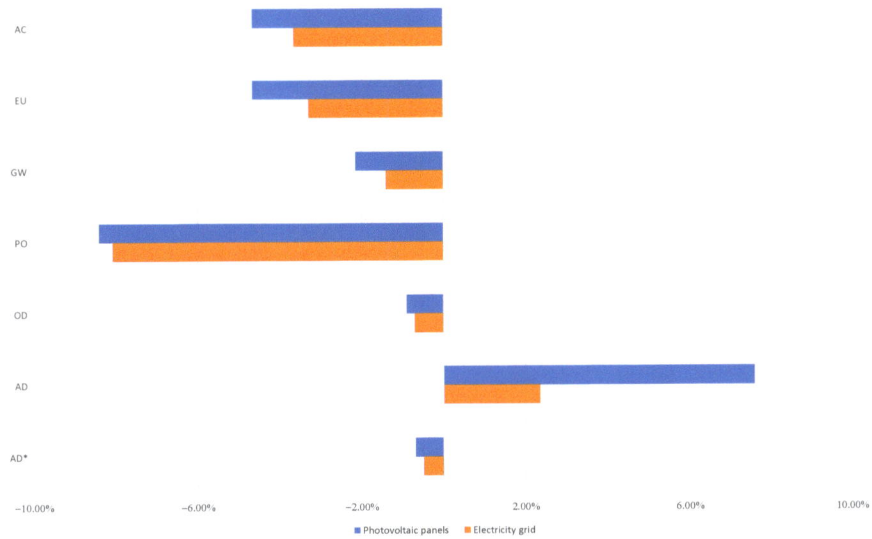

Figure 10. Variations of total impacts for different energy supply technologies.

4.4. *Scenario Analysis: Effects of Changes in the Mixing Temperature*

As underlined in Section 4.1, fuels have a huge impact on the sustainability of the asphalt mixture. In particular, LSF oil appeared to be the most affecting fuel (Figures 5 and 6). As already discussed, LSF oil is used in the plant for actually preparing aggregates—drying and heating, for allowing correct mixing with bitumen and, according to traditional technology of HMA, the mixture temperature is around 160 ÷ 170 °C. In recent years, alternative technologies have been studied with the aim of assuring equal adequate performance to the final product, without reaching similar temperatures [18,35,49]. Among these, the so-called warm technique (for producing WMA) guarantees mechanical levels in the field comparable with those of HMA, with significantly lower production temperature, around 130 ÷ 140 °C. From a sustainability perspective, such a solution may provide significant benefits, determining potential reduction in terms of impacts and emissions in plant and construction sites [15,28,48], even if the characteristics of the considered additive should be properly evaluated. For a preliminary quantification of the related improvements, according to the general approach of the paper, the impact values for all the considered categories were estimated, considering the production of a WMA, for comparison purposes with those obtained for the HMA. Calculation was based on the available data, considering just hypothetically the possibility of obtaining a DU of asphalt product using WMA technology, without variations in the mix-design. Such a hypothesis allowed useful simplifications in the calculation process and, further, may be accepted since influence of slight variations in the component proportions or the introduction of few quantities of additives do not strongly influence the final results. However, this is no longer valid when the additive percentages are high, and the additive nature and production processes determine such high direct impacts so as to offset the possible advantages assured by the lower production temperature. Regarding the energy consumption when using WMA technology, technical studies always agree on its reduction, although providing slightly different values [49]. Nevertheless, based on this literature, it is reasonable to consider as acceptable a 30% reduction in the energy needs for production of WMA compared to HMA, for a 20 °C reduction in the mixing temperature. This reduction has been applied to the LFS oil and LPG mass quantities provided in Section 3.2.3, passing respectively from 387.9 MJ and 25.9 MJ for HMA to 271.5 MJ and 18.1 MJ for WMA. The variations in the various impact categories due to the WMA technology compared to the original HMA are provided in Figure 11.

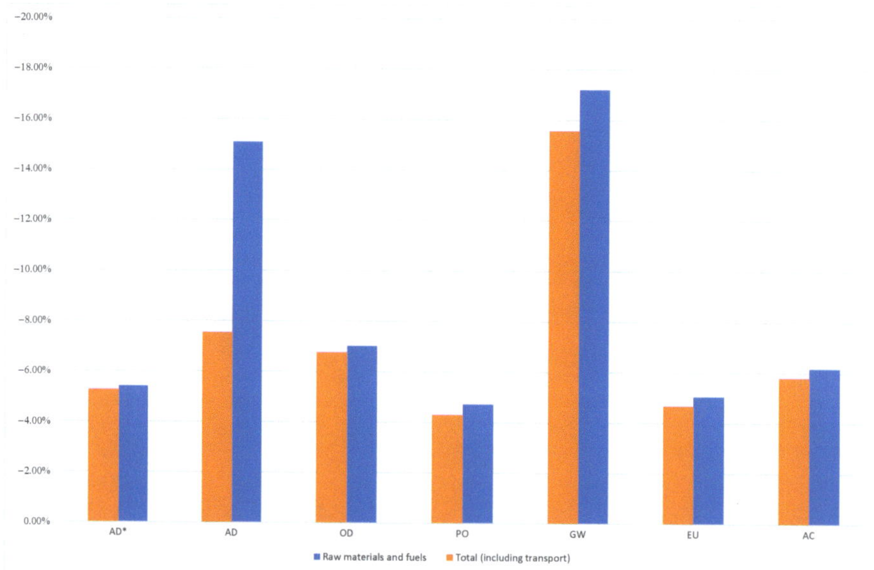

Figure 11. Impact variations (without and with transport) for WMA technology.

As expected, the effects of changes in the mixture temperature are relevant in practice. Relying on the previous hypotheses, considering 20 °C reduction in the mixing temperature, may assure saving 2.5 kg of LFS oil and 0.19 kg of LPG for each DU of asphalt mixture, with remarkable benefits on impact values. Considering the total values (including transport impacts), all the impact categories show more than 5% reduction, with a −15% peak for GW. From the analysis of the figure, it is evident how a similar technology should be boosted, for assuring more sustainable products for the pavement industry. Obviously, as already stated, similar considerations are not relevant if final performance in the field is excluded from the assessment. It would be not acceptable, in practical context, to compare two different solutions characterized by significantly different performance levels in exercise. In this regard, in a more exhaustive LCA of asphalt pavements for a more proper assessment of the numerical differences in impact values among different solutions, specific reference to the actual mix-design and the effective mechanical final responses must be included.

5. Conclusions

In this paper, the LCA of 1 ton of asphalt mixture was investigated, from raw materials extraction to product manufacturing. In particular, based on a detailed phase of primary data collection directly in an Italian production plant, all the inputs and outputs of traditional HMA production for road pavement construction were determined and analyzed. According to the LCA framework, for each productive phase, the data collection was performed within the inventory phase to identify and quantify material and fuel flows, determining their consumption for the DU, together with direct and indirect emissions. Then, all the available data, integrated with secondary reliable data, were used for estimating the environmental impacts of the examined declared unit.

The analysis evidenced the contribution of each component of the product and of fuels on the selected environmental impact categories, for understanding the most crucial phases of the process and the least sustainable materials. Numerical results showed a large influence of bitumen and fuels, in particular LFS oil, on every impact category, evidencing the possible benefits of boosting the development of alternative productive technologies, such as WMA and recycled mixtures. Indeed, similar alternatives require less fuel for drying and heating aggregates and may be equally adequate from a performance perspective in the field, with noticeable reduction of both virgin bitumen and aggregate consumption. In general, the study allowed to collect foreground data for the calculation of a HMA eco-profile and complete the LCA of the product according to the most common production procedure. The collected data and the results of this study may represent a reliable reference and numerical base for following steps of the LCA, considering all the phases of an asphalt pavement service life. LCAs concerning the overall construction process of asphalt pavements can rely on the actual values measured in the plant to provide accurate and reliable outcomes for stakeholders and decision makers in the reference geographical context/market. As preliminarily presented in this paper, the comparison of results of various scenarios may simplify the evaluation of the most convenient and sustainable solutions and approaches (also in terms of technological improvements) for driving improvement and innovation of the entire production industry. In fact, the possible benefits of alternative sources of electricity for the plant working and reduction in the mixture production temperature has been highlighted. In both cases, the considered modifications to the plant or to the product—assuring equal performance levels on field—may generally lower the impacts of the process, with significant environmental benefits, crucial for a novel sustainable approach.

Author Contributions: G.S.: resources, writing—original draft, formal analysis. S.L.: resources, validation, writing—original draft. M.C.: methodology, validation, supervision. C.C.: resources, conceptualization, writing—original draft, funding acquisition. All authors have read and agreed to the published version of the manuscript.

Funding: This work was financially supported by the Italian Ministry for University and research (research grant PRIN 2017 USR342).

Conflicts of Interest: The authors declare that they have no known competing financial interests or personal relationships that could have appeared to influence the work reported in this paper.

References

1. Butt, A.A.; Mirzadeh, I.; Toller, S.; Birgisson, B. Life cycle assessment framework for asphalt pavements: Methods to calculate and allocate energy of binder and additives. *Int. J. Pavement Eng.* **2012**, *15*, 290–302. [CrossRef]
2. Marzouk, M.; Abdelkader, E.M.; El-Zayat, M.; Aboushady, A. Assessing environmental impact indicators in road construction projects in developing countries. *Sustainability* **2017**, *9*, 843. [CrossRef]
3. Van Dam, T.J.; Harvey, J.T.; Muench, S.T.; Smith, K.D.; Snyder, M.B.; Al-Qadi, I.L.; Ozer, H.; Meijer, J.; Ram, P.V.; Roesler, J.R.; et al. *Towards Sustainable Pavement Systems: A Reference Document*; Technical Report Number: FHWA-HIF-15-002; Federal Highway Administration: Washington, DC, USA, 2015.
4. Lee, J.C.; Edil, T.B.; Tinjum, J.M.; Benson, C.H. Quantitative assessment of environmental and economic benefits of recycled materials in highway construction. *Transp. Res. Rec. J. Transp. Res. Board* **2010**, *2158*, 138–142. [CrossRef]
5. Farooq, M.A.; Mir, M.S. Use of reclaimed asphalt pavement (RAP) in warm mix asphalt (WMA) pavements: A review. *Innov. Infrastruct. Solut.* **2017**, *2*, 10. [CrossRef]
6. Plati, C. Sustainability factors in pavement materials, design, and preservation strategies: A literature review. *Constr. Build. Mater.* **2019**, *211*, 539–555. [CrossRef]

7. International Organization for Standardization. *Environmental Management-Life Cycle Assessment-Principles and Framework*; ISO 14040_2006(E); ISO: Geneva, Switzerland, 2006.
8. International Organization for Standardization. *Environmental Management-Life Cycle Assessment-Requirements and Guidelines*; ISO 14044_2006(E); ISO: Geneva, Switzerland, 2006.
9. Hakkinen, T.; Makela, K. Environmental adaptation of concrete-environmental impact of concrete and asphalt pavements. In Proceedings of the 1996 Espoo VTT Research Notes, Croatia, Balkans, 8 July 1996; Valtion Teknillinen Tuktumuskeskus (VTT): Helsinki, Finland, 1996.
10. Horvath, A.; Hendrickson, C. Comparison of environmental implications of asphalt and steel-reinforced concrete pavements. *Transp. Res. Rec. J. Transp. Res. Board* **1998**, *1626*, 105–113. [CrossRef]
11. Santero, N.J.; Masanet, E.; Horvart, A. Life-cycle assessment of pavements. Part I: Critical review. *Resour. Conserv. Recycl.* **2011**, *55*, 801–809. [CrossRef]
12. Celauro, C.; Corriere, F.; Guerrieri, M.; Casto, B.L. Environmentally appraising different pavement and construction scenarios: A comparative analysis for a typical local road. *Transp. Res. Part D Transp. Environ.* **2015**, *34*, 41–51. [CrossRef]
13. Santero, N.J.; Masanet, E.; Horvath, A. Life-cycle assessment of pavements Part II: Filling the research gaps. *Resour. Conserv. Recycl.* **2011**, *55*, 810–818. [CrossRef]
14. Celauro, C.; Corriere, F.; Guerrieri, M.; Casto, B.L.; Rizzo, A. Environmental analysis of different construction techniques and maintenance activities for a typical local road. *J. Clean. Prod.* **2017**, *142*, 3482–3489. [CrossRef]
15. Anthonissen, J.; Braet, J.; Bergh, W.V.D. Life cycle assessment of bituminous pavements produced at various temperatures in the Belgium context. *Transp. Res. Part D Transp. Environ.* **2015**, *41*, 306–317. [CrossRef]
16. Farina, A.; Zanetti, M.C.; Santagata, E.; Blengini, G.A. Life cycle assessment applied to bituminous mixtures containing recycled materials: Crumb rubber and reclaimed asphalt pavement. *Resour. Conserv. Recycl.* **2017**, *117*, 204–212. [CrossRef]
17. Balaguera, A.; Carvajal, G.I.; Jaramillo, Y.P.A.; Albertí, J.; Fullana-I-Palmer, P. Technical feasibility and life cycle assessment of an industrial waste as stabilizing product for unpaved roads, and influence of packaging. *Sci. Total Environ.* **2018**, *651*, 1272–1282. [CrossRef]
18. Gulotta, T.; Mistretta, M.; Praticò, F. A life cycle scenario analysis of different pavement technologies for urban roads. *Sci. Total Environ.* **2019**, *673*, 585–593. [CrossRef]
19. Hasan, U.; Whyte, A.; Al Jassmi, H. Critical review and methodological issues in integrated life-cycle analysis on road networks. *J. Clean. Prod.* **2019**, *206*, 541–558. [CrossRef]
20. Huang, Y.; Bird, R.; Bell, M. A comparative study of the emissions by road maintenance works and the disrupted traffic using life cycle assessment and micro-simulation. *Transp. Res. Part D Transp. Environ.* **2009**, *14*, 197–204. [CrossRef]
21. Ardente, F.; Beccali, M.; Cellura, M. FALCADE: A fuzzy software for the energy and environmental balances of products. *Ecol. Model.* **2004**, *176*, 359–379. [CrossRef]
22. Franzitta, V.; Longo, S.; Sollazzo, G.; Cellura, M.; Celauro, C. Primary data collection and environmental/energy audit of hot mix asphalt production. *Energies* **2020**, *13*, 2045. [CrossRef]
23. Stripple, H. *Life Cycle Inventory of Asphalt Pavements*; IVL Swedish Environmental Research Institute Ltd.: Gothenburg, Sweden, 2000.
24. Mukherjee, A. 2016 Life Cycle Assessment of Asphalt Mixtures in Support of an Environmental Product Declaration. ISO 14040 Compliant Life Cycle Assessment Study Supporting the National Asphalt Pavement Association, EPD Program for North American Asphalt Mixtures. Available online: https://www.asphaltpavement.org/PDFs/EPD_Program/LCA_final.pdf (accessed on 15 June 2020).
25. Mukherjee, A.; Dylla, H. Challenges to using environmental product declarations in communicating LCA results: The case of the asphalt industry; Transportation Research Board of the National Academies. *Transp. Res. Rec. J. Transp. Res. Rec.* **2017**, *2639*, 84–92. [CrossRef]
26. Mroueh, U.M.; Eskola, P.; Laine-Ylijoki, J.; Wellman, K.; Makela, E.; Juvankoski, M.; Ruotoistenmaki, A. Life cycle assessment oof road construction. *Tielaitoksen Selvityksia* 2000. Available online: https://cris.vtt.fi/en/publications/life-cycle-assessment-of-road-construction (accessed on 15 June 2020).
27. Yang, R.; Kang, S.; Ozer, H.; Al-Qadi, I. Environmental and economic analyses of recycled asphalt concrete mixtures based on material production and potential performance. *Resour. Conserv. Recycl.* **2015**, *104*, 141–151. [CrossRef]

28. Vidal, R.; Moliner, E.; Martínez, G.; Rubio, M.C. Life cycle assessment of hot mix asphalt and zeolite-based warm mix asphalt with reclaimed asphalt pavement. *Resour. Conserv. Recycl.* **2013**, *74*, 101–114. [CrossRef]
29. Giani, M.I.; Dotelli, G.; Brandini, N.; Zampori, L. Comparative life cycle assessment of asphalt pavements using reclaimed asphalt, warm mix technology and cold in-place recycling. *Resour. Conserv. Recycl.* **2015**, *104*, 224–238. [CrossRef]
30. Park, K.; Hwang, Y.W.; Seo, S.; Seo, H. Quantitative Assessment of environmental impacts on life cycle of highways. *J. Constr. Eng. Manag.* **2003**, *129*, 25–31. [CrossRef]
31. Ardente, F.; Beccali, G.; Cellura, M. Eco-sustainable energy and environmental strategies in design for recycling: The software "ENDLESS". *Ecol. Model.* **2003**, *163*, 101–118. [CrossRef]
32. SITEB. L'assemblea annuale di SITEB. *Rass. Bitume* **2019**, *92*, 19–25. (In Italian)
33. EAPA. *EAPA'S Position Statement on the Use of Secondary Materials, by-Products and Waste in Asphalt Mixtures*; European Asphalt Pavement Association: Brussel, Belgium, 2017. Available online: https://eapa.org/wp-content/uploads/2018/07/EAPA-position-statement-2.pdf (accessed on 15 June 2020).
34. NAPA; EAPA. *The Asphalt Paving Industry-A Global Perspective: Production, Use, Properties, and Occupational Exposure Reduction Technologies and Trends*, 2nd ed.; National Asphalt Paving Association and European Asphalt Pavement Association: Lanham, MD, USA, 2011; ISBN 0-914313-06-1.
35. D'Angelo, J.; Harm, E.; Bartoszek, J.; Baumgardner, G.; Corringan, M.; Cowsert, J.; Harman, T.; Jamshidi, M.; Jones, W.; Newcomb, D.; et al. *Warm-Mix Asphalt: European Practice*; Technical Report Number: FHWA-PL-08–007; Federal Highway Administration: Washington, DC, USA, 2008.
36. Ravaioli, S. Impianti d'asfalto: Considerazioni storico-economiche e statistiche del settore. *Rass. Bitume* **2010**, *66*, 33–39. (In Italian)
37. EAPA. *Asphalt in Figures 2018*; European Asphalt Pavement Association: Brussels, Belgium, 2020. Available online: https://eapa.org/asphalt-in-figures/ (accessed on 15 June 2020).
38. EPA. *Hot Mix Asphalt Plants Stakeholders Opinions Report*; EPA-454/R-00–030; U.S. Environmental Protection Agency: Research Triangle Park, NC, USA, 2001.
39. The International EPD®System. Asphalt mixtures (Europe, Australia). *Product Category Rules (PCR) of the Environmental Product Declaration (EPD)*. 2018. Version 1.03. Available online: https://www.battleco2.com/wp-content/uploads/2018/08/pcr2018-04-Asphalt-mixtures-BatteCO2.pdf (accessed on 16 June 2020).
40. European Commission, Joint Research Centre e Institute for Environment and Sustainability. *ILCD Handbook International Reference Life Cycle Data System. General Guide for Life Cycle Assessment Provisions and Action Steps*; Publications Office of the European Union: Luxembourg, 2010.
41. EPD International AB. *General Programme Instruction for the International EPD®System*; EPD INTERNATIONAL AB. 2019; Version 3.01. Available online: www.environdec.com (accessed on 13 June 2020).
42. Wernet, G.; Bauer, C.; Steubing, B.; Reinhard, J.; Moreno-Ruiz, E.; Weidema, B.P. The ecoinvent database version 3 (Part I): Overview and methodology. *Int. J. Life Cycle Assess.* **2016**, *21*, 1218–1230. [CrossRef]
43. MacLeay, I. *Digest of United Kingdom Energy Statistics*; The Stationery Office: London, UK, 2010. Available online: www.decc.gov.uk (accessed on 12 June 2020).
44. IEA. *Energy Statistics Manual*; OECD Publishing: Paris, France, 2004. Available online: https://doi.org/10.1787/9789264033986-en (accessed on 16 June 2020).
45. Google Maps. Available online: https://www.google.com/maps/place/Sicily (accessed on 30 November 2020).
46. De Brito, J.; Kurda, R. The past and future of sustainable concrete: A critical review and new strategies on cement-based materials. *J. Clean. Prod.* **2020**, *123558*, 123558. [CrossRef]
47. Hasan, U.; Whyte, A.; Al Jassmi, H. Life cycle assessment of roadworks in United Arab Emirates: Recycled construction waste, reclaimed asphalt pavement, warm-mix asphalt and blast furnace slag use against traditional approach. *J. Clean. Prod.* **2020**, *257*, 120531. [CrossRef]
48. Rubio, M.C.; Martínez, G.; Baena, L.; Moreno, F. Warm mix asphalt: An overview. *J. Clean. Prod.* **2012**, *24*, 76–84. [CrossRef]
49. Zaumanis, M. Warm Mix Asphalt. In *Climate Change, Energy, Sustainability and Pavements*; Gopalakrishnan, K., Steyn, W., Harvey, J., Eds.; Green Energy and Technology Springer: Berlin, Germany, 2014.
50. Ravaioli, S. DM 69/18-End of Waste per il fresato d'asfalto: Vera opportunità o occasione mancata? *Rass. Bitume* **2019**, *91*, 19–26. (In Italian)

51. Celauro, C.; Bosurgi, G.; Sollazzo, G.; Ranieri, M. Laboratory and in-situ tests for estimating improvements in asphalt concrete with the addition of an LDPE and EVA polymeric compound. *Constr. Build. Mater.* **2019**, *196*, 714–726. [CrossRef]
52. Cellura, M.; Longo, S.; Mistretta, M. Sensitivity analysis to quantify uncertainty in Life Cycle Assessment: The case study of an Italian tile. *Renew. Sustain. Energy Rev.* **2011**, *15*, 4697–4705. [CrossRef]
53. Muteri, V.; Cellura, M.; Curto, D.; Franzitta, V.; Longo, S.; Mistretta, M.; Parisi, M.L. Review on life cycle assessment of solar photovoltaic panels. *Energies* **2020**, *13*, 252. [CrossRef]

sustainability

MDPI

Article

The Bike-Sharing System as an Element of Enhancing Sustainable Mobility—A Case Study based on a City in Poland

Elżbieta Macioszek [1,*], Paulina Świerk [2] and Agata Kurek [1]

1 Transport Systems and Traffic Engineering Department, Faculty of Transport and Aviation Engineering, Silesian University of Technology, Krasińskiego 8 Street, 40-019 Katowice, Poland; agata.kurek@polsl.pl

2 Metropolitan Transport Board, Silesian Voivodeship, Barbary 21A Street, 40-053 Katowice, Poland; pauswierk@gmail.com

* Correspondence: elzbieta.macioszek@polsl.pl

Received: 20 March 2020; Accepted: 15 April 2020; Published: 17 April 2020

check for updates

Abstract: The bike-sharing system allows urban residents to rent a bike at one of the rental stations located in the city, use them for their journey, and return them to any other or the same station. In recent years, in many cities around the world, such systems were established to encourage their residents to use bikes as an element of enhancing sustainable mobility and as a good complement to travel made using other modes of transport. The main purpose of this article is to present the results of an analysis of the functioning of the bike-sharing system in Warsaw (Poland). Moreover, the article presents an analysis of the accessibility to individual stations. An important aspect is that the bike-sharing system has been popular among users and that more people use it. Therefore, the city should be provided with a dense network of conveniently located bike-sharing stations. Also, the quality of the bike-sharing system should be an adjustment to the user's expectations. In connection with the above, the article also presents the results of the analysis of factors affecting bike-sharing system usage as well as the level of satisfaction connected with bike-sharing system usage. The results of the analysis showed that there is a strong positive correlation between these variables. The obtained results can be helpful for carrying out activities whose purpose is to increase the bike-sharing system usage as well as to increase the level of satisfaction connected with bike-sharing system usage.

Keywords: bike-sharing system; bike; sustainable mobility; road traffic engineering; road transport

1. Introduction

Cities radically change into smart cities because of the increasing mobility of societies, urban growth, new products, and technological innovations [1–3]. Also, these factors enforce the need to properly shape travellers' behaviour. Currently, in many cities in urban areas, the biggest challenge is to overcome the problem of the prevailing use of private cars [4–7]. Bike-sharing systems are considered to be an effective tool for popularizing alternative ways of traveling, and can be used as an element of enhancing sustainable mobility in cities. Sustainable mobility is the communication behavior of users shaped in this way in the spatial structure and transport, in which the length of the travel route is rationalized, individual motorization does not degrade public transport and non-motorized transport (walking and cycling), and the functioning of the transport system makes it possible to maintain harmony with the environment. Sustainable transport is conducive to improving the city's image and spatial order, as well as creating good quality public space, and also reduces the diversity in the development and quality of life of individual city areas [8]. Contemporary system solutions that connect space and transport planning aim for greater use of the bike as a means of transport. Journeys made by private bikes as well as by bike with the bike-sharing system contribute to significant savings and benefits for both bike users and the natural environment (i.e., no negative impact on the quality of life in the city (no noise and no emissions) protection of monuments and flora; better use of space, both for movement and parking; less degradation of the road network; improvement of the attractiveness of the city center (business, culture, recreation, social life); reduction of congestion and economic losses; increased traffic flow; greater attractiveness of public transport; better access to urban services for the whole society; saving time and money of parents who are exempted from taking their children to school; significant saving of time for cyclists on short and medium distances; no possible necessity owning a second car by the household (budget increase) [9–11]).

However, travel by bike is not able to meet all mobility needs. In large cities and agglomerations, in highly urbanized areas, as well as in the outskirts and rural areas, the car will still be used. In these cases, solutions should be implemented that enable it to be used more effectively. D. Jankowska-Karpa and A. Wnuk in their work [9] proposed changes that may contribute to limiting traffic and at the same time promoting travels by bikes in cities. This may contribute to an increase in the role of travel by bikes in shaping sustainable mobility. These were the following changes:

- At short distances, residents should be encouraged to restrict car use, especially if the journey is short and begins and ends at home (e.g., home-shop-home). In this case, several solutions can be proposed: bike paths, facilities for pedestrians and cyclists to allow free movement, development of convenient infrastructure, e.g., parking spaces, etc.;
- Modernization of streets ensuring the safe and comfortable movement of pedestrians and cyclists at a distance of at least 1 km around each school and promotion of "pedestrian and bicycle buses" for children, which would contribute the reduction of parking lots and traffic flow around schools;
- Many trips by car taken from work during the lunch break. Initiatives should be taken to promote more environmentally friendly means of transport around jobs and planning public spaces in office districts to reduce the number of cars. Employers can support such initiatives, by creating bike parking, free bike rentals, installing lockers at workplaces for storing personal things, etc.;
- The so-called "time policy" can be developed to adapt the offer to the needs of mobility. For example, due to heavy traffic during peak hours, in consultation with employers and authorities, it is possible to synchronize activities and services. This would reduce congestion on some transport network at certain hours. These solutions have already been implemented in some European cities.

The bike-sharing system allows users to rent a bike in one of the automatic stations located in the city, use it during the journey, and return it at the same or any other station. In recent years, systems of this type have been created in many cities around the world to encourage residents to use the bike as an ecological mode of transport or as a part of the journey using other modes of transport. The basic idea of the bike-sharing concept is sustainable transportation. Such systems often operate as part of the city's public transport system. They provide fast and easy access, have different business models, and make use of applied technology like smart cards and/or mobile phones. Table 1 presents the classification of bike-sharing systems currently functioning in the world, due to their operating principles.

The area systems can be divided into four generations. The first generation (bike-sharing on the streets) is the Dutch "Witte Fietsplan" (White Bike Plan), which was announced in 1965 in Amsterdam. Its author was Luud Schimmelpenninck. The purpose of the plan was to provide a maximally simple to use, widely available communication alternative to growing motorization.

Table 1. The classification of bike-sharing systems currently functioning in the world, due to their operating principles. Source: Own research based on data presented in [12,13].

Operating Principles of Bike-Sharing Systems	Bike-Sharing System Description
Point systems	Bike rental and return take place at one or no more than a few points. These points can be automated or operated by staff. In this case, there are both typical commercial rentals and public systems that support cycling mobility by offering free access. A frequent limitation, in this case, is the limited time of renting and returning bikes within a day.
Area systems	Bike rental and return take place at numerous, any points located on a large area (usually in the whole area of dense urban development). In this case, the rental and return procedure is always automated and takes many different forms. This procedure is not usually hourly limited as in the case of the point systems. Due to the greatest potential for cycling mobility, this is the most popular form of the bike-sharing system. Its development can be grouped into subsequent generations.
Point-area systems	Developed based on a close connection between the bike-sharing system and public transport operators, most often based on railway stations. In these systems, bike rental usually takes place at a single point, i.e., at the railway station node. The return can take place at the rental point but also by leaving the bike anywhere in the designated area after completing the appropriate return procedure.
Long-term bike rental systems	These are systems completing the bike-sharing system in the area and mixed type. These systems based on renting bikes for a longer time (from months to years). The fee may be one-off or divided into installments in the amount of part of the costs of buying and utilization a private bike. In systems of this type, users may be required by the regulations, e.g., to use rented bikes for a specified number of journeys to nodes and continue the journey by public transport. These types of systems are a very attractive alternative to other (open) bike-sharing systems, especially in less-populated, peripheral urban areas and in isolated towns with smaller surface and with lower population density. A variant of this type of systems is the long-term rental of electric bikes.

At that time, it began to create huge spatial and environmental problems on the narrow streets of Amsterdam [14,15].

The second generation (bike-sharing systems on bail) started in 1995 in Copenhagen under the name "Bycyclen". It was the world's first area-based and automated bike-sharing system consisting of 1000 bikes and 110 bike rental and return stations. The principles of its functioning were a generational breakthrough: the bike could be rented by placing a coin (deposit) in a special slot. When the bike was returned, the coin was recovered. Apart from a returnable deposit, the system was free to use. The system was available 24 hours a day, 7 days a week, from mid-April to November. The bikes were also specially designed—they were very simplified, they had full wheels without spokes and consisted of parts of unusual sizes connected by custom screws (to prevent stealing of bikes for parts). The operating costs of the system were borne by the city of Copenhagen, private companies, e.g., Coca Cola, placing ads on bikes, and to some extent the Danish government. The system operator was a foundation established by the city authorities. The system only allowed movement around the city center. The users were punishable by a fine when they traveled outside the designated area. In this way, the supervision of bikes was improved. The problem was that in this way it prevented the main target group from using the bike-sharing system, i.e., people commuting to the center from neighboring districts. As a result, the system was mainly for tourists. It was a very recognizable and valued symbol of the city's tourist attractiveness [16,17].

The beginning of the third generation (smart docks) can be regarded the creation of the "Bikeabout" system in the city Portsmouth in England. At first, the system covered classic student rentals. In 1996 the system was expanded and fully automated by using personal magnetic cards. The introduction of full user identification for the first time in the history of the bike-sharing systems contributed to the creation of a system in which no bike has been stolen or damaged. The next step in the development of the third-generation bike-sharing systems was the appearance of systems characterized by a much greater range and number of stations. Additionally, the use of teleinformatics was the most important for supervision and management and introducing relocation to compensate the level of filling the bikes-sharing stations. France has become a significant area of the evolution of the third-generation bike-sharing system. In 1998, the "Velo a la Carte" bike-sharing system was launched in Rennes. It was the first computerized bike-sharing system in the world. This system included 200 bikes, 25 docking stations connected to the central management, three service employees, and a special car for transporting bikes. The system was available to all residents of the city and neighboring metropolitan areas and required a personal magnetic card. Using the vehicle was free within two hours of renting. The business model was based on the public-private partnership of the city authorities with the Clear Channel advertising group [18,19].

The fourth generation (smart bikes): a feature that distinguishes the fourth-generation bike-sharing systems is the departure from the "smart dock" idea that allows rental and return bikes in docking stations, towards "smart bikes". These bikes have built-in identification, rental, closing, and return of bikes. The role of the dock is taken over by a bike, which is equipped with an electric lock and multi-system electronic supervision in real-time (thanks to the GPS (Global Positioning System) and/or GSM (Global System for Mobile Communications) tracking module). The role of the rental terminal in the dock was known from the third generation—partly taken over by the smartphone with Internet access, partly the active rental panel built into the fourth-generation system bike and powered by dynamos (battery, photovoltaic panel). The second characteristic of the fourth generation, presented in the literature [20,21], is much greater integration of the bike-sharing systems with public transport systems and car-sharing systems. It is important to create common channels of remote access to information and services, integrated tariffs and fees for systems operating in the same area, and the implementation of integrated electronic access cards (agglomeration city cards, communication, etc.) [22].

The bike-sharing systems developed around the world in recent years are mainly based on third-generation systems. In Warsaw (Poland), the Warsaw Bike-Sharing System operates as part of Warsaw Public Transport, which is organized by the Municipal Road Management. This system began to operate in 2012 and is a third-generation system. The system's operator and a contractor is NextBike Polska sp. z o.o. According to the ranking prepared by the consumer service ShopAlike.pl [23], Warsaw ranks fourth in Europe in terms of transport-sharing systems. Only Paris, Brussels, and Berlin are more friendly to bike-sharing, scooter-sharing, and car-sharing systems. In Poland, the first bike-sharing system was launched in 2011. Since then, the total number of rentals in all systems in Poland amounted to less than 60 million. Table 2 presents the bike-sharing systems in Poland with the most number of bike rentals in the years 2011–2019.

Table 2. The bike-sharing systems in Poland with the most number of bike rentals in the years 2011–2019. Source own research based on data presented in [24].

Name of the Bike-Sharing System	City in which the System Occurs	System Operation Start Date	The Number of Rentals (millions)
Veturilo—Warsaw Bike-Sharing System	Warsaw	1 August 2012	25.6
Wrocław Bike-Sharing System	Wrocław	8 June 2011	6.3
Łódź Bike-Sharing System	Łódź	30 April 2016	6.0
Poznań Bike-Sharing System	Poznań	15 April 2012	4.5
Lublin Bike-Sharing System	Lublin	19 September 2014	3.4
BiKeR—Białystok Bike-Sharing System	Białystok	31 May 2014	3.1

The main purpose of this article is to present the results of the analysis of the functioning of the bike-sharing system in Warsaw. Moreover, this paper presents an analysis of the accessibility to individual stations, which is made by using isochrones. An important aspect is that the bike-sharing system has been popular among users and that more people use it. Therefore, a dense network of conveniently located bike-sharing stations should be provided in the city. Also, the quality of the bike-sharing system should be an adjustment to the user's expectations. In connection with the above, the article also presents the results of the analysis of factors affecting the bike-sharing system usage, as well as the level of satisfaction connected with the bike-sharing system usage.

2. Literature Review on the Bike-Sharing Systems in the Cities

In Poland, the most important works related to the bike-sharing system belong to the Warsaw Traffic Study 2015 [25–28]. A traffic model was developed for the Warsaw agglomerations as part of the Warsaw Traffic Study. The traffic model was divided into three independent models. The third model is for cycling traffic. The cycling model was prepared according to the principles of the classic four-stage transport model [29–32]. Part of the first stage was determining the generations and absorption of individual origins and destinations of the bike traffic stream, for the identified travel motivations. Then, as part of the second stage, the streams were allocated between individual origins and destinations of the traffic stream. The last stage was the distribution of traffic flow over individual transport roads. This model also mapped the bike-sharing stations Vertuilo which allowed the analysis of travel behaviors of both private bike users and the bike-sharing system Vertuilo users [33]. The research showed that the flow of bike travel is evenly distributed over the Warsaw transport network. The cyclists choose the main communication routes in both the morning and afternoon peak hours. Comparing the morning and afternoon peak hours, an increase in traffic can be observed on selected sections of the network and a decrease in traffic on selected parts of it. In turn, perceived enablers and barriers of the bike-sharing system in Warsaw according to cyclists and non-cyclists presented [34]. The results of a public opinion survey about the bike-sharing system in Warsaw in the first years of its functioning were included in the work of B. Klepackiego and P. Sakowskiego [35].

Due to the fact that the bike-sharing systems became popular in the world only a few decades ago, scientific literature dedicated bike-sharing systems is relatively new. The scientific papers published so far, first of all, relate to the establishment, functioning, analysis, and modeling of bike-sharing systems. The works available in the literature can be divided into several distinct research areas. These are empirical research on the functioning and use of a bike-sharing system—e.g., M.Z. Austwick et al. [36], P. De Maio [37], A. Faghih-Imani et al. [38]—or problems of location and capacity of bike-sharing stations—e.g., I. Frade and A. Ribero [39], J.R Lin and T.H. Yang [40]. Moreover, one can find research works dedicated to modeling and optimizing the repositioning problem—e.g., T. Raviv et al. [41], M. Benchimol et al. [42], D. Chemla et al. [43], T. Raviv and O. Kolka [44]—as well as the functioning of the bike-sharing system and its efficiency in various conditions—e.g., C. Ficker and N. Gast [45], P. Vogel and D.C. Mattfeld [46]—or bike-sharing usage—e.g., C. Morency et al. [47]. An important group of works are works dedicated to bike-sharing system usage prediction—e.g., Y. Li and Y. Zheng [48], Y. Li et al. [49], most often using time series models (e.g., A. Kaltenbrunner et al. [50]), and factors determining the use of the bike-sharing system—e.g., A. Faghih-Imani and N. Eluru [51], T.D. Tran et al., [52], or also the positive influence of the bike-sharing system on people's health and life as well as the natural environment—e.g., B. Alberts et al. [53], R. Jurdak [54], Y. Guo et al. [55].

Also, in the related literature one can find works that show that the location of stations in densely populated places and places of concentration of workplaces increases the number of rentals (e.g., R. Rixey [56]). A similar pattern can occur in the situation of stations near to schools, universities, places of concentration of services, and places of entertainment such as restaurants, bars, cinemas, and shops (e.g., A. Faghih-Imani et al. [57]).

In turn, F. Gonzalez et al. [58], R. Nair et al. [59], A. Faghih-Imani et al. [60] analyzed the influence of the occurrence of bike-sharing stations at public transport systems such as bus stations, train stations, metro stations, and bus stops. The results of these analyses show that the location of the bike-sharing stations at public transport systems also increases the bike-sharing stations usage.

Analyzing articles on bike-sharing systems, they indicate that the analyses contained in these works are most often based on data obtained from surveying bike-sharing system users or managers—e.g., E. Fishman et al. [61], Y. Tang et al. [62], S. Kaplan et al. [63], or surveying the bike-sharing system operator—e.g., T. Raviv et al. [41], P. Świerk [64], A. Kurek [65], or online sources of usage at stations—e.g., P. Jimenez et al. [66], R. Hampshire [67], X. Wang et al. [68].

In cities where bike-sharing systems have been operating for a long time, research has been conducted to obtain information from bike-sharing system users about the quality of system functioning. These data can be used by decision-makers to improve the offer and functioning of the bike-sharing system so that it meets users' expectations as much as possible. Studies on factors affecting on bike-sharing system usage as well as the level of satisfaction connected with bike-sharing system usage have been conducted so far, e.g., Y. Guo et al. [69]. The purpose of this research was to understand the factors affecting the low bike-sharing system usage in Ningbo (China). Based on their research a few conclusions connected with planning, engineering, and public advocacy were discussed in order to increase the bike-sharing system usage in the mentioned city. A similar survey was performed by L. Caggiani et al. [70]. They proposed an optimization model able to determine how to employ a given budget to enhancing a bike-sharing system, maximizing global user satisfaction. In turn, C. Etienne and O. Latifa proposed a model for the bike-sharing system in Paris [71], which identifies the latent factors that shape the geography of trips, and the results offer insights into the relationships between station neighborhood type and the generated mobility patterns. G. Manzi and G. Saibene [72] analyzed the level of satisfaction connected with bike-sharing service usage through surveys carried out in Milan.

In the literature, there are many different scientific papers about the attempt to determine the level of satisfaction with the bike-sharing system usage, which were conducted in various cities around the world, e.g., F. Xin et al. [73] conducted this type of research for Shanghai, D. Efthymiou et al. [74] for Greece, J. Shi et al. [75] for China. Analyzing these works, it can be stated that the approaches to conducting research were very different in terms of the method of obtaining data for analysis, the number of research samples, the manner of conducting analyses, as well as the conclusions obtained.

3. Veturilo—Warsaw Bike-Sharing System

The operator and contractor of the Warsaw Bike-Sharing System (Veturilo) is the company Nextbike Polska z o.o., which operates under the license of Nextbike GmbH [76]. The Nextbike system is popular in the world. Currently, it operates in over 200 cities in the world, in 16 countries on 4 continents. In Poland, Nextbike operates in dozens of cities. According to the data from the system operator, in 2019 in Poland had access for over 8 million people to the bike-sharing system. In Warsaw, it operates in 17 districts (no system in the Wesoła district) and consists of 377 stations (which 335 are city stations and 42 are sponsor stations, i.e., financed by private parties) (see Figure 1).

Figure 1 shows statistical data characterizing the number of stations, bikes, users and rentals in the Veturilo system in Warsaw in 2012–2019.

Figure 2 shows the schematic location of the Veturilo bike-sharing stations in Warsaw. At the stations are 5337 bikes, which are available in the following types [78]:

- standard;
- children 4+ (dedicated to children over 4 years and 110 cm tall, these bikes have 18-inch wheels);
- children 6+ (designed for children over 6 years and 120 cm tall, these bikes have 20-inch wheels);
- tandem (designed for two people who sit one behind the other. The tandem has one frame and two wheels, while each person has its own steering wheel, saddle and pedals. The maximum load of a tandem is 227 kg);

- electric (equipped with an electric motor, which covers up to 82% of the effort associated with driving. The battery allows travel from 60 to 130 km. Electric support works at speeds up to 25 km/h);
- bike with cardan shaft (it is not equipped with a chain but with a set of joints that drive the bike. This type of bike is definitely less emergency than a standard bike with a chain).

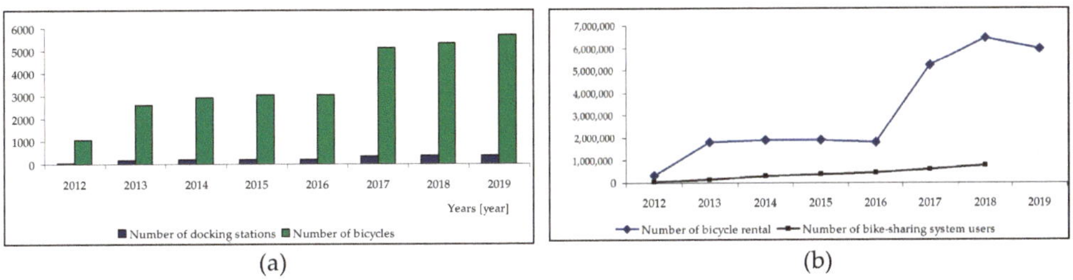

Figure 1. Characteristics of the Veturilo system in Warsaw in 2012–2019; (**a**) number of docking stations and bikes; (**b**) number of bike-sharing systems users and bikes rental. Source: Own research based on data presented in [76–79].

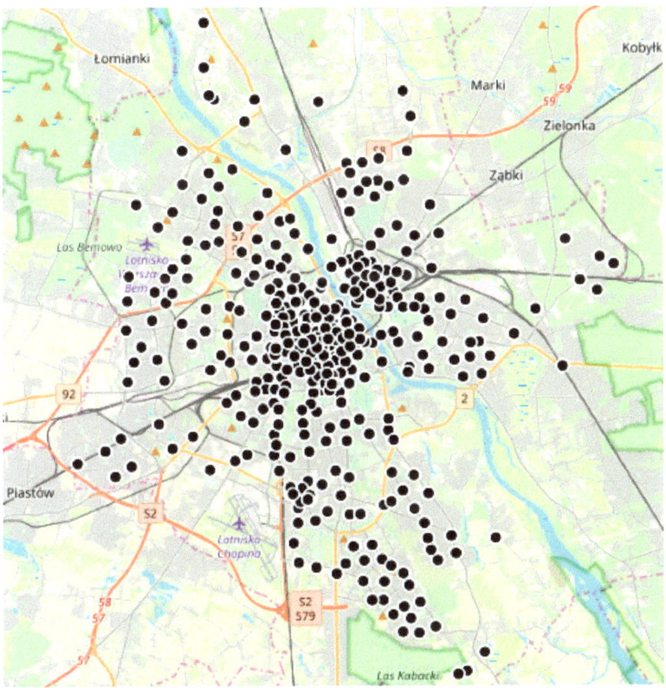

Figure 2. Locations of the Veturilo system stations in Warsaw. Source: Own research based on data presented in [64,76,78,79].

The children's bikes are available at selected stations. The children may use these bikes only in the care of an adult. In turn, the tandem bikes increase the attractiveness of the system in terms of recreation journey and give the possibility of a joint journey of people with the same origin and destination. In addition, a tandem is a great solution, e.g., for people with disabilities who can use this mode of transport only in this form. In Warsaw, the electric bike has been available for rent since August 2017 at one of the eleven selected stations. The electric bikes have a separate price list. Renting an electric bike is more expensive than renting other types of bikes.

To start using the Veturilo system, it is necessary to register as a user in the system. This can be done in four ways, i.e., by completing the registration form on the website www.veturilo.waw.pl, using the terminal located at each station, using the Veturilo mobile application, or by contacting the 24-hour call center.

Then, the newly created account must be topped up with an initial payment of 10 PLN or a credit card added as a payment method. Account activation is done by clicking on the verification link, which is sent to the e-mail address provided in the registration. After registering, the user receives an SMS and an e-mail containing the PIN code. The PIN code together with the user's phone number are identifiers in the Veturilo system.

To rent a bike, the "Rental" button must be selected at the bike-sharing station terminal and the instructions that will be shown on the display must be followed. After completing the rental procedure, the bike is automatically released from the electric lock. The user can also rent a bike via the Veturilo mobile application or by contacting the call center. If the bike is secured with a combination lock, the code displayed on the terminal should be entered. This code is also provided in the mobile application throughout the bike rental time. The return of the bike is done by connecting the bike to the electric lock. This is confirmed by a sound signal and a green diode. On the terminal or in the mobile application the user should check that the return has been made correctly. When all electric locks are occupied at the station, the user can secure the bike with a combination lock. Table 3 presents the price list for renting a standard, children, tandem, or electric bike from the bike-sharing system in Warsaw. The price list also contains information about the number of fines for theft, loss, or destruction of bikes (Table 4).

Table 3. The price list for renting a standard, children, tandem, or electric bike from the bike-sharing system in Warsaw system (assuming exchange rate 1 = 4.26 PLN). Source: Own research based on data presented in [79].

Action/Rental Time	Price List for Renting a Standard, Child, or Tandem Bike	Price List for Renting an Electric Bike
During registration to the system	10 PLN (2.35)	10 PLN (2.35)
The first 20 min	0 PLN (0.00)	0 PLN (0.00)
From 21 to 60 min	1 PLN (0.23)	6 PLN (1.41)
Second hour	3 PLN (0.70)	14 PLN (3.27)
Third hour	5 PLN (1.17)	14 PLN (3.27)
The fourth and subsequent hours	7 PLN (1.64)	14 PLN (3.27)
Over 12 h of using the bike	200 PLN (46.95)	300 PLN (70.42)

Table 4. Information about the fines for theft, loss, or destruction of bikes from the bike-sharing system in Warsaw. Source: Own research based on data presented in [79].

Types of Bicycles	The Price List
Standard bike	2000 PLN (469.48)
Tandem bike	7000 PLN (1643.19)
Children's bike	1900 PLN (446.01)
Electric bike	12000 PLN (2816.90)

4. Research Methodology

The main purpose of this work was:

- analysis of the functioning of the bike-sharing system in Warsaw;
- analysis of access to stations;
- analysis of factors affecting bike-sharing system usage as well as the level of satisfaction connected with bike-sharing system usage;
- surveying preferences among users of the bike-sharing systems.

Data were obtained from the operator Veturilo Nextbike to analyze the functioning of the bike-sharing system in Warsaw. These data covered the functioning of the system in 2018, i.e., from March to November 2018. In order, QGIS (Quantum Geographic Information System)and Open Street Maps were used to perform the availability analysis. In turn, a survey questionnaire was used for the analysis of factors affecting the bike-sharing system usage as well as the level of satisfaction connected with the bike-sharing system. This type of research has been used commonly in transport engineering. The form and content of the survey questionnaire was developed based on a broad review of the literature. The reliability of the questionnaire and its internal consistency were checked using the Cronbach's alpha coefficient (α) ([80–85]). The value of this factor was $\alpha = 0.743$, which confirmed that the constructed questionnaire was a tool that could be used to measure the bike-sharing system usage as well as the level of satisfaction connected with bike-sharing usage. The final form of the questionnaire was discussed, assessed, and corrected by an expert team (four independent experts assessed its layout, readability, style, and transparency of questions). To assess respondents' answers a five-point Likert scale was used—from 1 (bad) to 5 (excellent) [86–88]. The average rating for the questionnaire was 4.5, which confirmed the satisfactory validity of the questionnaire.

The data obtained from the questionnaire were analyzed. The bivariate ordered probit modeling technique was used to identify the significant factors affecting bike-sharing usage and users' level of satisfaction. This modeling technique took into account the correlation between independent variables. As a result, a conceptual model was obtained consisting of factors related to the bike-sharing system usage and users' level of satisfaction. The questionnaire consisted of several parts. In the survey, respondents answered questions about:

- data that characterized the respondent's profile: age, gender, occupation, education level, monthly income, information about having in the household a bike, e-bike, car (and number of them), etc.;
- use of the bike-sharing system (the answers used a three-point Likert scale, where: 1—rarely, 2—medium, 3—often);
- the level of satisfaction of bike-sharing system usage (the answers used a three-point Likert scale, where: 1—lack of satisfaction, 2—average satisfaction, 3—fully satisfaction);
- travel patterns, including such information as: trip purpose (during weekdays and weekends), trip mode, travel time, bike-sharing station availability within 500 m near home and workplace/university, bus stops/tram stops/railway station availability within 500 m near home and workplace/university;
- perception of the bike-sharing system, including such information as: familiarity with the bike-sharing system, satisfaction with bike-sharing fees, easiness to check-in and check-out, saving travel costs by bike-sharing, etc.

The surveys were conducted among users of the bike-sharing system in Warsaw, around bike-sharing stations, from 2016 to 2019. All persons participating in the survey had previously agreed to participate in the survey. The survey was conducted on weekdays and weekends to obtain a large size and varied answers of the respondents. The random sampling technique was used in order to select the bike-sharing system users. People over 18 years took part in the survey. A total of 1600 questionnaires were collected. The questionnaires were examined in terms of data correctness and completeness (e.g., the questionnaires were rejected in which: respondents indicated that they were using the system for the first time; respondents said that they contained incomplete information; respondents could not indicate their assessment, e.g., "I do not know", "I have no opinion of my own", "I have not such as situation"). A total of 1584 survey questionnaires remained in the database as a result of the selection. These data were the basis for analysis.

5. Analysis of the Functioning of Warsaw Bike-Sharing System in 2018

5.1. The Characteristics of System Users and the Structure of Using Bike-Sharing Stations

According to [89], the number of users of the bike-sharing system in Warsaw was 785,000. Figure 3a shows the age structure of users. The obtained data indicated that most users are between 25 and 35 years old (which is 37.13% of the total, i.e., about 291,500 people). Another numerous age group of people using Veturilo is the range from 19 to 25 years old. This range is 29.47% of all users (i.e., around 231,500 people). The least numerous group of the bike-sharing system users are elderly people over the age of 66 (6240 users). The average age of people using the system in 2018 was 32 years.

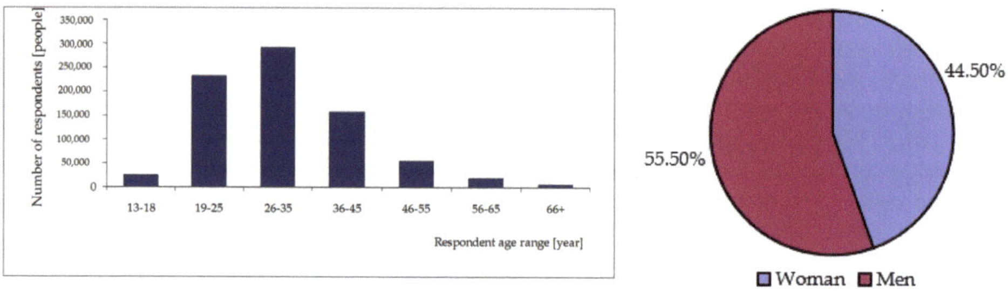

(a) (b)

Figure 3. (**a**) Age structure; (**b**) sex structure of people using Veturilo in Warsaw in 2018. Source: Own research based on data presented in [77].

Figure 3b shows the gender structure of the bike-sharing system users. The data indicate that slightly more men (55.5%) than women (44.5%) use the bike-sharing system in Warsaw.

The obtained data also made it possible to state that in 2018 standard bikes were the most often rented (95.5%) and the least often rented were bikes for children 4+ (which was 0.1% of all rentals—i.e., 5821 rentals)—see Figure 4a. In turn, Figure 4b presents data on bike rentals on weekdays and weekends. A total of 72% of all rentals took place on weekdays. In the analyzed period, this was the majority of all rentals. On weekdays, the bikes were rented on average 22,150 times a day. Weekend rentals were 28% of all rentals.

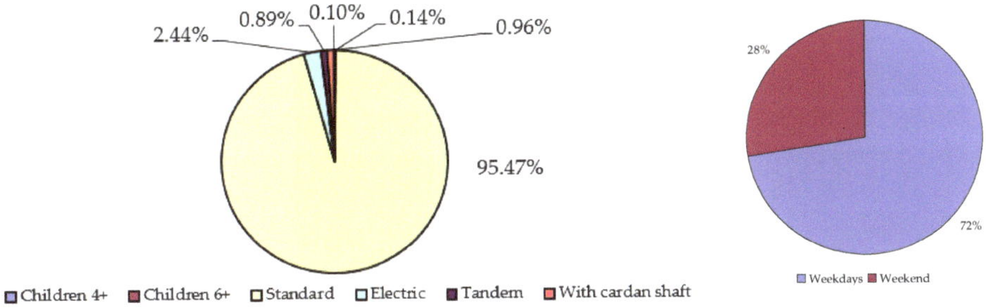

(a) (b)

Figure 4. The number of bike rentals from the Veturilo system in Warsaw in 2018 (**a**) considering types of bikes; (**b**) on weekdays and weekends. Source: Own research based on data presented in [77].

Moreover, Figure 5a shows the number of bike rentals from March to November on weekdays and weekends. The most rentals were in May. In April, June, July, and August, the number of rentals were at a similar level. From August to November a decrease could be observed in the number of Veturilo bike rentals. The least rentals were in March and November. It is possible that this was caused by weather conditions in Poland at this time of the year. The most rentals on weekends took place in April. The number of rentals decreased in each subsequent month. The least rentals on weekends were in March and November, as on weekdays.

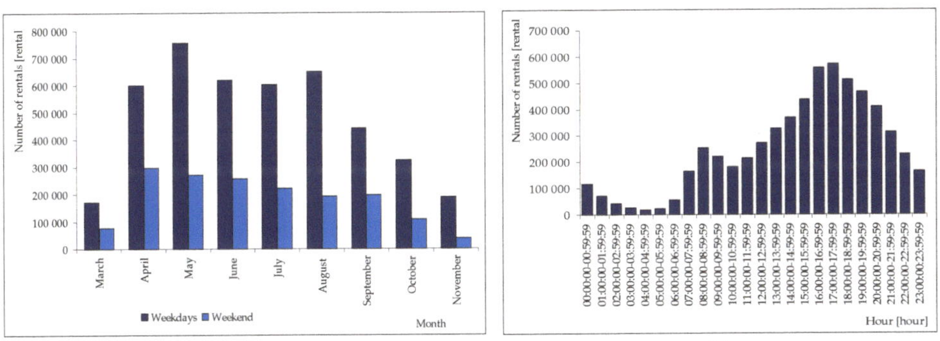

(a) (b)

Figure 5. The number of bike rentals from the Veturilo system in Warsaw in 2018; (**a**) from March to November on weekdays and weekends; (**b**) at particular hours of the day. Source: Own research based on data presented in [77].

Figure 5b shows the number of bike rentals at particular hours of the day. The largest number of rentals were between 17:00 and 17:59 and between 16:00 and 16:59. In the following hours, the number of rentals decreased, up to hours between 04:00 and 04:59, when the number of rentals was the lowest. Between 07:00 and 11:59, the number of rentals was around 200,000 rentals per hour. In the afternoon this number increased with each subsequent hour until it was maximum value between 17:00 and 17:59.

Figures 6 and 7 present the number of bike rentals and returns on particular stations. The most bikes were rented from the stations Arkadia, Plac Wileński, Metro Centrum Nauki Kopernik, and Al. Niepodległości-Batorego, whereas the least were from the stations Komandosów-Niedziałkowskiego, Paderewskiego-Dziewanowska, Frontowa—Czerwonych Beretów, PKN Orlen-Wydawnicza, and Niepołomicka-Chełmżyńska.

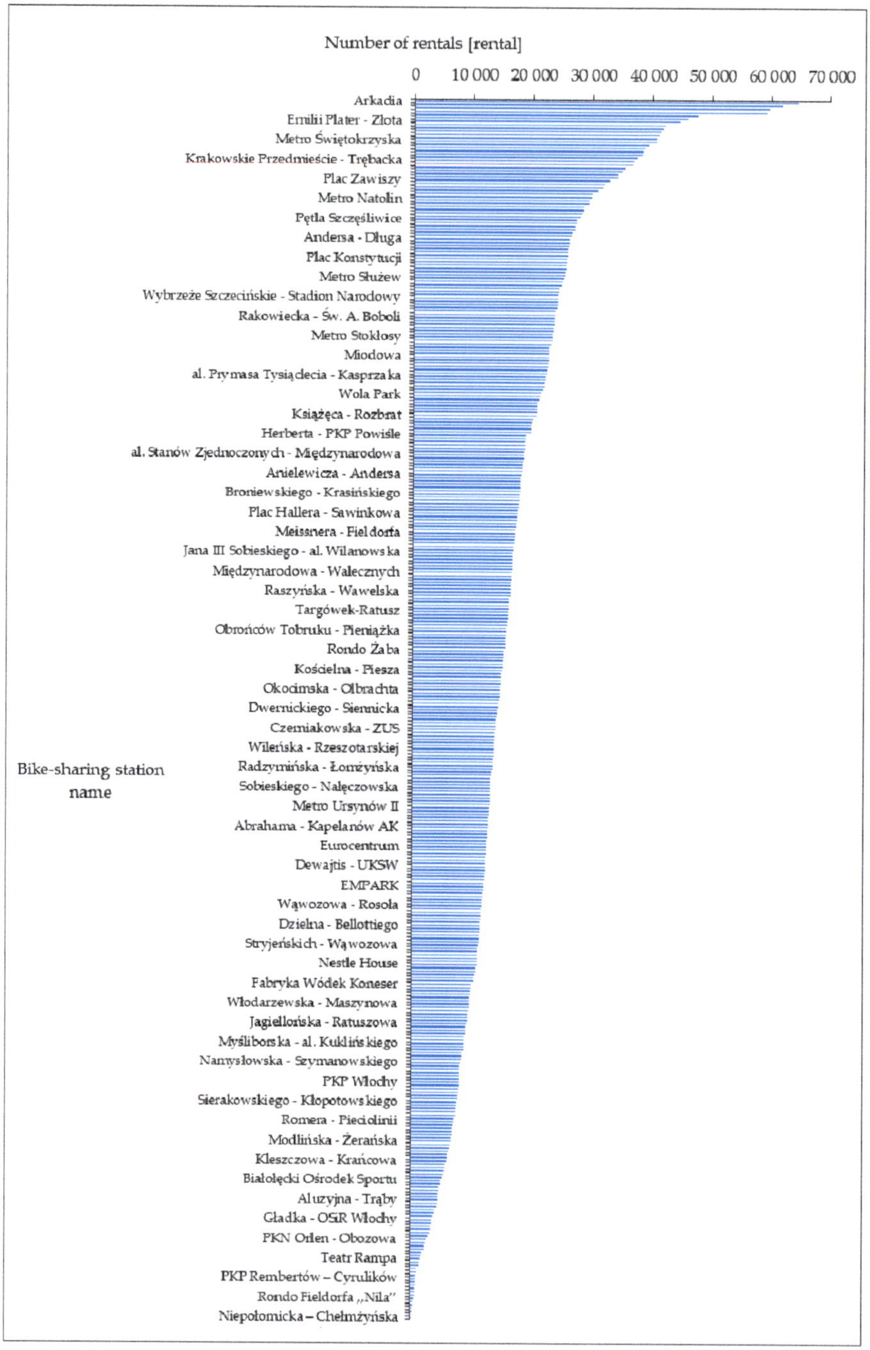

Figure 6. The number of bike rentals from individual stations in the Veturilo system in Warsaw in 2018. Source: Own research based on data presented in [77].

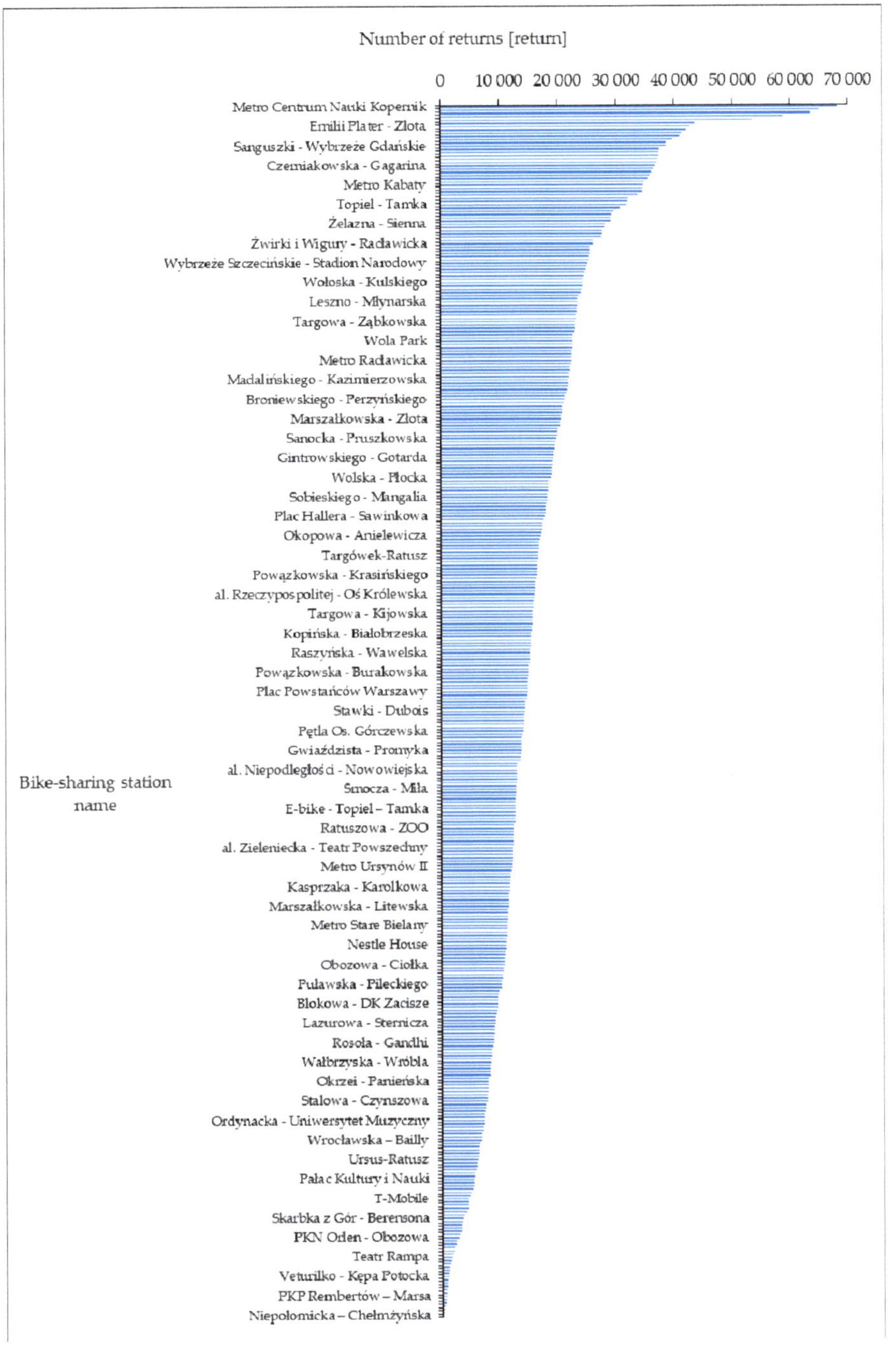

Figure 7. The number of bike returns to individual stations in the Veturilo system in Warsaw in 2018. Source: Own research based on data presented in [77].

In the case of bike returns, the most bikes were returned to the stations Arkadia, Plac Wileński, Metro Centrum Nauki Kopernik, and Al. Niepodległości-Batorego. In turn, the least returns were to the stations Komandosów-Niedziałkowskiego, PKN Orlen-Wersalska, Frontowa-Czerwonych Beretów, PKN Orlen-Wydawnicza, and Niepołomicka-Chełmżyńska.

The average number of rentals and returns from and to each station was around 16,000 rentals/stations in 2018.

5.2. The Bike-Sharing Stations in the Criterion of Popularity among Users

Figure 8 shows the bike-sharing stations which were the most and the least popular among system users in 2018. The most popular stations included: Metro Centrum Nauki Kopernik, Arkadia, Plac Wileński, Al. Niepodległości-Batorego, Port Czerniakowski, Rondo Jazdy Polskiej, Emilii Plater-Złota, Plac Unii Lubelskiej, Stefana Banacha-UW, Metro Nowy Świat-Uniwersytet, Metro Świętokrzyska, Rondo ONZ, Sanguszki-Wybrzeże Gdańskie, Al. Jerozolimskie-Emilii Plater, Al. Jana Pawła II- Plac Mirowski, Metro Dworzec Gdański, Krakowskie Przedmieście-Trębacka, Al. Jana Pawła II-Grzybowska, Czerniakowska-Gagarina, Plac na Rozdrożu, Rondo Waszyngtona-Stadion Narodowy, Al. Jana Pawła II-Al. Solidarności, Belwederska-Gagarina, Grójecka-Bitwy Warszawskiej 1920 roku, Metro Kabaty, Rondo Wiatraczna, Sobieskiego-Chełmska, and Metro Imielin. In turn, the least popular stations included: Niepołomicka-Chełmżyńska, PKN Orlen-Wydawnicza, PKN Orlen-Wersalska, Frontowa-Czerwonych Beretów, Paderewskiego-Dziewanowska, Komandosów-Niedziałkowskiego, Rondo Fieldorfa "Nila", PKN Orlen-Migdałowa, CH Marywilska 44, PKP Rembertów-Marsa, and PKN Orlen-Powsińska.

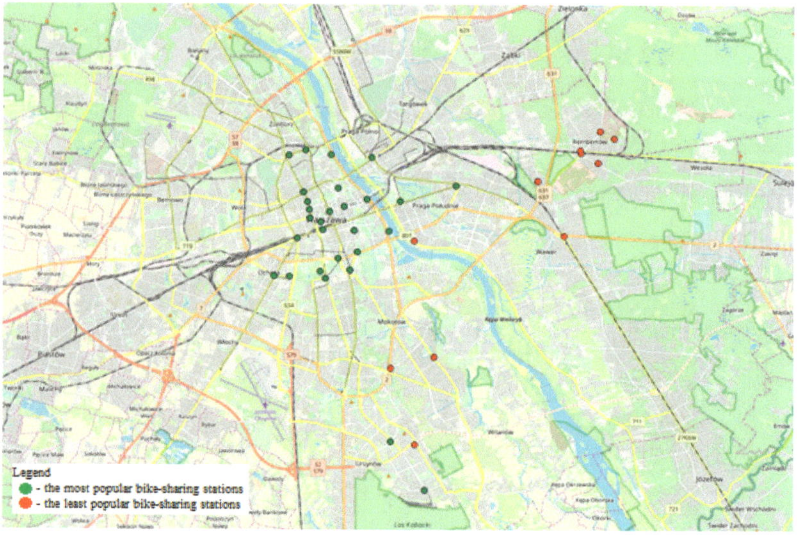

Figure 8. The bike-sharing stations; (**green**) the most popular among users; (**red**) the least popular among users in 2018. Source: Own research based on data presented in [77].

As can be seen in Figure 8, in the area of two districts of Warsaw, the bike-sharing stations which were the least popular among users are located near the bike-sharing stations which were the most popular among users. In the case of bike-sharing stations located in the Ursynów district, the reason may be that both the bike-sharing stations which were the most popular among users are located near the metro line stations. However, the bike-sharing station which was the least popular among users is located at the Orlen petrol station. Another district in which this is the case is Praga Południe. The bike-sharing station which was the least popular among users, also as in the previous case, is located at the Orlen petrol station. However, the bike-sharing stations which were the most popular among users located in the same district and on the same side of the Vistula River are located near public transport stops, which are served by a large number of bus and tram lines. The above analyses indicate that probably in the case of the most popular among users bike-sharing stations which are located at metro line stations and public transport stops, users are people who use bike-sharing stations as a continuation of their journey.

5.3. Pedestrian Accessibility to the Bike-Sharing Stations

The quality of bike-sharing stations functioning depends on, e.g., time of arrival at the nearest station, the accessible bikes at the station, the technical condition of the bikes, prices for using the bike as well as cleanliness and equipment of the bike, weather conditions. ArcGIS (Geographic Information Systems) technologies with the Network Analyst extension were used to analyze pedestrian accessibility to the bike-sharing stations in Warsaw. It allows realizing various network analyses, e.g., modeling pedestrian or car movements using existing roads. The areas were designated from which it is possible to get to the bike-sharing station at the assumed travel time (time distance of 5 and 10 min) only on foot. Then, the obtained information was compared with data on the location of residential buildings in Warsaw. This was transformed into indicators showing the spatial extent of accessibility areas and the number of inhabitants for which access to particular stations is the best. This made it possible to make a general assessment of the location of stations in terms of the time of arrival using different methods.

The necessary information was used to model the accessibility of the bike-sharing stations (i.e., vector data illustrating the location of single-family and multi-family buildings accessible in the Topographic Objects Database, vector data regarding the streets in Warsaw, data illustrating the location of bike-sharing stations, obtained as a result of a query of the information in the OpenStreetMap [90] database). The obtained data about the bike-sharing stations were checked and corrected using maps accessible on Google Maps [91]. Then, the information was converted to GTFS (General Transit Feed Specification) format by exporting to text. Figure 9 presents the final results of the analysis of pedestrian access to individual bike-sharing stations in Warsaw in the time category (time distance of 5 and 10 min). This accessibility is marked on the map with isochrones.

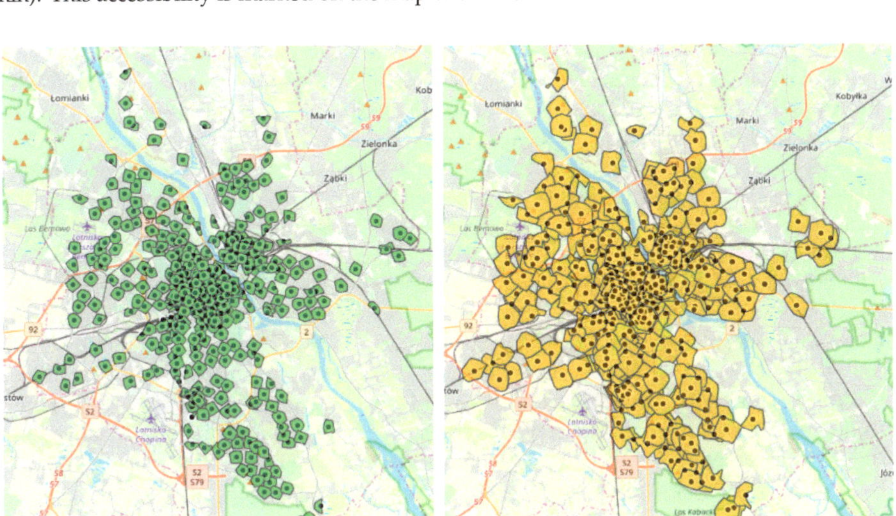

(a) (b)

Figure 9. Pedestrian access to the bike-sharing station in Warsaw; (**a**) on the assumption of an average travel time of 5 min to the station; (**b**) on the assumption of an average travel time of 10 min to the station. Source: Own research based on data presented in [77].

On the basis of data presented on Figure 9, it can be stated that pedestrian accessibility to bike-sharing stations is very good, assuming travel times to the station of 5 and 10 min for stations located in the city center and some districts distant from the center, e.g., Wilanów, Bródno, and Natolin. Pedestrian accessibility deteriorates in districts away from the city center, close to the administrative borders of Warsaw, which is due to the small number of bike-sharing stations in these districts.

6. Analysis of Factors Determining the Bike-Sharing System Usage as Well as the Level of Satisfaction Connected with Bike-Sharing Usage

6.1. Features of Respondents, Household of Respondents and Travel Patterns

Table 5 presents the results of the cross-tabulation of the respondents in the field bike-sharing system usage and level of satisfaction connected with bike sharing-system usage. The bike-sharing system usage was defined as a typical ordinal variable according to three groups: (1) rarely—one–two days per week, (2) average—three–four days per week, and (3) often—five–seven days per week. In turn, the level of satisfaction was also defined as a typical ordinal variable according to three groups: (1) lack of, (2) average, (3) fully.

Table 5. Respondents' bike-sharing system usage and level of satisfaction connected with using the bike-sharing system.

Usage	**Satisfaction**			**Total**
	Fully	**Average**	**Lack of**	
Often	229	102	7	338 (21.34%)
Medium	385	212	11	608 (38.38%)
Rarely	325	286	27	638 (40.28%)
Total	939 (59.28%)	600 (37.88%)	45 (2.84%)	1584 (100%)

Respondents who often used the bike-sharing system were 21.34% of the total, while 38.38% of all respondents were people who used the bike-sharing system an average amount. This value is comparable to those who rarely use the bike-sharing system (40.28%). The results confirmed that the bike-sharing system is not used often by users (most of the answers were "average" and "rarely").

In the case of the level of satisfaction connected with the bike-sharing system usage, it can be stated that the majority of respondents were fully satisfied with the bike-sharing system and its functions and they assessed the bike-sharing system at a very high level (59.28%). A total of 37.88% of all respondents assessed the level of satisfaction connected with the bike-sharing usage at an average level without indicating the disadvantages or advantages of the system. Meanwhile, 2.84% of all respondents expressed dissatisfaction with the system. The conclusion is that most of all respondents positively assessed the functioning of the bike-sharing system.

The questions included in the survey allowed the selection of many explanatory variables (independent, endogenous). Then, these variables were used to determine their impact (or no impact) on bike-sharing system usage and users' level of satisfaction connected with using the bike-sharing system. Tables 6–9 present information about the characteristics of households of respondents, travel patterns, and information on the environment in which respondents live (bike-sharing station availability close to home/workplace/school/university and bus station/tram stop/railway station availability near home/workplace/school/university (within 500 m)).

Table 6. Characteristics of respondents.

Variable	Characteristics	Symbol	Frequency	Percentage (%)
Gender	Male	1	945	59.66
	Female	0	639	40.34
Age group	Young (18–29)	2	682	43.06
	Middle aged (30–45)	1	802	50.63
	Old (50–70)	0	100	6.31
Level of education	Primary	3	12	0.76
	Secondary	2	472	29.80
	Bachelor/Engineer	1	693	43.75
	Higher	0	407	25.69
Occupation	Student	6	529	33.40
	Employee in company or enterprise	5	493	31.12
	Self-employed	4	172	10.86
	Military	3	12	0.76
	Freelance	2	258	16.29
	Retired	1	81	5.11
	Others	0	39	2.46
Monthly income (Gross) (PLN)	>5000	3	251	15.85
	4000–5000	2	532	33.59
	3000–4000	1	633	39.96
	<3000 *	0	168	10.61

* In 2019, the minimum wage in Poland was PLN 2250 gross.

Table 7. Characteristics of respondents' households.

Variable	Characteristics	Symbol	Frequency	Percentage (%)
Owning a car	Yes	1	1359	85.80
	No	0	225	14.20
Owning bicycle/e-bike in house	Yes	1	112	7.07
	No	0	1472	92.93

Table 8. Characteristics of travel patterns.

Variable	Characteristics	Symbol	Frequency	Percentage (%)
Type of means of transport used for travel	Walking	5	108	6.82
	Bicycle	4	257	16.22
	E-bike/motorcycle/Electric scooter	3	21	1.33
	Public bus or rail transit	2	567	35.80
	Public transport and bike	1	113	7.13
	Passenger car	0	518	32.70
Travel time [min]	<30	2	479	30.24
	30–60	1	733	46.28
	>60	0	372	23.48
Trip purpose during weekdays	Work	5	429	27.08
	School/University	4	684	43.18
	Shopping	3	57	3.60
	Spending free time	2	278	17.55
	Visiting friends	1	102	6.44
	Other	0	34	2.15
Trip purpose during weekends	Work	5	387	24.43
	School/University	4	437	27.59
	Shopping	3	162	10.23
	Spending free time	2	327	20.64
	Visiting friends	1	257	16.22
	Other	0	14	0.88

Table 9. Characteristics of the environment in which the respondents operate (i.e., bike-sharing station availability close to home/workplace/school/university and bus station/tram stop/railway station availability near home/workplace/school/university (within 500 m)).

Variable	Characteristics	Symbol	Frequency	Percentage (%)
Bike-sharing station availability close to home/workplace/school/university (within 500m)	Yes	1	484	30.56
	No	0	1100	69.44
Bus station/tram stop/railway station availability near home/workplace/school/university (within 500m)	Yes	1	1208	76.26
	No	0	376	23.74

Based on the data contained in Table 6, it can be concluded that the group of respondents was diverse due to different characteristics such as age, gender, level of education, occupation, and monthly income. On the other hand, analysis of the data contained in Table 7 shows that the majority of respondents had their own car and did not have a bike in the household.

In turn, travel patterns included in Table 8 indicated that a significant proportion of respondents were public transport users or people using a passenger car in their daily travels. These were the respondents whose travel time usually ranged from 30 to 60 min.

The data in Table 9 indicate the positive characteristics of the environment in which the respondents operate. The majority of respondents indicated that within 500 m from their place of residence, workplace, school, university, they could find a bike-sharing station and bus station (or tram stop or railway station). The above indicates the positive features of public transport in Warsaw.

Table 10 presents the potential independent variables affecting the perception of the bike-sharing system. These are the variables:

- knowledge of the rules of using bike-sharing systems, i.e., rules of use, fees for use, rules of technical maintenance when renting and returning a bike;
- acceptance of the fee for using the bike-sharing system;
- the existence of incentives for travel in the city using an ecological form of transport like bikes, e-bikes, public transport, walking;
- the ability to reduce travel costs by using the bike-sharing system;
- extension of travel time (loss of time) by using the bike-sharing system;
- the ability to adapt the travel route to user needs thanks to the use of the bike-sharing system;
- knowledge about the bike-sharing system obtained from extensive social campaigns, e.g., explaining to society the advantages and disadvantages of using the system, transport policy, potential benefits;
- easy to use the bike-sharing system in terms of registering in the system, paying fees, using a credit card;
- easy to use the bike-sharing system during bike rental—for the bike rental process up to 5 minutes (for the process of renting for more than 5 minutes, it was assumed that technical service is not easy);
- easy to use the bike-sharing system during bike return for the bike return process up to 5 minutes (for the process of returning for more than 5 minutes, it was assumed that technical maintenance is not easy).

Table 10. Potential variables affecting the perception of the bike-sharing system.

Variable	Characteristics	Symbol	Frequency	Percentage (%)
Knowledge of the principles of using bike-sharing systems (i.e., rules of use, fees for use, rules of maintenance when renting and returning a bike)	Yes	1	1557	98.30
	No	0	27	1.70
The existence of incentives for travel in the city using an ecological form of transport (like bikes, e-bikes, public transport, walking)	Yes	1	251	15.85
	No	0	1333	84.15
Acceptance of fees for using a bike-sharing system	Yes	1	749	47.29
	No	0	835	52.71
Extending travel time (loss of time) by using a bike-sharing system	Yes	1	301	19.00
	No	0	1283	81.00
The possibility of reducing travel costs by using a bike-sharing system	Yes	1	489	30.87
	No	0	1095	69.13
Knowledge of the bike-sharing system obtained from extensive social campaigns (regarding, among others, explaining to the public the advantages and disadvantages of using the system, transport policy, potential benefits)	Yes	1	61	3.85
	No	0	1523	96.15
The option of adapting the travel route to your needs thanks to the use of a bike-sharing system	Yes	1	692	43.69
	No	0	892	56.31
Easy to use of the bike-sharing system in terms of registration in the system, payment of fees, use of a credit card	Yes	1	990	62.50
	No	0	594	37.50
Easy to use the bike-sharing system during the bike rental process up to 5 minutes	Yes	1	1428	90.15
	No	0	156	9.85
Easy to use the bike-sharing system during bike return measured by the time of the bike return process up to 5 minutes	Yes	1	1390	87.75
	No	0	194	12.25

6.2. *Bivariate Ordered Probit Model Formulation*

A bivariate ordered probit model (BOP) is a hierarchical structure of two equations that can be used to model the simultaneous relationship of two variables. BOP makes it posisble to solve potential endogenous problems, e.g., the correlation between the explained variable and the explanatory variable in the model [92]. The bivariate ordered probit model has been used many times in modeling the variability of the phenomenon occurring in road transport [93–96].

This study aimed to identify factors that simultaneously determine the bike-sharing system usage as well as the level of satisfaction connected with bike-sharing system usage. Discrete modeling techniques were used as the dependent variables consist of category variables. At the beginning, for each observation ordinal data were defined [94–98]:

$$\begin{cases} y^*_{i,1} = \beta_1 X_{i,1} + e_{i,1},\ y_{i,1} = j\ if\ \lambda_{j-1} < y^*_{i,1} < \lambda_j,\ j = 0,\ldots,J \\ y^*_{i,2} = \beta_2 X_{i,2} + e_{i,2},\ y_{i,2} = k\ if\ \gamma_{k-1} < y^*_{i,2} < \gamma_k,\ k = 0,\ldots,K \end{cases}, \tag{1}$$

where y is the integer ordering, $y^*_{i,1}$, $y^*_{i,2}$ are the hidden dependent variables, $y_{i,1}$, $y_{i,2}$ are the bike-sharing system usage results according to the ordinal scale (1, 2, 3) and according to the ordinal scale for level of satisfaction connected with bike-sharing system usage (1, 2, 3), $X_{i,1}$, $X_{i,2}$ are vectors consisting of explanatory variables in two models, β_1, β_2 are the parameter vectors for estimation, $e_{i,1}$, $e_{i,2}$ are the random error for the models with normal distribution with mean = 0 and variance = 1, i is individual observations, j is bike-sharing system usage, k is the level of satisfaction connected with bike sharing usage, λ, γ are the estimated threshold parameters, which define $y_{i,1}$ and $y_{i,2}$.

The cross-equation of correlated errors in the BOP model can be written as:

$$\begin{bmatrix} e_{i,1} \\ e_{i,2} \end{bmatrix} \approx N\left(\begin{bmatrix} 0 \\ 0 \end{bmatrix}, \begin{bmatrix} 1 & r \\ r & 1 \end{bmatrix}\right), \tag{2}$$

where r is the correlation coefficient between $e_{i,1}$ and $e_{i,2}$.

If random errors have the bivariate normal distribution, a common probability for $y_{i,1} = j$ and $y_{i,2} = k$ can be written as:

$$\begin{aligned} P\left(y_{i,1} = j, y_{i,2} = k \middle| X_{i,1}, X_{i,2}\right) &= \Pr\left(\lambda_{j-1} < y^*_{i,1} < \lambda_j; \gamma_{k-1} < y^*_{i,2} < \gamma_k\right) \\ &= \Pr\left(\lambda_{j-1} < \beta_1 X_{i,1} + e_{i,1} < \lambda_j; \gamma_{k-1} < \beta_2 X_{i,2} + e_{i,2} < \gamma_k\right) \\ &= \Pr\left(\lambda_{j-1} - \beta_1 X_{i,1} < e_{i,1} < \lambda_j - \beta_1 X_{i,1}; \gamma_{k-1} - \beta_2 X_{i,2} < e_{i,2} < \gamma_k - \beta_2 X_{i,2}\right) \\ &= \Psi_2\left[\left(\lambda_j - \beta_1 X_{i,1}\right), \left(\gamma_k - \beta_2 X_{i,2}\right), r\right] - \Psi_2\left[\left(\lambda_{j-1} - \beta_1 X_{i,1}\right), \left(\gamma_k - \beta_2 X_{i,2}\right), r\right] \\ &\quad - \Psi_2\left[\left(\lambda_j - \beta_1 X_{i,1}\right), \left(\gamma_{k-1} - \beta_2 X_{i,2}\right), r\right] + \Psi_2\left[\left(\lambda_{j-1} - \beta_1 X_{i,1}\right), \left(\gamma_{k-1} - \beta_2 X_{i,2}\right), r\right], \end{aligned} \tag{3}$$

where Ψ is the cumulative distribution function.

Parameters in the BOP model can be calculated by maximizing the log-likelihood function and are given by:

$$LL = \sum_{i=1}^{n} \sum_{j=0}^{J} \sum_{k=0}^{K} \delta_{jk} \begin{bmatrix} \Psi_2\left[\left(\lambda_j - \beta_1 X_{i,1}\right), \left(\gamma_k - \beta_2 X_{i,2}\right), r\right] - \Psi_2\left[\left(\lambda_{j-1} - \beta_1 X_{i,1}\right), \left(\gamma_k - \beta_2 X_{i,2}\right), r\right] \\ -\Psi_2\left[\left(\lambda_j - \beta_1 X_{i,1}\right), \left(\gamma_{k-1} - \beta_2 X_{i,2}\right), r\right] + \Psi_2\left[\left(\lambda_{j-1} - \beta_1 X_{i,1}\right), \left(\gamma_{k-1} - \beta_2 X_{i,2}\right), r\right] \end{bmatrix}, \tag{4}$$

where i = 1,2,3, . . . , and n is the size of the sample.

The signs of the factors connected with explanatory variables indicate the positive and negative effects of the variable on the results. The positive sign means the probability increases of the most strong result and the probability decreases of the least strong result. In the case of the negative sign, the opposite is true. These coefficients do not determine the quantitative impact of these variables. Also, they cannot be interpreted, in particular for intermediate categories. To determine the influence on intermediate categories, the marginal effects are calculated at the average sample for each category of explanatory variable $X_{i,1}$ for $y_{i,1}$ as [94,99]:

$$\frac{P(y_{i,1} = j)}{\partial X_{i,1}} = \left[\tau\left(\lambda_{j-1} - \beta_1 X_{i1}\right) - \tau\left(\lambda_j - \beta_1 X_{i,1}\right)\right]\beta_1, \tag{5}$$

where $P(y_{i,1} = j)$ is the probability that bike-sharing system usage is on level j, $\tau(\ldots)$ is the probability mass function of the standard normal distribution, j is the levels of importance ordered by integers, λ is the thresholds.

Thus, the marginal effect of explanatory variable $X_{i,2}$ for $y_{i,2}$ is given as:

$$\frac{P(y_{i,2} = k)}{\partial X_{i,2}} = [\tau(\gamma_{k-1} - \beta_2 X_{i2}) - \tau(\gamma_k - \beta_2 X_{i,2})]\beta_2. \tag{6}$$

6.3. Bivariate Ordered Probit Model Results

In order to determine the factors affecting the bike-sharing system usage as well as the level of satisfaction connected with the bike-sharing system usage the bivariate ordered probit model was determined. Table 6 to Table 8 present the explanatory variables and descriptive statistics. In turn, Table 9 presents bivariate ordered probit model results. The obtained results show a significant correlation between the bike-sharing system usage as well as the level of satisfaction connected with the bike-sharing system usage. The obtained results seem logical and find confirmation in reality. Regular users of the bike-sharing system were satisfied with the offered services. In contrast to them, people who rarely used the system indicated its disadvantages. In the final form of the model, variables were only included which were significant at the 95% confidence level. The correlation parameter was positive ($p = 0.128$, p-value = 0.027) which allows us to state that a higher level of satisfaction connected with bike-sharing system usage could increase the likelihood of the bike-sharing system usage.

Based on Table 11, it can be concluded that there were 12 significant variables in the model of the bike-sharing system usage and 10 significant variables in the model of the level of satisfaction connected with the bike-sharing system usage. The bivariate ordered probit model results presented in Table 11 represent the general direction of influence of factors determining bike-sharing system usage as well as the level of satisfaction connected with bike-sharing system usage. In turn, Tables 12 and 13 show the numerical values for those variables that had marginal effects for the bivariate ordered probit model. The most desirable group of respondents were those who were characterized by "often" bike-sharing system usage and a "fully" level of satisfaction connected with bike-sharing system usage.

Table 11. Bivariate ordered probit model results.

Variable	Bike-Sharing System Usage			Level of Satisfaction Connected with Bike-Sharing Usage		
	β	Standard Error	p-Value	β	Standard Error	p-Value
Gender	0.315	0.062	0.001	-	-	-
Monthly income (Gross) 2000–4000 (PLN)	-	-	-	0.462	0.215	0.036
Monthly income (Gross) 4001–∞ (PLN)	-	-	-	0.594	0.273	0.002
Travel time(< 30 min)	0.405	0.123	0.006	-	-	-
Trip mode (bicycle)	0.701	0.164	0.001	-	-	-
Trip mode (public transport and bicycle)	0.704	0.178	0.002	-	-	-
Bicycle/e-bike in household	0.11	0.082	0.023	-	-	-
Satisfaction with bike-sharing fees	0.312	0.097	0.018	0.572	0.093	0.000
Bike-sharing station close to home/workplace	0.629	0.201	0.003	0.342	0.171	0.022
Familiarity with bike-sharing system	0.736	0.101	0.001	0.638	0.112	0.034
Saving travel cost by bike-sharing	-	-	-	0.407	0.173	0.022
Encouragement of green travel	0.301	0.250	0.00	-	-	-
Flexible route by bike-sharing system	0.328	0.152	0.010	0.418	0.162	0.002
Wasting travel time by bike-sharing	−0.159	0.101	0.009	-	-	-
Great effort on the introduction to the public	0.482	0.146	0.000	0.389	0.099	0.001
Easy to check-in	-	-	-	0.378	0.141	0.002
Easy to check-out	-	-	-	0.302	0.159	0.001
ρ	0.128	0.036	0.027	-	-	-
μ_1	2.789	-	-	-	-	-
μ_2	4.276	-	-	-	-	-
θ_1	−0.904	-	-	-	-	-
θ_2	1.538	-	-	-	-	-
Number of observations	1584					
Log-likelihood convergence	−1622.03					

Table 12. Numeric values for variables that had marginal effects for the bivariate ordered probit model in the case of bike-sharing system usage.

Bike-Sharing System Usage	Rarely	Medium	Often
Gender	−0.1172	0.0463	0.0709
Trip mode (bicycle)	−0.2358	0.0250	0.2108
Trip mode (public transport and bicycle)	−0.3718	0.0246	0.3472
Bicycle, e-bike in household	−0.0983	0.0681	0.0302
Travel time (< 30 min)	−0.1426	0.0471	0.0955
Familiarity with bike-sharing system	−0.2199	0.1082	0.1117
Bike-sharing station close to home/workplace	−0.3018	0.0967	0.2051
Satisfaction with bike-sharing fees	−0.0938	0.0429	0.0509
Wasting travel time by bike-sharing	0.0892	−0.0572	−0.0320
Encouragement of green travel	−0.1702	0.0857	0.0845
Flexible route by bike-sharing	−0.0798	0.0610	0.0188
Great effort on the introduction to the public	−0.3219	0.0529	0.2690

Table 13. Numeric values for variables that had marginal effects for the bivariate ordered probit model in the case of level of satisfaction connected with bike-sharing system usage.

Level of Satisfaction Connected with Bike-Sharing System Usage	Lack of	Average	Fully
Familiarity with bike-sharing	−0.3081	−0.2138	0.5219
Monthly income (Gross) 2000–4000 (PLN)	−0.0078	−0.2394	0.2472
Monthly income (Gross) 4001−∞ (PLN)	−0.0143	−0.3192	0.3335
Bike-sharing station close to home/workplace	−0.0074	−0.0903	0.0977
Satisfaction with bike-sharing fees	−0.0328	−0.2790	0.3118
Flexible route by bike-sharing	−0.0085	−0.2014	0.2099
Saving cost by bike-sharing	−0.0261	−0.1839	0.2100
Great effort on the introduction to the public	−0.0326	−0.2106	0.2432
Easy to check-in	−0.0267	−0.1523	0.1790
Easy to check-out	−0.0084	−0.0623	0.0707

7. Discussion

The analysis presented in the article makes it possible to state that the bike-sharing system plays the role of an element enhancing sustainable mobility in the city. The data presented in the paper for Warsaw (Poland) indicated that from year to year the number of bike-sharing system users increases. These users are mainly young people in a group of age 19–45, using standard bikes, most often in the afternoon (from 3:00 p.m. to 9:00 p.m.) and mostly on working days, in the months from April to September. An important aspect is that the bike-sharing system has been popular among users and that more people use it. Therefore, a dense network of conveniently located bike-sharing stations should be provided in the city. Also, the quality of the bike-sharing system should be an adjustment to the users' expectations. Hence, this article presented an analysis of the factors the bike-sharing system usage as well as the level of satisfaction connected with bike-sharing system usage. The results of the analysis showed that there is a strong positive correlation between these variables. The obtained results can be helpful for carrying out activities to increase bike-sharing system usage as well as the level of satisfaction connected with bike-sharing system usage. The analysis presented in this article allowed us to formulate the following conclusion that respondents do not use often the bike-sharing system (they asked about the frequency of using the system, the answers "average" and "rarely" were, respectively, 38.38% and 40.28%), nevertheless, most of the respondents were fully satisfied with the functioning of the bike-sharing system and assessed it at a very high level (59.28%). A total of 37.88% of respondents rated the level of satisfaction connected with bike-sharing usage at an average level without indicating the disadvantages or advantages of the system. Meanwhile, respondents who were not satisfied with the functioning of the system represented 2.84% of the total. Conducted analyses allowed us to establish that there is a significant correlation between bike-sharing system usage as well as the level of satisfaction connected with bike-sharing system usage. The obtained results seem logical and find confirmation in reality. Regular users of the bike-sharing system are satisfied with the offered services. In contrast to them, people who rarely use the system indicate its disadvantages. The correlation parameter was positive ($p = 0.128$, p-value = 0.027) which makes it possible to state that a higher level of satisfaction connected with bike-sharing system usage could increase the likelihood of bike-sharing system usage. Analysis has shown that men use the bike-sharing system 7% more often than women. The obtained conclusion is consistent with the results of previous research conducted by E. Fishman [100], J. Pucher et al. [101], C.C. Hsu et al. [102]. Moreover, respondents who indicated that they have a bike or e-bike in their households more often used the bike-sharing system. This result can be explained by the fact that those respondents are familiarized with cycling, so they willingly decide to use a bike during their journey. The marginal effect indicated that the probability of bike-sharing system usage for this group is about 3% higher than respondents without a bike or e-bike in their household. This conclusion is consistent with the obtained results by other authors, e.g., by Y. Guo et al. [69].

Respondents used the bike as well as public transport and bike on their daily journey, they often used the bike-sharing system. Marginal effects indicated that the probability of bike-sharing system usage increases by about 21%–35% for people who travel with these two modes of transport. Similar research results are presented in the paper R.B. Noland et al. [103]—metro stations located a short distance from the bike-sharing stations contributed to increasing the number of trips. In turn, respondents who indicated that their travel time was less than thirty minutes were 9.55% more likely to use the bike-sharing system than respondents whose travel time was longer. This conclusion is also confirmed by earlier scientific papers in which travel time using a bike was usually about 30 minutes [104–106].

In the case that the availability of the bike-sharing stations is in the distance up to 500 m of home, workplace, school, university, then the probability of "often" bike-sharing system usage increases by 20.51%. Additionally, the probability of full satisfaction connected with bike-sharing system usage increases by 9.77%. M. Austwick et al. [36], Y. Guo et al. [69], and E. Eren et al. [107] obtained similar results in their works. They indicated that an increasing number of bike-sharing stations cause increasing bike-sharing system usage, whereas the authors W. El-Assi et al. [108] indicated that the distance between stations affects the number of trips using a bike-sharing system.

A total of 11.17% of respondents who knew the rules of the bike-sharing system well used the bike-sharing system more often. In addition, the probability of their full satisfaction increased by as much as 52.19%. This conclusion is important because it indicates the need to familiarize the society with the principles of the functioning and operation of the bike-sharing system. Activities in this area may be conducted by information campaigns or social consultations. These types of activities should contribute to increasing the number of bike-sharing system users. Respondents who accepted and who expressed their satisfaction with the current fees for using the bike-sharing system were characterized by 5.09% higher probability of often using the bike-sharing system. Additionally, they were characterized by 31.18% higher probability of full satisfaction connected with bike-sharing system usage. This conclusion is consistent with the results of the works of A. Faghih-Imani and N. Eluru [51], and E. Fishman [100]. The results from these scientific studies showed the relationship between the bike-sharing system usage and respondents financial savings. As shown in the work of Z. Li et al. [109], people usually minimize travel costs.

Respondents who did not want to incur a loss of time in connection with a cycling trip were 3.20% less likely to choose this form of transport. In turn, a great effort connected with the introduction of the bike-sharing system to the city in order to enhance to sustainable mobility also increased the bike-sharing system usage and level of satisfaction respectively by 26.90% and 24.32%. This conclusion confirms the effectiveness of existing social campaigns and all other activities aimed at encouraging the local community to travel using pro-ecological forms of transport and promotes sustainable transport development. Respondents who were environmentally friendly and who support ecological forms of transport were 8.45% more likely to often use bike-sharing systems.

Respondents who recognized the flexible route offered to them by the bike-sharing system tended to use it often and were fully satisfied with bike-sharing system usage. Marginal effects in these cases show that the probability of often using bike-sharing systems and full satisfaction connected with bike-sharing system usage increased in this case by 1.88% and 20.99%, respectively. Respondents also indicated that easy check-in and easy check-out increase the probability of full satisfaction connected with bike-sharing system usage. In this case, the marginal effects show that the probability of full satisfaction increased by 17.90% and 7.07%, respectively. Results of analysis also indicated that monthly income gross was positively correlated with the level of satisfaction connected with bike-sharing system usage. For the respondents with a monthly gross income of 2000–4000 PLN, marginal effects showed that the probability of full satisfaction increase by 24.72%. In the case of respondents with monthly income gross over 4000 PLN the probability of full satisfaction increased by 33.35%. A similar conclusion was found from research carried out by K.M. Gámez-Pérez et al. [110]. In this work it was indicated that as the socioeconomic level increases, the likelihood that a person will be a user of a bike-sharing system decreases.

However, the presented system is not without defects. Its main disadvantages include an insufficient number of bikes for the disabled, tricycles that could be used by older users, or bikes with manual drive. In addition, it should consider extending the season by winter, especially in mild winter. It is also critical to approach children's bikes and the security level offered by the system for this type of user. This disadvantage of the system is also the malfunction of bikes and stations, which the respondents pointed out in their answers. As reported in [111], there were 59,380 reports on bike defects (which accounted for 0.92% of all rentals) and 22,861 reports of station defects in 2018. There was also a decrease in the popularity of the bike-sharing system associated with the increase in the popularity of electric scooters.

In addition, future analysis would have to be done of the city's configuration which might contribute to strengthening the obtained results about the most integrated bike-sharing routes, as well as the level of connectivity within the bike-sharing network. In Warsaw agglomeration, where there are several transport subsystems (pedestrian, bike, individual, and public transport), ensuring a high degree of their integration is of particular importance. This integration must consist of the continuous modernization of transport nodes so that users of the bike-sharing system in Warsaw can get anywhere and that they have places to park the bike. The customer service systems should be improved that sometimes does not work well. Particular importance will be attributed to the qualitative change in the functioning of the main transport nodes related to the railway system and the increase in the number of stations and bikes at stations. To encourage new customers to use these solutions, the cost of use should not be forgotten, which must be at a level achievable by every resident. It can certainly be stated that the Vertuilo system is integrated with other modes of urban transport. It makes it faster and healthier to arrive at the previously selected destination. Warsaw is perceived as a city which is cyclist-friendly and choosing innovative solutions that are friendly to the natural environment, thanks to investments and implementation of new projects in the bike-sharing system.

Author Contributions: Conceptualization, E.M., and P.Ś.; methodology, E.M.; software, A.K.; validation, E.M., P.Ś. and A.K.; formal analysis, E.M., and P.Ś; investigation, P.Ś. and A.K; resources, P.Ś. and A.K.; data curation, E.M.; writing—original draft preparation, E.M.; writing—review and editing, E.M.; visualization, A.K., and P.Ś.; supervision, E.M. All authors have read and agreed to the published version of the manuscript.

Funding: This research received no external funding.

Acknowledgments: The authors wish to acknowledge Nextbike Poland Company for providing access to data for research and analysis purposes as well as for their contributions in the understanding of the system. The Authors would like to also thanks the Students, PhD Student, Graduate from Transport Systems and Traffic Engineering Department, Faculty of Transport and Aviation Engineering, Silesian University of Technology in Katowice who take part in field questionnaires survey collection.

Conflicts of Interest: The authors declare no conflict of interest.

References

1. Mrówczynska, B.; Cieśla, M.; Król, A.; Sładkowski, A. Application of artificial intelligence in prediction of road freight transportation. *Promet Traffic Transp.* **2017**, *29*, 363–370. [CrossRef]
2. Sładkowski, A.; Cieśla, M. Analysis and Development Perspective Scenarios of Transport Corridors Supporting Eurasian Trade. In *Transport Systems and Delivery of Cargo on East–West Routes*; Studies in Systems, Decision and Control; Sładkowski, A., Ed.; Springer: Cham, Switzerland, 2018; Volume 155, pp. 70–120.
3. Cieśla, M. Outsourcing strategy selection for transportation services based on the make-or-buy decision. *Transp. Probl.* **2015**, *10*, 91–98. [CrossRef]
4. Macioszek, E. Roundabout entry capacity calculation—A case study based on roundabouts in Tokyo, Japan, and Tokyo surroundings. *Sustainability* **2020**, *12*, 1533. [CrossRef]
5. Macioszek, E. The Passenger Car Equivalent Factors for Heavy Vehicles on Turbo Roundabouts. Frontiers in Built Environment. *Sect. Transp. Transit Syst.* **2019**, *5*, 1–13. [CrossRef]
6. Macioszek, E. Changes in Values of Traffic Volume—Case Study Based on General Traffic Measurements in Opolskie Voivodeship (Poland). In *Directions of Development of Transport Networks and Traffic Engineering*; Lecture Notes in Networks and Systems 51; Macioszek, E., Sierpiński, G., Eds.; Springer: Cham, Switzerland, 2019; pp. 66–76.
7. Macioszek, E. Models of Critical Gaps and Follow-up Headways for turbo Roundabouts. In *Roundabouts as Safe and Modern Solutions in Transport Networks and Systems*; Lecture Notes in Networks and Systems 52; Macioszek, E., Akçelik, R., Sierpiński, G., Eds.; Springer: Cham, Switzerland, 2019; pp. 124–134.
8. Rudnicki, A. Zrównoważona mobilność a rozwój przestrzenny miasta (Sustainable mobility and spatial development of a city). *Archit. Czas. Tech.* **2010**, *1-A*, 57–74.
9. Jankowska-Karpa, D.; Wnuk, A. System roweru publicznego (Bike Sharing System—BBS) jako element polityki zrównoważonej mobilności na przykładzie Francji i Polski. *Logistyka - Nauka* **2014**, *6*, 4803–4812.
10. Kwiatkowski, M. Bike-sharing boom—Development of new forms of sustainable transport in Poland on the example of a public bicycle. *Transp. Geogr. Pap. Polich Geogr. Soc.* **2018**, *21*, 60–69.
11. Drynda, P. Influence of Bike-share system for sustainable transport on the example of the Opole on bikes. *Studia Miej.* **2012**, *6*, 105–115.
12. *Metropolitan Area of Gdańsk, Gdynia, Sopot: Conceptual study of the Bicycle System Metropolitan Area for the Metropolitan Area Gdańsk—Gdynia—Sopot*; Metropolitan Area of Gdansk, Gdynia, Sopot: Gdańsk, Poland, 2016.
13. Ważna, A.; Bieliński, T. New Generation of Bike-Sharing Systems in China: Lessons for European Cities. *J. Manag. Financ. Sci.* **2018**, *11*, 25–42.
14. Shaheen, S.A.; Guzman, S.; Zhang, H. Bikesharing in Europe, the Americas and Asia: Past, Present and Future. Transportation Research Record. *J. Transp. Res. Board* **2010**, *2143*, 159–167. [CrossRef]
15. Patel, S.J.; Patel, C.R. An Infrastructure Review of Public Bicycle Sharing System (PBSS): Global and Indian Scenario. In *Innovative Research in Transportation Infrastructure*; Deb, D., Balas, V.E., Dey, R., Shah, J., Eds.; Springer: Singapore, 2019; pp. 111–120.
16. Shaheen, S.; Guzman, S. Worldwide bikesharing. *Access Mag.* **2011**, *1*, 22–27.
17. Parkes, S.D.; Marsden, G.; Shaheen, S.A.; Cohen, A.P. Understanding the diffusion of public bikesharing systems: Evidence from Europe and North America. *J. Transp. Geogr.* **2013**, *31*, 94–103. [CrossRef]
18. Frade, I.; Ribeiro, A. Bicycle sharing systems demand. *Procedia-Soc. Behav. Sci.* **2014**, *111*, 518–527. [CrossRef]
19. Midgley, P. Bicycle-sharing schemes: Enhancing sustainable mobility in urban areas. *United Nations Dep. Econ. Soc. Aff.* **2011**, *8*, 1–12.
20. Mátrai, T.; Tóth, J. Comparative assessment of public bike sharing systems. *Transp. Res. Procedia* **2016**, *14*, 2344–2351. [CrossRef]
21. Dell'Olio, L.; Ibeas, A.; Moura, J.L. Implementing bike-sharing systems. *Proc. Inst. Civ. Eng. Munic. Eng.* **2011**, *164*, 89–101. [CrossRef]
22. Shen, Y.; Zhang, X.; Zhao, J. Understanding the usage of dockless bike sharing in Singapore. *Int. J. Sustain. Transp.* **2018**, *12*, 686–700. [CrossRef]
23. Competition for Veturilo You Borrow a Bike and Take it Where You Need. Available online: https://warszawa.wyborcza.pl/warszawa/7,54420,22279830,konkurencja-dla-veturilo-pozyczasz-rower-i-oddajesz-go-gdzie.html (accessed on 21 March 2019).

24. Wrocławski Rower Miejski. Available online: https://wroclawskirower.pl/miliony-kilometrow-przejechanych-na-rowerach-miejskich-nextbike-polska/ (accessed on 3 January 2020).
25. Szarata, A., with the Team, Jacyna, M., with the Team: WBR 2005, Warsaw Traffic Study along with the Development of a Traffic Model: Synthesis. BPRW S.A. Available online: http://transport.um.warszawa.pl/wbr-2015 (accessed on 2 February 2020).
26. Szarata, A., with the Team, Jacyna, M., with the Team: WBR, 2015, Warsaw Traffic Study along with the Development of a Traffic Model. Roads and Public Transportation Department of the Capital City of Warsaw. Available online: http://transport.um.warszawa.pl/warszawskie-badanie-ruchu-2015/model-ruchu (accessed on 15 January 2020).
27. Szarata, A., with the Team, Jacyna, M., with the Team: WRR, 2015, Warsaw Bike Report. Roads and Public Transportation Department of the Capital City of Warsaw. Available online: http://www.transport.um.warszawa.pl/sites/default/files/Raport%20rowerowy%202015.pdf (accessed on 10 January 2020).
28. Jacyna, M.; Wasiak, M.; Gołębiowski, P. Model ruchu rowerowego dla Warszawy według Warszawskiego Badania Ruchu 2015. *Transp. Miej. I Reg.* **2016**, *10*, 5–11.
29. Jacyna, M. *Modelowanie i Ocena Systemów Transportowych*; Oficyna Wydawnicza Politechniki: Warszawskiej, Warszawa, 2009.
30. Jacyna, M.; Wasiak, M. Modelowanie podziału zadań przewozowych w segmencie przewozów pasażerskich. In *Zeszyty Naukowo-Techniczne Stowarzyszenia Inżynierów i Techników Komunikacji w Krakowie*; Materiały Konferencyjne: Seria, Brunei, 2014; Volume 1, pp. 123–135, Association of Polish Engineers and Communication Technicians, Kracow, Poland.
31. Jacyna-Gołda, I.; Żak, J.; Gołębiowski, P. Models of traffic flow distribution for various scenarios of the development of proecological transport system. *Arch. Transp.* **2014**, *32*, 17–28. [CrossRef]
32. Jacyna, M.; Wasiak, M. Data Exploration for Determining the Parameters of Volume-Delay Function for Sections in the Traffic Models for Heavily Urbanized Areas. In Proceedings of the 20th International Scientific Conference Transport Means, Juodkrantė, Lithuania, 5–7 October 2016; Technologija: Kaunas, Lithuania, 2016; pp. 866–871.
33. Jacyna, M.; Wasiak, M.; Kłodawski, M.; Gołębiowski, P. Modelling of bicycle traffic in the Cities using Visum. *Procedia Eng.* **2017**, *187*, 435–441. [CrossRef]
34. Iwańska, K.; Blicharska, M.; Pierotti, L.; Tainio, M.; deNazelle, A. Cycling in Warsaw, Poland—Perceived enablers and barriers according to cyclists and non-cyclists. *Transp. Res. Part A* **2018**, *113*, 291–301. [CrossRef] [PubMed]
35. Klepacki, B.; Sakowski, P. Warsaw public bicycles network in its users opinion. *Logist. Sci.* **2014**, *4*, 3562–3569.
36. Austwick, M.Z.; O'Brien, O.; Strano, E.; Viana, M. The structure of spatial networks and communities in bicycle sharing systems. *PLoS ONE* **2013**, *8*, 1–17.
37. De Maio, P. Bike-sharing: History, impacts, models of provision, and future. *J. Public Transp.* **2009**, *12*, 41–56. [CrossRef]
38. Faghih-Imani, A.; Hampshire, R.; Marla, L.; Eluru, N. An empirical analysis of bike sharing usage and rebalancing: Evidence from Barcelona and Seville. *Transp. Res. Part A* **2017**, *97*, 177–191. [CrossRef]
39. Frade, I.; Ribeiro, A. Bike-sharing stations: A maximal covering location approach. *Transp. Res. Part A* **2015**, *82*, 216–227. [CrossRef]
40. Lin, J.R.; Yang, T.H. Strategic design of public bicycle sharing systems with service level constrains. *Transp. Res. Part E* **2011**, *47*, 284–294. [CrossRef]
41. Raviv, T.; Tzur, M.; Forma, I.A. Static repositoring in a bike-sharing system: Models and solution approaches. *Eur. J. Transp. Logist.* **2013**, *2*, 187–229. [CrossRef]
42. Benchimol, M.; Benchimol, P.; Chappert, B.; Taille, A.D.L.; Laroche, F.; Meunier, F.; Robinet, L. Balancing the stations of a self service "bike hire" system. *Rairo Oper. Res.* **2011**, *45*, 37–61. [CrossRef]
43. Chemla, D.; Meunier, F.; Calvo, R.W. Bike sharing systems: Solving the static rebalancing problem. *Discret. Optim.* **2013**, *10*, 120–146. [CrossRef]
44. Raviv, T.; Kolka, O. Optimal inventory management of a bike-sharing station. *IIE Trans.* **2013**, *45*, 1077–1093. [CrossRef]
45. Fricker, C.; Gast, N. Incentives and Regulations in Bike Sharing Systems with Stations of Finite Capacity. Available online: https://arxiv.org/pdf/1201.1178.pdf (accessed on 6 July 2019).

46. Vogel, P.; Mattfeld, D.C. Modeling of repositioning activities in bike-sharing systems. In Proceeding of the 12th world conference on transport research, Transport Research Conference, Lisbon, Portugal, 11–15 July 2010.
47. Morency, C.; Trepanier, M.; Paez, A.; Verreault, H.; Faucher, J. *Modelling Bikesharing Usage in Montreal over 6 Years*; Cirrelt, Interuniversity Research Centre on Enterprise Networks, Logistics and Transportation: Montreal, QC, Canada, 2017; pp. 1–16.
48. Li, Y.; Zheng, Y. Citywide Bike Usage Prediction in a Bike-Sharing System. In *IEEE Transactions on Knowledge and Data Engineering*; IEEE: Piscataway, NJ, USA, 2019; pp. 1–12.
49. Li, Y.; Zheng, Y.; Zhang, H.; Chen, L. Traffic Prediction in a Bike-Sharing System. In Proceedings of the 23rd ACM International Conference on Advances in Geographical Information System, Seattle, DC, USA, 3–6 November 2015.
50. Kaltenbrunner, A.; Meza, R.; Grivolla, J.; Codina, J.; Banchs, R. Urban cycles and mobility patterns: Exploring and predicting trends in a bicycle-based public transport system. *Pervasive Mob. Comput.* **2010**, *6*, 455–466. [CrossRef]
51. Faghih-Imani, A.; Eluru, N. Incorporating the impact of spatio-temporal interactions on bicycle sharing system demand: A case study of New York CityBike system. *J. Transp. Geogr.* **2016**, *54*, 218–227. [CrossRef]
52. Tran, T.D.; Ovtracht, N.; d'Arcier, B.F. Modeling bike sharing system using built envinronment factors. *Procedia Cirp* **2015**, *30*, 293–298. [CrossRef]
53. Alberts, B.; Palumbo, J.; Perce, E. *Vehicle 4 Change: Health Implications of the Capital Bikeshare Program*; The George Washington University: Washington, DC, USA, 2012; pp. 1–37.
54. Jurdak, R. The impact of cost and network topology on urban mobility: A study of public bicycle usage in 2 US cities. *PLoS ONE* **2013**, *8*, 1–6. [CrossRef] [PubMed]
55. Guo, Y.Y.; Liu, P.; Bai, L.; Xu, C.C.; Chen, J. Red light running behavior of electric bicycles at signalized intersections in China. *Transp. Res. Rec. J. Transp. Res. Board* **2014**, *2468*, 28–37. [CrossRef]
56. Rixey, R. Station-level forecasting of bikesharing ridership: Station network effects in three US systems. *Transp. Res. Rec. J. Transp. Res. Board* **2013**, *2387*, 46–55. [CrossRef]
57. Faghih-Imani, A.; Eluru, N.; El-Geneidy, A.; Rabbat, M.; Haq, U. How does land-use and urban form impact bicycle flows: Evidence from the bicycle-sharing system (BIXI) in Montreal. *J. Transp. Geogr.* **2014**, *41*, 306–314. [CrossRef]
58. Gonzalez, F.; Melo-Riquelme, C.; De Grange, L. A combined destination and route choice model for a bicycle sharing system. *Transportation* **2015**, *43*, 1–17. [CrossRef]
59. Nair, R.; Miller-Hooks, E.; Hampshire, R.; Busic, A. Large-scale vehicle sharing systems: Analysis of Velib. *J. Sustain. Transp.* **2013**, *7*, 85–106. [CrossRef]
60. Faghih-Imani, A.; Eluru, N. Analyzing Bicycle Sharing System User Destination Choice Preferences: An Investigation of Chicago's Divvy System. Available online: https://www.semanticscholar.org/paper/Analysing-Bicycle-Sharing-System-User-Destination-%3A-Faghih-Imani/fac01190574742c25ac366035ebb0aa2e22e6e23 (accessed on 2 January 2019).
61. Fishman, E.; Washington, S.; Haworth, N.; Watson, A. Factors influencing bike share membership: An analysis of Melbourne and Brisbane. *Transp. Res. Part A* **2015**, *71*, 17–30. [CrossRef]
62. Tang, Y.; Pan, H.; Shen, Q. Bike-sharing systems in Beijing, Shanghai, and Hanghou and their impact on travel behavior. In Proceedings of the Transportation Research Board 90th Annual Meeting, Transportation Research Board, Washington, DC, USA, 24–27 January 2011; pp. 1–12.
63. Kaplan, S.; Manca, F.; Nielsen, T.; Prato, C. Intentions to use bike-sharing for holiday cycling: An application of the theory of planned behavior. *Tour. Manag.* **2015**, *47*, 34–46. [CrossRef]
64. Świerk, P. The Analysis of the Functioning of Bike-Sharing Systems in Warsaw and Cracow. Master's Thesis, Silesian University of Technology, Faculty of Transport and Aviation Engineering, Transport Systems and Traffic Engineering Department, Katowice, Poland, September 2019.
65. Kurek, A. Analysis of Transport Availability of Warsaw City. Master's Thesis, Silesian University of Technology, Faculty of Transport and Aviation Engineering, Transport Systems and Traffic Engineering Department, Katowice, Poland, June 2019.
66. Jimenez, P.; Nogal, M.; Caulfield, B.; Pilla, F. Perceptually important points of mobility patterns to characterise bike sharing systems: The Dublin case. *J. Transp. Geogr.* **2016**, *54*, 228–239. [CrossRef]

67. Hampshire, R.C.; Marla, L. An analysis of bike sharing usage: Explanation trip generation and attraction from observed demand. In Proceedings of the Transportation Research Board 91st Annual Meeting, Washington, DC, USA, 22–26 January 2012; pp. 1–17.
68. Wang, X.; Lindsey, G.; Schoner, J.E.; Harrinson, A. Modeling bike share station activity: The effects of nearby businesses and jobs on trips to and from stations. In Proceedings of the Transportation Research Board 92nd Annual Meeting, Washington, DC, USA, 13–17 January 2013; pp. 1–14.
69. Guo, Y.; Zhou, J.; Wu, Y.; Li, Z. Identifying factors affecting bike-sharing usage and degree of satisfaction in Ningbo, China. *PLoS ONE* **2017**, *12*, 1–19. [CrossRef]
70. Caggiani, L.; Camporeale, R.; Marinelli, M.; Ottomanelli, M. User satisfaction based model for resource allocation in bike-sharing systems. *Transp. Policy* **2019**, *80*, 117–126. [CrossRef]
71. Etienne, C.; Latifa, O. Model-Based Count series clustering for bike sharing system usage mining: A case study with the Velib'System of Paris. *Acm Trans. Intell. Syst. Technol.* **2014**, *5*, 39-1–39-21. [CrossRef]
72. Manzi, G.; Saibene, G. Are they telling the truth? Revealing hidden traits of satisfaction with a public bike-sharing service. *Int. J. Sustain. Transp.* **2018**, *12*, 2532–2570. [CrossRef]
73. Xin, F.; Chen, Y.; Wang, X. Cyclist satisfaction evaluation model for free-floating bike-sharing system: A case study on Shanghai. *Transp. Res. Rec. J. Transp. Res. Board* **2018**, *2672*, 21–32. [CrossRef]
74. Efthymiou, D.; Antoniou, C.; Wadell, P. Factors affecting the adoption of vehicle sharing systems by young drivers. *Transp. Policy* **2013**, *29*, 64–73. [CrossRef]
75. Shi, J.; Si, H.; Wu, G.; Su, Y.; Lan, J. Critical factors to achieve dockless bike-sharing sustainability in China: A stakeholder-oriented network perspective. *Sustainability* **2018**, *10*, 90. [CrossRef]
76. Systemy Rowerów Miejskich. Nextbike. Available online: https://nextbike.pl/ (accessed on 29 December 2019).
77. Nextbike Poland Company: Data on Bike-Sharing System in Warsaw. *Warsaw* **2019**, unpublished data.
78. Miasto Stołeczne Warszawa. To Był Dobry Rok—Rowerowe Podsumowanie. 2018. Available online: http://www.um.warszawa.pl/aktualnosci/dobry-rok-rowerowe-podsumowanie-2018 (accessed on 29 December 2019).
79. Veturilo. Warszawski Rower Publiczny. Available online: https://www.veturilo.waw.pl/cennik/ (accessed on 11 July 2019).
80. Carmines, E.G.; Zeller, R.A. *Reliability and Validity Assessment*, 5th ed.; Sage Publications Inc.: Beverly Hills, CA, USA, 1982.
81. Cronbach, L.J. Coefficient alpha and the internal structure of tests. *Psychometrica* **1951**, *16*, 297–334. [CrossRef]
82. Yurdugul, H. Minimum sample size for Cronbach's coefficient alpha: A Monte-Carlo study. *Hacet. Univ. J. Educ.* **2008**, *35*, 397–405.
83. Cortina, J.M. What is Coefficient Alpha? An Examination of Theory and Applications. *J. Appl. Psychol.* **1993**, *104*, 78–98. [CrossRef]
84. Streiner, D. Starting at the beginning: An introduction to coefficient alpha and internal consistency. *J. Personal. Assess.* **2003**, *80*, 99–103. [CrossRef]
85. Tavakol, M.; Dennick, R. Making sense of Cronbach's alpha. *Int. J. Med Educ.* **2011**, *2*, 53–55. [CrossRef]
86. Allen, E.; Seaman, C.A. Likert Scales and Data Analyses. Quality Progress 2007. Available online: http://rube.asq.org/quality-progress/2007/07/statistics/likert-scales-and-data-analyses.html (accessed on 1 February 2019).
87. Boone, H.N.; Boone, D.A. Analyzing Likert Data. *J. Ext.* **2012**, *50*, 1–5.
88. Jamieson, S. Likert scales: How to (ab) use them. *Med. Educ.* **2004**, *38*, 1212–1218. [CrossRef]
89. Podsumowanie Sezonu Roweru Publicznego w Aglomeracji Warszawskiej. Available online: https://www.veturilo.waw.pl/podsumowanie-sezonu-roweru-publicznego-w-aglomeracji-warszawskiej/ (accessed on 29 December 2019).
90. OpenStreetMap. Available online: https://www.openstreetmap.org/#map=6/51.507/21.560 (accessed on 8 August 2019).
91. Google Maps. Available online: https://www.google.pl/maps (accessed on 6 June 2019).
92. Greene, W. *Estimating Econometric Models with Fixes Effects*; Department of Economics, Stern School of Business, New York University: New York, NY, USA, 2001; pp. 1–14.
93. Chiou, Y.C.; Hwang, C.C.; Chang, C.C.; Fu, C. Modeling two-vehicle crash severity by a bivariate generalized ordered probit approach. *Accid. Anal. Prev.* **2013**, *51*, 175–184. [CrossRef]

94. Anastasopoulos, P.; Karlaftis, M.; Haddock, J.; Mannering, F. Household automobile and motorcycle ownership analyzed with random parameters bivariate ordered probit model. *Transp. Res. Rec.* **2012**, *2279*, 12–20. [CrossRef]
95. Russo, B.J.; Savolainen, P.T.; Schneider, W.H.; Anastasopoulos, P.C. Comparison of factors affecting injury severity in angle collisions by fault status using a random parameters bivariate ordered probit model. *Anal. Methods Accid. Res.* **2014**, *2*, 21–29. [CrossRef]
96. Weiss, A.A. A bivariate ordered probit model with truncation: Helmet use and motorcycle injuries. *J. R. Stat. Soc. Ser. C (Appl. Stat.)* **1993**, *42*, 487–499. [CrossRef]
97. Greene, W.H.; Hensher, D.A. *Modeling Ordered Choices*, 1st ed.; Cambridge University Press: Cambridge, UK, 2009.
98. Sajaia, Z. Maximum likelihood estimation of a bivariate ordered probit model: Implementation and Monte Carlo simulations. *Stata J.* **2009**, *2*, 1–18.
99. Washington, S.; Karlaftis, M.; Mannering, F. *Statistical and Econometric Methods for Transportation Data Analysis*, 2nd ed.; Chapman and Hall/CRC: Boca Raton, FL, USA, 2011.
100. Fishman, E. Bikeshare: A review of recent literature. *Transp. Rev.* **2016**, *36*, 92–113. [CrossRef]
101. Pucher, J.; Buehler, R.; Seinen, M. Bicycling renaissance in North America? An update and re-appraisal of cycling trends and policies. *Transp. Res. Part A Policy Pract.* **2011**, *45*, 451–475. [CrossRef]
102. Hsu, C.C.; Liou, J.J.; Lo, H.W.; Wang, Y.C. Using a hybrid method for evaluating and improving the service quality of public bike-sharing systems. *J. Clean. Prod.* **2018**, *202*, 1131–1144. [CrossRef]
103. Noland, R.B.; Smart, M.J.; Guo, Z. Bikeshare trip generation in New York city. *Transp. Res. Part A Policy Pract.* **2016**, *94*, 164–181. [CrossRef]
104. Mei, Z.; Wang, D.; Chen, J.; Wang, W. Investigation of bicycle travel time estimation using bluetooth sensors for low sampling rates. *Promet Traffic Transp.* **2014**, *26*, 383–391. [CrossRef]
105. Shafizadeh, K.; Niemeier, D. Bicycle journey-to-work: Travel behavior characteristics and spatial attributes. *Transp. Res. Rec. J. Transp. Res. Board* **1997**, *1578*, 84–90. [CrossRef]
106. Caulfield, B.; O'Mahony, M.; Brazil, W.; Weldon, P. Examining usage patterns of a bike-sharing scheme in a medium sized city. *Transp. Res. Part A Policy Pract.* **2017**, *100*, 152–161. [CrossRef]
107. Eren, E.; Uz, V.E. A Review on Bike-Sharing: The Factors Affecting Bike-Sharing Demand. *Sustain. Cities Soc.* **2019**, *54*, 101882. [CrossRef]
108. El-Assi, W.; Mahmoud, M.S.; Habib, K.N. Effects of built environment and weather on bike sharing demand: A station level analysis of commercial bike sharing in Toronto. *Transportation* **2017**, *44*, 589–613. [CrossRef]
109. Li, Z.; Wang, W.; Yang, C.; Ragland, D.R. Bicycle commuting market analysis using attitudinal market segmentation approach. *Transp. Res. Part A Policy Pract.* **2013**, *47*, 56–68. [CrossRef]
110. Gámez-Pérez, K.M.; Arroyo-López, P.; Cherry, C.R. Defining a primary market for bikesharing programs: A study of habits and usage intentions in León, Mexico. *Transp. Res. Rec.* **2017**, *2634*, 50–56. [CrossRef]
111. Veturilo do garażu. A drogowcy szykują zmiany. Available online: https://tvn24.pl/tvnwarszawa/najnowsze/warszawa-koniec-sezonu-veturilo-2416326 (accessed on 3 April 2020).

sustainability

MDPI

Article

Sustainable Urban Street Comprising Permeable Pavement and Bioretention Facilities: A Practice

Yiqing Dai [1,2], Jiwang Jiang [1,3,*], Xingyu Gu [1], Yanjing Zhao [1] and Fujian Ni [1]

[1] Department of Highway and Railway Engineering, Southeast University, Nanjing 210096, China; yiqing.dai@monash.edu (Y.D.); guxingyu1976@163.com (X.G.); benbenzhao@gmail.com (Y.Z.); nfj@seu.edu.cn (F.N.)

[2] Department of Civil Engineering, Monash University, Clayton, VIC 3800, Australia

[3] Department of Civil and Environmental Engineering, the Hong Kong Polytechnic University, Hung Hom, Kowloon, Hong Kong

* Correspondence: jiang_jiwang@hotmail.com

Received: 1 September 2020; Accepted: 3 October 2020; Published: 8 October 2020

Abstract: Roadside bioretention and permeable pavements have proven effectiveness in rainwater filtration and waterlogging mitigation, but conventional street design approach could not accommodate their work in conjunction. In this research, possible roadside facilities allowing water transmission from permeable pavements and bioretention to the pipe system are proposed. Hydraulic properties of the comprised elements were analyzed, including rainfall intensity, permeable pavements, soil layers and pipe systems. A transformation method was formulated to obtain a successive time-intensity formula from conventional design parameters to describe the rainfall behavior, and therefore the water retention capacity of the bioretention could be considered. A test section of 1.6 km combining permeable pavements and roadside bioretention was constructed, and its hydraulic performance was predicted based on the proposed design method and Storm Water Management Model (SWMM). The research results suggest that the bioretention facilities and permeable pavements cooperate well in the test section. In a light rain event, the proposed street has favorable performance in rainwater collection and filtration. In a relatively intense rainstorm event, the street collects and filters water in the initial stage, but will have similar hydraulic performance to a conventional street once the retention facilities are saturated. Thus, no reduction in diameters of drainage pipes from conventional designs is suggested in similar projects.

Keywords: permeable pavement; bioretention; simulation; rainwater; hydraulic performance

1. Introduction

Runoff from streets is a main source of water contamination in urban areas, as dozens of contaminants, typically TSS (total suspended solid), heavy metals and organic contaminants, have been reported to exist with a considerable amount in runoff from urban roads [1–5]. Characteristics of the road runoff in the early stage of rainfall events are mainly investigated in related research due to high density of contaminants [6,7]. Besides, waterlogging is another typical hazard caused by runoff due to insufficient processing capacity of pipeline system in extreme storm events [8,9]. Several urban hydrological planning concepts have been introduced for runoff purification and waterlogging mitigation [10], among which "Best Management Practices, BMPs" [3] and "Low Impact Development, LID" [1,5] in the United States, "Sustainable Urban Drainage System, SUDS" in Europe and "Sponge City, SPC" in China [11] have been widely reported and proven to be efficient. In various design concepts, permeable pavement and roadside bioretention facilities are recommended since they are more practical and cost-efficient compared to some other facilities; for example, the rain gardens are limited by space shortage in urban areas, and the green roofs require labor-consuming construction and maintenance.

The permeable pavement was originally introduced to address rainwater membrane on pavement surface and therefore improve skid resistance for safety while driving [12]; however, its performance in runoff purification and waterlogging mitigation have been noticed and evaluated [13,14], and several additives [15,16] have been developed to improve the purification performance of permeable asphalt concrete. In existing permeable pavement projects, the rainwater in the pavement structure is collected and discharged to the pipe system despite of the contamination, while it would be favorable if the collected rainwater could be transmitted to the bioretention for further purification before the discharge. The roadside bioretention is reported to have the most favorable performance among all LID practices, and it is usually comprised of designed plants, pipe systems and engineering soil layers [1,17]. However, the retention capacity of roadside bioretention is not included in existing codes for street design, therefore it needs to be estimated individually in the design of a street. Several tools have been developed to predict the runoff treatment and retention performance of bioretention, such as "National Storm-water Management Calculator, SWC", "System for Urban Storm-water Treatment and Analysis Integration, SUSTAIN" and "Storm Water Management Model, SWMM", which has developed an LID module in the latest version [18,19]. However, these hydraulic simulation tools are not specially developed for road engineering and they demand hydraulic backgrounds.

In existing projects, conjunction work of permeable pavement and roadside bioretention is limited, possibly because their connections need special design to allow rainwater transmission, which is not included in the present design code, and a practical tool is needed for road engineers to evaluate the hydraulic performance conveniently and systematically. This paper investigates the possible solutions for cooperative work of permeable pavement and bioretention facilities. Hydraulics of relevant elements including the rainfall, pavement, soil and bioretention are analyzed. A transformation method is formulated to obtain a successive time-intensity formula from conventional design parameters to describe the rainfall behavior, and therefore the water retention capacity of the bioretention could be considered. A convenient and practical method is developed to predict the hydraulic performance of such streets. A test section was also constructed with its hydraulic performance evaluated.

2. Street Structure

The structure and facilities of a typical roadside bioretention are presented in Figure 1. The ponding volume of the ponding area is controlled by the height of the overflow inlet as shown in Figure 1a. The vegetation needs special selection since ponding may happen frequently and the soil depth is limited by the underlying structures. Several herbaceous plants and undershrub such as *Zinnia* and *Nandina domestica* are generally planted instead of traditional street trees. In addition, the resistance of plants to salt should be considered as salt is usually used as the deicing agent for winter maintenance of pavement. The bridging layer is composed by medium-sized aggregates to create a separation between the planting soil and the underdrain facilities. An alternative geotextile as also shown in Figure 1a may provide a better separation but a potential clogging risk exists.

The washed stone and underdrain (perforated pipes) as shown in Figure 1b are for harvesting the filtered rainwater which will then be discharged, but in some cases, the water is stored for reuse [20]. When cooperating with compact pavement, the bioretention collects runoff through curbs with holes or gaps and no additional facilities are required. Permeable pavement has proven efficiency in road safety improvement, rainwater filtration and increasing the time of concentration, however the connection with bioretention needs specific design.

As shown in Figure 2a, one practical solution for permeable pavement with roadside bioretention is setting rain grates and catch-basins beneath the pavement, and introducing the curbs with gaps for bioretention. For street stability concerns, the permeable base and the permeable subbase are not suggested for streets although they may be used in sidewalks, parks or other pedestrian occasions; therefore, the base and the subbase in Figure 2a,b are impermeable. The drainage process for the street system in Figure 2a with the increasing intensity of a rain event has three stages. In the first stage, where the rainfall intensity is within the processing capacity of the permeable pavement, rainwater is filtered by pavement and collected through the rain grate to the catch basin. In the second stage, where the rainfall intensity is beyond the capacity of the pavement, the surface runoff develops and enters the bioretention through gaps or holes among the curbs. The runoff will then be filtered by the functional layers and enter the underdrain. In the last stage, where the processing capacity of the underdrain is saturated, the overflow develops and the rainwater will be collected through the overflow inlet.

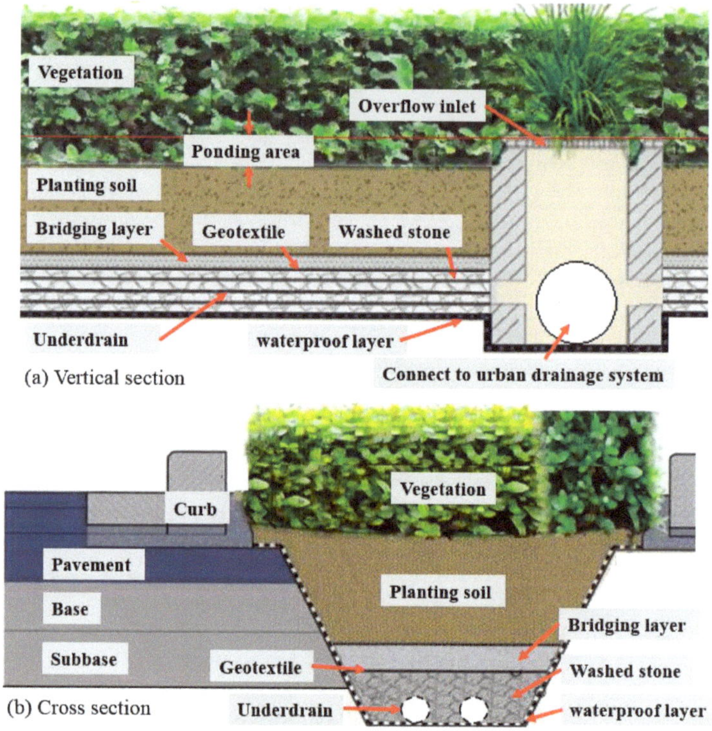

Figure 1. Typical structure of roadside bioretention.

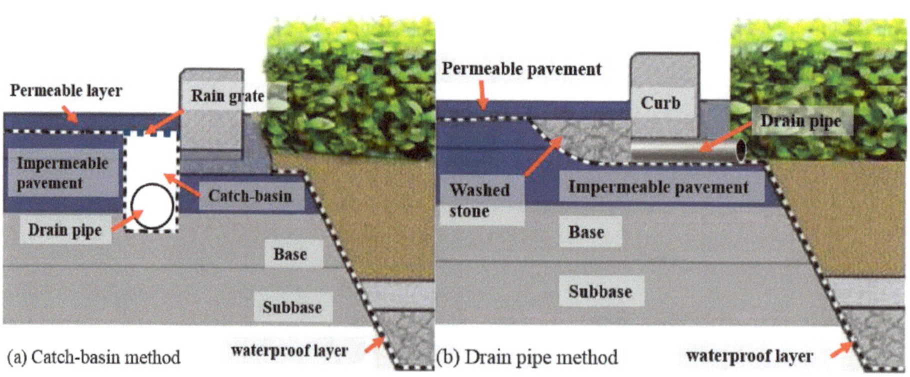

Figure 2. Connections between permeable pavement and roadside bioretention.

Another practical method is presented in Figure 2b where a discontinuous washed stone layer is introduced to collect the filtered water from the permeable pavement, and the collected water is transported to bioretention through a drain pipe. A longitudinally continuous washed stone strip or perforated pipe would have more favorable water-collecting performance, but would lead to compaction problems in the surface pavement; therefore, the traffic loads need to be considered before utilization. This method is not applicable for thick permeable pavement as a backward flow from the bioretention to the pavement structure may happen. However, in an urban street with a thin permeable upper layer, this connection method is suggested since the rainwater from pavement structures could be filtered further by the bioretention and this method is implemented in the test section.

3. Method

A hydraulic analysis was conducted on the element concerned in the street design, including rainfall intensity, pavement and soil permeability, and ponding and overflow in the bioretention. The computational logic comprising these elements are presented and a calculation procedure based on C# was developed to help in the calculation.

3.1. Rainfall

In a conventional urban drainage design process, the required processing capacity or flow capacity (Q, m^3/s) of the pipeline system is determined by the expected runoff as follows [21]:

$$Q = 16.67 A q \psi \tag{1}$$

where A is the catchment area in m^2; q is the average rainfall intensity in mm/min; and ψ is the runoff coefficient, a dimensionless factor taken as the ratio between the runoff and the precipitation. However, apart from drainage through the pipeline system, the pavement and bioretention can retain a certain amount of rainwater and mitigate the waterlogging risks. In order to take the retention capacity of bioretention into consideration, the time-intensity characteristic of the rainfall needs to be obtained, and a method is therefore formulated to build such time-intensity characteristic based on the average rainfall intensity (q). In the conventional street design method, the average rainfall intensity (q) is calculated by an empirical intensity-duration formula as seen in Equation (2) [22].

$$q = \frac{a}{(t+b)^c} \tag{2}$$

where t is the duration of rainfall in min; a, b and c are rainfall parameters, varying with meteorological localities and the return period. If the rain intensity at the moment of time 't' (min) is assumed to be $i(t)$, the average rainfall intensity, q (mm/min), can be written as:

$$q = \frac{1}{t}\int_0^t i(t)dt \tag{3}$$

Taking the derivative of both sides of Equation (3) gives:

$$q't + q = i(t) \tag{4}$$

The rain intensity at the moment of time 't' can therefore written as:

$$i(t) = \frac{a[(1-c)t+b]}{(t+b)^{c+1}} \tag{5}$$

In this model, the moment with the maximum rainfall intensity needs to be assigned, and a parameter r % is defined as the percentage of time before the maximum rainfall intensity appears, thus rainfall intensity before and after the maximum can be described as Equation (6) and Equation (7).

$$i_{tb} = \frac{a\left[\frac{(1-c)t_b}{r\%}+b\right]}{\left(\frac{t_b}{r\%}+b\right)^{1+c}} \tag{6}$$

$$i_{ta} = \frac{a\left[\frac{(1-c)t_a}{1-r\%}+b\right]}{\left(\frac{t_a}{1-r\%}+b\right)^{1+c}} \tag{7}$$

where the variable t_b is the time before the moment with the maximum rainfall intensity, min; i_{tb} is the rainfall intensity at t_b, mm/min; t_a is the time after maximum rainfall intensity, min; and i_{ta} is the rainfall intensity at t_a, mm/min.

3.2. Pavement

Hydraulic and mechanical performance of a permeable pavement has been well investigated in existing literatures to help in design, construction, maintenance and rainwater quality prediction [23] of permeable pavement projects. The main purpose of permeable pavement is to increase the skip resistance and filter the rainwater, while the retention capacity of the pavement is relatively limited. A runoff coefficient (namely the ratio between the volume of discharged rainwater and the precipitation received by the pavement [24,25]) is used to describe the retention capacity of pavement as shown in Equation (8). Considering structural differences among driveways, bicycle lanes and sidewalks, different runoff coefficients can be assigned to individual lanes in the calculation procedure.

$$f = \frac{\sum f_n w_n}{\sum w_n} \tag{8}$$

where f is the weighted average runoff coefficient, dimensionless; f_n is the runoff coefficient of the nth pavement, dimensionless; and w_n is the width of the nth pavement, m. It should be mentioned that porous asphalt may gradually lose permeability due to void blocking and compaction, which could be reflected by an increased runoff coefficient. Therefore, Equation (8) is also valid in characterizing the hydraulic performance of pavement after long-time services.

3.3. Soil and Ponding Area

Soil layers are the main media for water filtration in the bioretention and several researches have been conducted for the filtration improvement. Aggregates including the bridging layer and the washed stone are also treated as soil in the design. Ponding happens when the soil layers are saturated. The relationships between the soil permeability, rainfall intensity and ponding are presented in Figure 3. The soil permeability will decrease with water content until saturation. Several theories have been developed to describe the relationship, such as Horton's infiltration model [26]. The soil permeability is presented by a decreasing dashed line in Figure 3 and the saturated permeability is indicated by a horizontal dashed line. A storm event with constant rainfall intensity is presented in the figure by a horizontal line.

Three stages can be observed from Figure 3. In stage 1, the permeability of soil layers exceeds the rainfall intensity. In this stage, no ponding develops and the volume of water infiltrated is determined by the rainfall intensity. All water is absorbed by soil particles and no infiltration happens [27]. In stage 2, the rainfall intensity exceeds the soil permeability but ponding does not happen immediately and the excessive water will fill the voids in the soil. In stage 3, the voids are filled gradually and a bottom-top saturation happens in soil layers. Ponding develops gradually and the retention capacity of the ponding area is calculated as the volume of a quadrangular frustum pyramid considering the longitudinal slope of a street. Concrete dams could be constructed to contain more water if the slope is relatively sharp [1]. In the calculation procedure, two individual infiltration rates (before and after saturation) can be assigned to each soil layer [26].

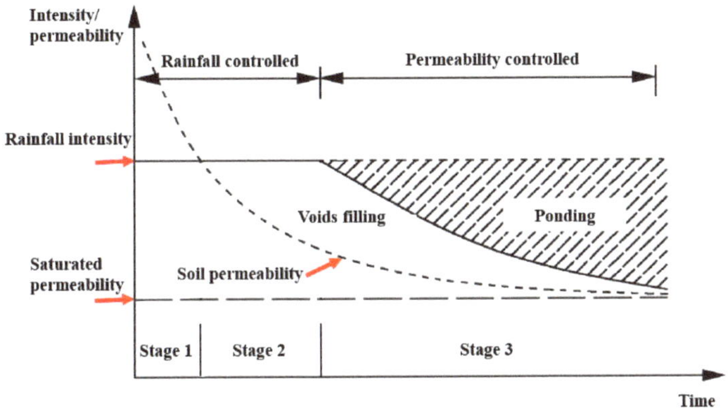

Figure 3. Relationships between soil permeability, rainfall intensity and ponding.

3.4. Overflow

When the retention capacity of the ponding area is exceeded, overflow enters the drainage system through the overflow inlet without soil filtration. A hanging basket is usually installed at the overflow inlet for retention of leaves and major solid objects. Overflow capacity is determined by the discharge capacity of the pipe outlets, which can be predicted by the full pipe flow equation, an adapted version from Manning's formula, as shown in Equation (9) [28]. The discharge capacity of the perforated pipes collecting in the underdrain is also calculated by Equation (9), while the Hazen-Williams equation [29] need to be introduced if pumps are utilized in drainage.

$$C = \frac{\pi d^2 \sqrt{S}}{4n}(\frac{d}{4})^{\frac{2}{3}} \tag{9}$$

where C is the discharge capacity, m^3/s; d is the pipe diameter, m; n is the Manning's roughness coefficient, dimensionless; and S is the pipe slop, dimensionless.

3.5. Computational Logic

The computational logic of the calculation procedure is presented in Figure 4. Rainfall parameters (i.e., the average rainfall intensity (q) and the percentage of time before the maximum rainfall intensity (r)) are input at the beginning of the procedure, and these parameters are turned into a time-intensity series (i.e., a series of rainwater volume by minutes) according to Equations (6) and (7). The rainwater enters the bioretention after a reduction of pavement structures by weighted average runoff coefficient (see Equation (8)). In the computational logic, the rainwater is then absorbed by the soil layers and the status of the soil layers is judged after absorption of the rainwater produced in one minute, where the depths of different layers, permeability (i.e., infiltration rate) and water content for saturation need to be provided. The perforated pipe starts to work if all soil layers reach their field water capacity, and the discharge capacity of the pipe is determined by Equation (9). Ponding takes place when the soil layers are saturated or the infiltration rate is insufficient. Excessive water enters the drainage system directly through overflow inlet if the total ponding capacity is saturated. Also, the discharge capacity of the outlet pipes is determined by Equation (9). The calculation is realized by a software based on C#.

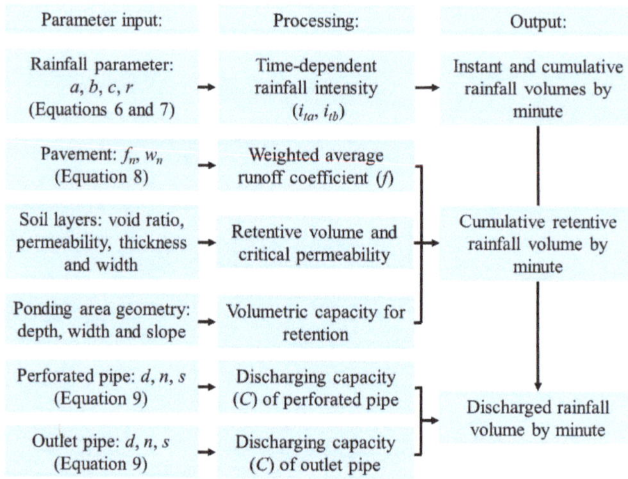

Figure 4. Computational logic of the calculation procedure.

4. Shi-Yang Street Project

4.1. Project Description

The test section is located at Nanjing (32.06° N, 118.80° E), Jiangsu Province, China. The city has a humid subtropical climate with damp conditions throughout the year and the waterlogging risk is a major concern in its urban planning. For example, several severe waterlogging events have been witnessed in recent years [9]. Shi-yang Street project is adjacent to the Qin-huai River, an important water system and scenery spot, as presented in Figure 5. Runoff from the street is collected and discharged to the Qin-huai River, which may intensify the waterlogging and bring contaminations to the river. Thus, a bioretention was constructed to improve the water quality and mitigate waterlogging problem.

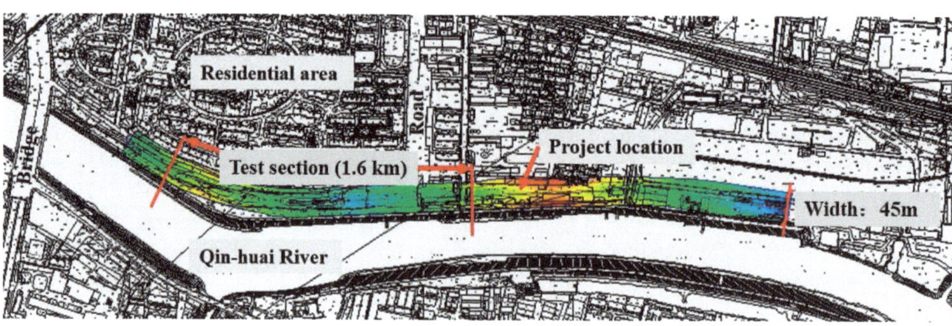

Figure 5. Project location.

This project had a total length of 3.1 km including a test section of 1.6 km that combined the permeable pavement and roadside bioretention. The total width of the street was 45 m, including sidewalks of 3.5 × 2 m, bicycle lanes of 3.5 × 2 m, driveways of 11.5 × 2 m, roadside bioretention of 2.5 × 2 m. The median divider with a width of 3 m was also used as a bioretention and was referred to as median bioretention. Permeable asphalt (PA) concrete with a nominal maximum aggregate size (NMAS) of 13.2 mm (i.e., PA-13) was utilized in bicycle lanes. High-viscosity asphalt was used for the PA-13 pavement and had a kinetic viscosity (60 °C) of 1.43×10^5 Pa·s and a softening point of 91 °C.

Underlying drainage pipes were installed inside the PA pavement to collect water. Dense graded asphalt concrete (AC) pavement was used in driveways considering heavy traffic loads. The NMAS for the AC was also 13.2 mm and the concrete was therefore referred as AC-13. The asphalt used for the AC concrete was styrene-butadiene-styrene (SBS) modified asphalt (PG 76–22). The air void ratios for the PA and AC concretes were 18.5% and 4.2%, respectively. Rutting resistance of the PA and AC concretes used in this test section in dry or saturated conditions was evaluated by repeated loading tests as described in literature [30], where cyclic pressure up to 700 kPa was applied to PA and AC concrete specimens in dry and saturated conditions and the increase in the strain of the specimens per load cycle was used as the indicator for rutting resistance, a slow increase in strain indicated superior rutting resistance. The results showed that for PA specimens, the ratio was 1.99 $\mu\varepsilon$/cycle in a dry condition and 3.41 $\mu\varepsilon$/cycle in a saturated condition, while for AC specimens, it was 0.08 $\mu\varepsilon$/cycle and 0.13 $\mu\varepsilon$/cycle in dry and saturated conditions respectively. The results suggest that AC concrete has much superior resistance to rutting than PA concrete, and rainwater is unfavorable for both types of concrete in terms of rutting resistance. Therefore, pavement structures were not suggested for rainwater retention and the rainwater needs to be drained out. PA concrete was not used in driveways for heavy traffic loads.

Traditional rainwater grates were replaced by gaps (0.3 m wide) between curbs to collect runoff from driveways and bicycle lanes. Ponding areas had a maximum depth of 15 cm while the overflow inlets were 5 cm above the planting soil. Charcoal (2%, by volume) was added to planting soil for water filtration. The total depth of the soil layers was about 70 cm including a geotextile layer and a 15 cm washed stone layer that contained a perforated pipe with a diameter of 10 cm. It should be mentioned that the proposed design increased the construction cost by only about 3%, mainly due to the added charcoal.

Construction details of the test section are presented in Figure 6. The main differences between the test and conventional street constructions lies in the permeable bicycle lanes and the roadside bioretention. Figure 6a shows the overflow inlet and general structure of the under layer of soil layers. The washed stone layer lies between a permeable geotextile and a waterproof layer. Connection between the overflow inlet and the drainage system is presented in Figure 6b, and the perforated pipe in washed stone layer can be observed. The inlet lid and filtering basket, as shown in Figure 6c, were set for retention of solid wastes in relatively large sizes. In street section with a sharp longitudinal slope, concrete dams as shown in Figure 6d were constructed to accommodate more rainwater. Relative narrow curb gaps, as shown in Figure 6e, were used since the longitudinal slope was small in the test section. Underlying pipes collecting water filtered by the permeable pavement were set as shown in Figure 6f.

Figure 6. Details in construction: (**a**) soil layers; (**b**) connection to drainage system; (**c**) outflow inlet; (**d**) concrete dams; (**e**) curb gap; (**f**) pipe under permeable pavement.

4.2. Systematic Hydraulic Evaluation Based on SWMM

To show the working process of Storm Water Management Model (SWMM) and compare the hydraulic performance of conventional street and the proposed street design, models for the test section and a conventional street (i.e., a street without bioretention or permeable pavement) were established. The rainfall intensity formula of Nanjing issued by Nanjing Urban Administration Bureau is shown as Equation (10). It should be noted that Equation (10) is a specific case of Equation (2).

$$q = \frac{64.3(1 + 0.8367\lg P)}{(t + 32.9)^{1.011}} \tag{10}$$

where q is the rainfall intensity, mm/min; P is the return period; and t is the time duration of rainfall, min, taken as 120 min in the models. The return period is 5 years and the duration of the rainfall is 120 min.

The model based on SWMM for the street region, including catchment area, nodes and partition, is shown in Figure 7. Eight nodes were selected for analysis, where nodes Y9, Y13, Y26 and Y27′ were overflow inlets in the test section or drainage system inlet in a conventional street; nodes Y19′J and Y29J were rainwater grate locations on driveways; and nodes Y29F and Y30′F were curb gaps on bicycle lanes. Flow volumes of the selected nodes were simulated to evaluate the waterlogging mitigation effect of the test section. In the modelling of the conventional streets, the pavement and the roadside strip were modelled as impermeable and the rainwater was all collected by the pipe system.

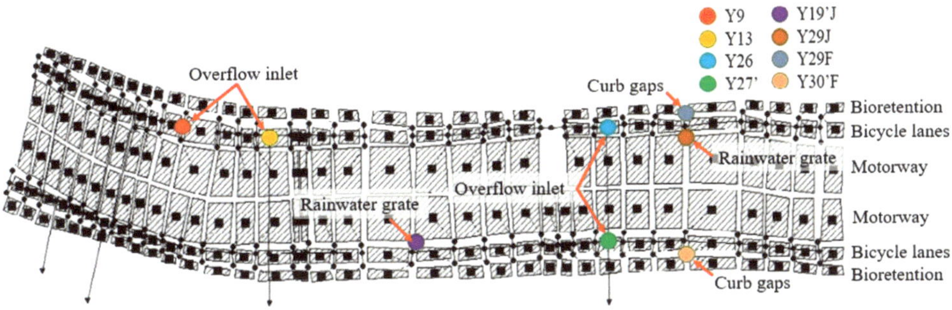

Figure 7. Storm Water Management Model (SWMM) model for the test section.

According to the results from SWMM, flow volumes of all selected nodes in the conventional street and the test section during a typical rainstorm with a return period of 5 years and 120-min duration (peak rainfall intensity 2.98 mm/min) are shown in Figure 8. As shown in Figure 8a–d, flow volumes of overflow inlets or the drainage system were only slightly reduced in the test section, indicating the effectiveness of these LID infrastructures in waterlogging mitigation during a relatively intense storm (for example with a 5-year return period) is limited. Thus, the same requirements for pipe diameters and slope gradient are still necessary. While curb gaps in driveways, as shown in Figure 8e,f, will have higher loads than conventional street due to less favorable drainage efficiency of curb gaps than rainwater grates, curb gaps have less blocking risks, thus they have less performance degradation in use. Flow volumes of curb gaps on bicycle lanes, as shown in Figure 8g,h, were obviously reduced as permeable materials are introduced in sidewalks and bicycle lanes. Both peak flow volume was reduced by about 52%.

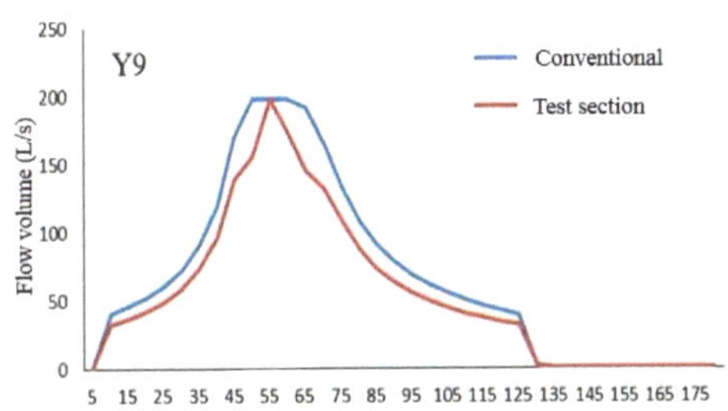

(**a**) Node Y9 with overflow inlets

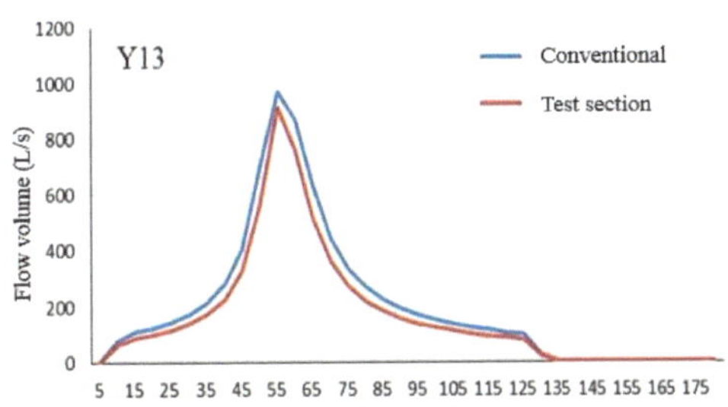

(**b**) Node Y13 with overflow inlets

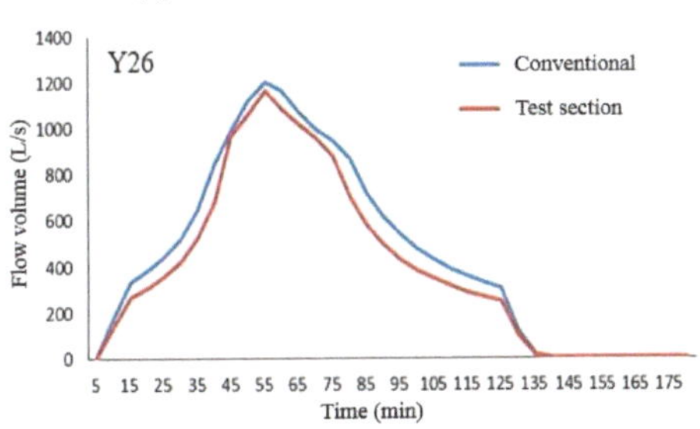

(**c**) Node Y26 with overflow inlets

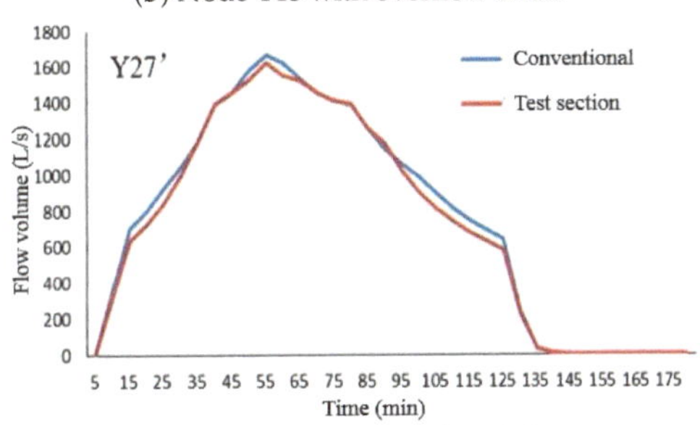

(**d**) Node Y27′ with overflow inlets

Figure 8. *Cont.*

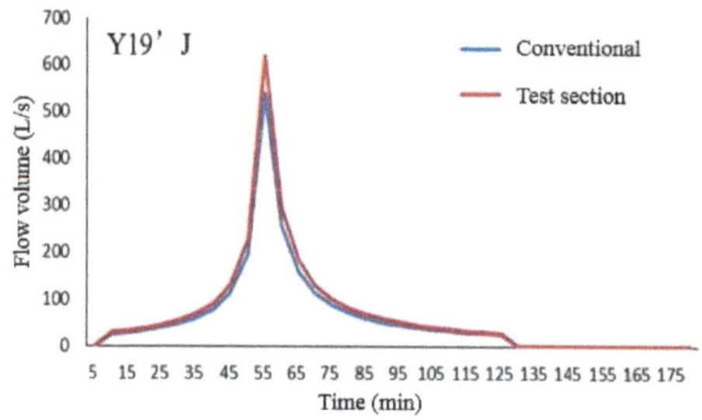

(e) Node Y19′J with rainwater grate location in driveways

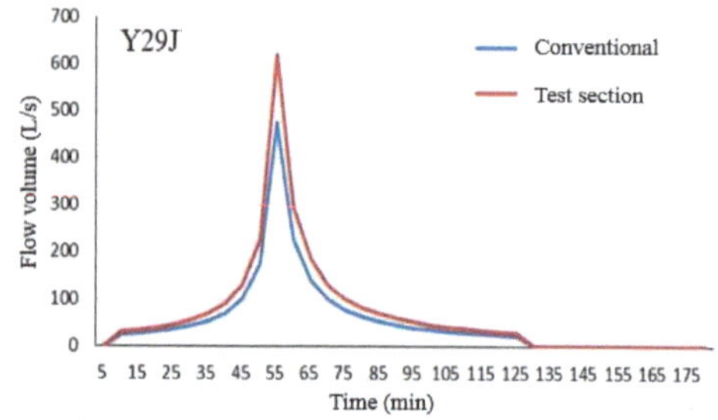

(f) Node Y29J with rainwater grate location in driveways

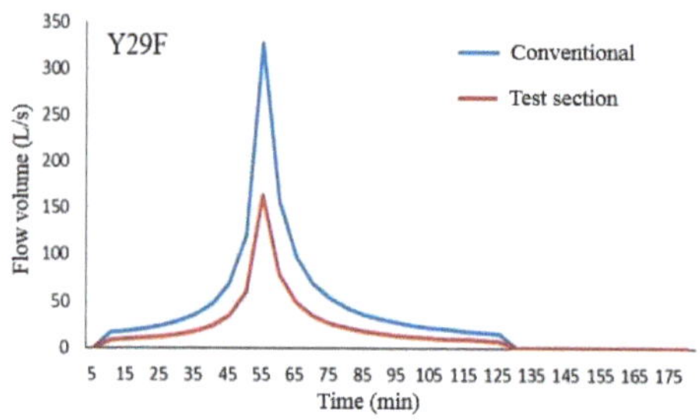

(g) Node Y29F with curb gaps on bicycle lanes

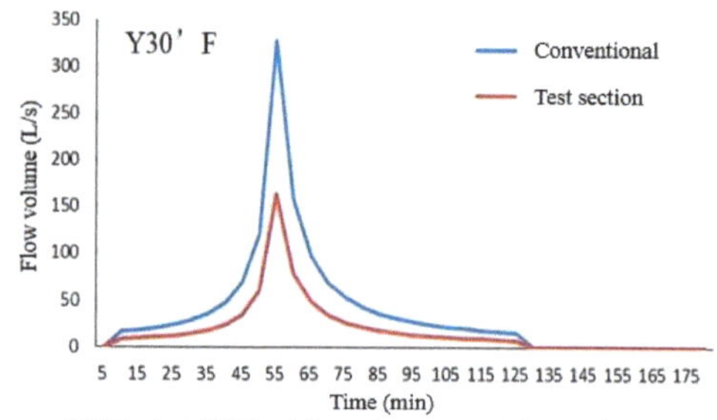

(h) Node Y30′F with curb gaps on bicycle lanes

Figure 8. Flow volume of selected nodes.

4.3. Evaluation Based on the Proposed Calculation Procedure

Based on the proposed calculation procedure, a conventional street and a test section of 250 m were considered. The 0.55-m thick planting soil layer was divided into a 0.25 m layer and a 0.3 m layer to show the permeation process in soil layers. The street slope and pipe slope were taken as 0.3%. The diameter of the drainage pipe was taken as 0.6 m and the roughness coefficient was 0.2. A relatively intense rainstorm and a light rain were generated by Equations (6) and (7) with return periods of 5 years and 0.2 years respectively. For conventional streets, the pavement and the roadside strip were set as impermeable and the rainwater was all collected by the pipe system. The maximum intensity of the rainfall events were set at the 48th min.

According to the results from the proposed calculation procedure, flow volumes of the drainage pipes at a 5-year return period are shown in Figure 9a. During a 120 min rainfall, the soil layers and ponding area absorbed all rainwater in the first 34 min and no water entered the drainage system, but after the retention infrastructures were saturated, the hydraulic performance of the test section was almost the same as a conventional street. The result accords with the simulation results from the SWMM model. The slight flow reduction in the later period can be attributed to the runoff reduction by permeable pavement. Pavement runoff in the first several minutes are highly contaminative, thus, from the point of water filtration, the facilities are meaningful.

However, the test section had much better performance in the 0.2-year return period rainfall as depicted in Figure 9b. The drainage pipes began to collect rainwater at the 51st min and the peak flow was reduced from 7.1 m^3/min to 4.0 m^3/min. Meanwhile, the time when peak flow happens was delayed by 4 min from the 48th min to the 52nd min. Details in soil layers under the 120 min rainfall with 0.2-year return period are presented in Figure 9c. The upper layer of the planting soil reaches field water capacity at the 36th min and is saturated in the 50th min. Ponding area was filled up within 2 min after all soil layers were saturated and the facilities lose further retention capacity.

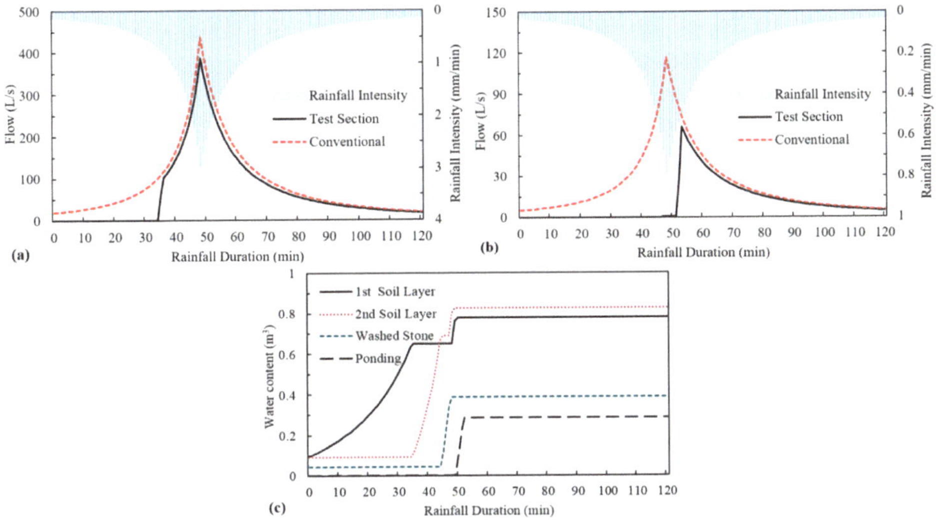

Figure 9. Details of street hydraulics: (**a**) 5-year return period; (**b**) 0.2-year return period; (**c**) details in soil layers.

5. Conclusions

This paper proposes a street structure with permeable pavement and roadside bioretention. Detailed design for connection facilities between the pavement and bioretention is presented. A convenient calculation procedure was developed to predict the hydraulic performance of such street structures. A test section was also built. Simulation through the SWMM model and a calculation by the proposed procedure were conducted to compare the performance of the test section with a conventional street. Several conclusions can be drawn as follows:

1. The bioretention facilities are able to collect the runoff at the initial stage of rainfall with different intensities. Therefore, the runoff at the initial stage containing a relatively higher amount of contaminants can be filtered and retained.
2. The test section has favorable performance in rainwater collection for light rainfall (with short return period or duration). For example, in a 120-min rainfall event with 0.2-year return period, the peak flow is reduced from 7.1 m^3/min to 4.0 m^3/min, and no overflow develops in the first 50 min.
3. For a relatively intense rainstorm (e.g., with a return period of 5 years), the proposed structure has only water filtration function while the effectiveness of waterlogging mitigation is quite limited. No reduction in diameter of drainage pipes is suggested. While the volume and quality of the rainwater including influx and outflow in the test section would be monitored after construction to evaluate the actual effectiveness of the facilities in retention and purification of rainwater.

Author Contributions: Conceptualization, Y.D. and X.G.; data curation, Y.D. and J.J.; formal analysis, Y.D. and J.J.; funding acquisition, Y.D.; investigation, Y.D. and Y.Z.; methodology, Y.D. and J.J.; project administration, X.G. and F.N.; resources, X.G.; software, X.G.; supervision, X.G. and F.N.; validation, Y.Z. and F.N.; visualization, Y.D.; writing—original draft, Y.D.; writing—review & editing, J.J. All authors have read and agreed to the published version of the manuscript.

Funding: This work was supported by the program of China Scholarships Council (No. 201606090204).

Acknowledgments: The authors acknowledge the financial support from the program of China Scholarships Council (No. 201606090204).

Conflicts of Interest: The authors declare no conflict of interest.

References

1. Dietz, M.E. Low Impact Development Practices: A Review of Current Research and Recommendations for Future Directions. *Water Air Soil Pollut.* **2007**, *186*, 351–363. [CrossRef]
2. Wei, Q.; Zhu, G.; Wu, P.; Cui, L.; Zhang, K.; Zhou, J.; Zhang, W. Distributions of typical contaminant species in urban short-term storm runoff and their fates during rain events: A case of Xiamen City. *J. Environ. Sci.* **2010**, *22*, 533–539. [CrossRef]
3. Fassman, E.; Fassman-Beck, E.A. Stormwater BMP treatment performance variability for sediment and heavy metals. *Sep. Purif. Technol.* **2012**, *84*, 95–103. [CrossRef]
4. Caschetto, M.; Barbieri, M.; Galassi, D.M.P.; Mastrorillo, L.; Rusi, S.; Stoch, F.; Di Cioccio, A.; Petitta, M. Human alteration of groundwater–surface water interactions (Sagittario River, Central Italy): Implication for flow regime, contaminant fate and invertebrate response. *Environ. Earth Sci.* **2013**, *71*, 1791–1807. [CrossRef]
5. Shafique, M.; Kim, R. Low Impact Development Practices: A Review of Current Research and Recommendations for Future Directions. *Ecol. Chem. Eng. S* **2015**, *22*, 543–563. [CrossRef]
6. Berndtsson, J.C. Storm water quality of first flush urban runoff in relation to different traffic characteristics. *Urban Water J.* **2013**, *11*, 284–296. [CrossRef]
7. Cheng, J.; Yuan, Q.; YoungChul, K. Evaluation of a first-flush capture and detention tank receiving runoff from an asphalt-paved road. *Water Environ. J.* **2017**, *31*, 410–417. [CrossRef]
8. Quan, R.-S.; Liu, M.; Lu, M.; Zhang, L.-J.; Wang, J.; Xu, S.-Y. Waterlogging risk assessment based on land use/cover change: A case study in Pudong New Area, Shanghai. *Environ. Earth Sci.* **2010**, *61*, 1113–1121. [CrossRef]
9. Zhang, X.; Hu, M.; Chen, G.; Xu, Y. Urban Rainwater Utilization and its Role in Mitigating Urban Waterlogging Problems—A Case Study in Nanjing, China. *Water Resour. Manag.* **2012**, *26*, 3757–3766. [CrossRef]
10. Fletcher, T.D.; Shuster, W.; Hunt, W.F.; Ashley, R.; Butler, D.; Arthur, S.; Trowsdale, S.; Barraud, S.; Semadeni-Davies, A.; Bertrand-Krajewski, J.-L.; et al. SUDS, LID, BMPs, WSUD and more – The evolution and application of terminology surrounding urban drainage. *Urban Water J.* **2014**, *12*, 525–542. [CrossRef]
11. Li, X.; Li, J.; Fang, X.; Gong, Y.; Wang, W. Case Studies of the Sponge City Program in China. In Proceedings of the World Environmental and Water Resources Congress 2016, West Palm Beach, FL, USA, 22–26 May 2016; pp. 295–308.
12. Dell'Acqua, G.; De Luca, M.; Lamberti, R. Indirect Skid Resistance Measurement for Porous Asphalt Pavement Management. *Transp. Res. Rec. J. Transp. Res. Board* **2011**, *2205*, 147–154. [CrossRef]
13. Fassman, E.A.; Blackbourn, S.; Fassman-Beck, E.A. Urban Runoff Mitigation by a Permeable Pavement System over Impermeable Soils. *J. Hydrol. Eng.* **2010**, *15*, 475–485. [CrossRef]
14. Bentarzi, Y.; Ghenaim, A.; Terfous, A.; Wanko, A.; Feugeas, F.; Poulet, J.; Mosé, R. Hydrodynamic behaviour of a new permeable pavement material under high rainfall conditions. *Urban Water J.* **2015**, *13*, 687–696. [CrossRef]
15. Scholz, M. Water Quality Improvement Performance of Geotextiles within Permeable Pavement Systems: A Critical Review. *Water* **2013**, *5*, 462–479. [CrossRef]
16. Barrett, M.; Katz, L.; Taylor, S. Removal of Dissolved Heavy Metals in Highway Runoff. *Transp. Res. Rec. J. Transp. Res. Board* **2014**, *2436*, 131–138. [CrossRef]
17. Chen, X.; Peltier, E.; Sturm, B.S.; Young, C.B. Nitrogen removal and nitrifying and denitrifying bacteria quantification in a stormwater bioretention system. *Water Res.* **2013**, *47*, 1691–1700. [CrossRef]
18. Jayasooriya, V.; Ng, A.W.M. Tools for Modeling of Stormwater Management and Economics of Green Infrastructure Practices: A Review. *Water Air Soil Pollut.* **2014**, *225*, 2055. [CrossRef]
19. Rosa, D.J.; Clausen, J.C.; Dietz, M.E. Calibration and Verification of SWMM for Low Impact Development. *JAWRA J. Am. Water Resour. Assoc.* **2015**, *51*, 746–757. [CrossRef]
20. Cheng, Y.; Wang, R. A novel stormwater management system for urban roads in China based on local conditions. *Sustain. Cities Soc.* **2018**, *39*, 163–171. [CrossRef]

21. Dierkes, C.; Lucke, T.; Helmreich, B. General Technical Approvals for Decentralised Sustainable Urban Drainage Systems (SUDS)—The Current Situation in Germany. *Sustainability* **2015**, *7*, 3031–3051. [CrossRef]
22. Chen, C. Rainfall Intensity-Duration-Frequency Formulas. *J. Hydraul. Eng.* **1983**, *109*, 1603–1621. [CrossRef]
23. Drake, J.; Bradford, A.; Van Seters, T. Stormwater quality of spring–summer-fall effluent from three partial-infiltration permeable pavement systems and conventional asphalt pavement. *J. Environ. Manag.* **2014**, *139*, 69–79. [CrossRef] [PubMed]
24. Ranieri, V. Runoff Control in Porous Pavements. *Transp. Res. Rec. J. Transp. Res. Board* **2002**, *1789*, 46–55. [CrossRef]
25. Blume, T.; Zehe, E.; Bronstert, A. Rainfall—Runoff response, event-based runoff coefficients and hydrograph separation. *Hydrol. Sci. J.* **2007**, *52*, 843–862. [CrossRef]
26. Verma, S. Modified Horton's infiltration equation. *J. Hydrol.* **1982**, *58*, 383–388. [CrossRef]
27. Jabro, J.; Evans, R.G.; Kim, Y.; Iversen, W.M. Estimating in situ soil–water retention and field water capacity in two contrasting soil textures. *Irrig. Sci.* **2008**, *27*, 223–229. [CrossRef]
28. Elger, D.F.; Roberson, J.A. *Engineering Fluid Mechanics*; Wiley: Hoboken, NJ, USA, 2013.
29. Liou, C.P. Limitations and Proper Use of the Hazen-Williams Equation. *J. Hydraul. Eng.* **1998**, *124*, 951–954. [CrossRef]
30. Wang, X.; Gu, X.; Ni, F.; Deng, H.; Dong, Q. Rutting resistance of porous asphalt mixture under coupled conditions of high temperature and rainfall. *Constr. Build. Mater.* **2018**, *174*, 293–301. [CrossRef]

sustainability

MDPI

Article

Identifying the Long-Term Thermal Storage Stability of SBS-Polymer-Modified Asphalt, including Physical Indexes, Rheological Properties, and Micro-Structures Characteristics

Peng Wang *, **Hong-Rui Wei, Xi-Yin Liu, Rui-Bo Ren and Li-Zhi Wang**

School of Transportation Engineering, Shandong Jianzhu University, Jinan 250101, China; whr50081858@126.com (H.-R.W.); liuxiyin123@163.com (X.-Y.L.); rrbgq@sdjzu.edu.cn (R.-B.R.); wlz85503@sdjzu.edu.cn (L.-Z.W.)

* Correspondence: peng0462@126.com or wangpeng0462@sdjzu.edu.cn; Tel.: +86-18560027486

Abstract: The thermal storage stability of styrene–butadiene–styrene tri-block copolymer modified bitumen (SBSPMB) is the key to avoid performance attenuation during storage and transportation in pavement engineering. However, existing evaluation index softening point difference within 48 h (ΔSP_{48}) cannot effectively distinguish this attenuation of SBSPMB. Thus, conventional physical indexes, rheological properties, and micro-structure characteristics of SBSPMB during a 10-day storage were investigated in this research. Results showed that during long-term thermal storage under 163 °C for 10 days, penetration, ductility, softening point, recovery rate (R%), and anti-rutting factor ($G^*/\sin\delta$) were decayed with storage time increasing. This outcome was ascribed to the phase separation of SBS, which mainly occurred after a 4-day storage. However, ΔSP_{48} after a 6-day storage met the specification requirements (i.e., below 2.5 °C). Thus, the attenuation degree of asphalt performance in field storage was not effectively characterized by ΔSP_{48} alone. Results from network strength (I) and SBS swelling degree tests revealed that the primary cause was SBS degradation and base asphalt aging. Moreover, conventional indexes, including penetration, ductility, and softening point, were used to build a prediction model for rheological properties after long-term storage using partial least squares regression model, which can effectively predict I, R, J_{nr}, $G^*/\sin\delta$, and SBS amount. Correlation coefficient is above 0.8. $G^*/\sin\delta$ and I at the top and bottom storage locations had high coefficient with SBS amount. Thus, phase separation of SBSPMB should be evaluated during thermal storage.

Keywords: polymer-modified asphalt; thermal storage stability; relationship between macro performance and micro structures; partial least squares regression

check for updates

Citation: Wang, P.; Wei, H.-R.; Liu, X.-Y.; Ren, R.-B.; Wang, L.-Z. Identifying the Long-Term Thermal Storage Stability of SBS-Polymer-Modified Asphalt, including Physical Indexes, Rheological Properties, and Micro-Structures Characteristics. *Sustainability* **2021**, *13*, 10582. https://doi.org/10.3390/su131910582

Academic Editors: Joel R.M. Oliveira, Hugo Silva, R. Christopher Williams and Zejiao Dong

Received: 24 August 2021
Accepted: 17 September 2021
Published: 24 September 2021

1. Introduction

Styrene–butadiene–styrene tri-block (SBS) polymer-modified bitumen (PMB) (i.e., SBSPMB) has been applied globally in road pavements owing to its superior performance. Compared with polyethylene (PE), styrene butadiene rubber (SBR), and ethylene-vinyl acetate (EVA), SBS can provide outstanding stiffness and elastic recovery characteristics over a wide temperature range to conventional asphalt [1,2]. SBSPMB also serves in porous pavement for sustainability of urban environment [3,4], thus, this binder has dominated the PMB market. However, SBS copolymers are far from perfect. The poor storage stability of SBSPMB remains a crucial but difficult challenge to be addressed.

The poor storage stability of SBSPMB often results from poor compatibility between SBS polymers and bitumen, which is controlled by the different properties of polymers and bitumen, such as density, molecular weight, polarity, and solubility [5]. Poor compatibility leads to phase separation between SBS and asphalt [6–8]. Actually, SBS has a biphasic morphology composed of polystyrene (PS) and polybutadiene (PB), in which rigid PS and flexible PB are the dispersed and continuous phases, respectively [9,10]. When a suitable amount of SBS is added into asphalt, PB can be swollen by the light components in asphalt but PS is pure without change. The combination of the hardness of PS and softness of PB could form a rubbery supporting network in the asphalt, and increase the anti-rutting, anti-cracking, and elastic response to the asphalt matrix. The strength of this physical crosslink network comes from intermolecular forces [11]. Thus, it is easily destroyed by heat and oxidation because of the presence of double bonds and hydrogen in the α- position (α-H) in PB blocks [12] and loss of the light components of asphalt [5]. Zhu believed that density difference was one of the causes of PMB instability in the gravity field [13]. Consequently, an extensive rule for SBSPMB is that SBS is partially swollen by the light components of bitumen to form a thermodynamically unstable but kinetically stable system. Nevertheless, this kinetically stable system is a type of metastable state in the gravity field. Lastly, phase separation resulted in a worse performance for SBSPMB during thermal storage, which is called segregation in road engineering.

Therefore, segregation is a crucial issue in PMB. Numerous studies have concentrated on preventing the segregation of SBSPMB, including optimizing the production process, grafting modifier, or adding stabilizers. Ali found that the mixing time during the blending process of the polymer and asphalt had a significant influence on the polymer particle distribution [14]. Fu reported that SBS-g-M grafted with vinyl monomer under c-rays irradiation could significantly improve the storage stability of SBS-modified asphalt [15]. Ren adopted SBS latex to avoid the segregation of SBSPMB [16]. Among many technologies, sulfur or polymerized sulfur is a common stabilizer for SBSPMB [17–19]. Sulfur is a chemically coupling polymer and bitumen through sulfide or polysulfide bonds, that leads to polymer molecule crosslinking, and provides considerably stronger interactions to form a stable polymer network compared with the physical ones [5]. Aromatics-rich oil (light components) is also a frequently used way to improve SBS swelling that decreases the segregation of SBSPMB [20,21]. Also, there is a side effect of the use of sulfur, as some studies have indicated that there was some element of insoluble sulfur in the vapour produced during the process of mixing and dumping of the asphalt mixture which could have bad effects on the safety of workers, particularly effects on the eyes and skin. Thus, antioxidants [22], nanomaterials [10,23–25], and other functionalization of polymers [24,26] are proposed. Overall, sulfur stabilizers remain the most widely used.

Extensive effort has been focused on the effective evaluation indexes on segregation. This method aims to extensively use closed aluminum toothpaste tubes at 163 °C to mimic the storage and transportation conditions of SBSPMB [19]. The softening point difference after a 48-h storage period (ΔSP_{48}) of SBSPMB in the top and bottom 1/3 part of the tube is the most popular control index used to evaluate PMB stability. However, this conventional index is insufficient to identify whether or not PMB is stable, particularly for long-term storage. Thus, Zhu introduced a phase-field method to capture the phase separation of SBSPMB [27]. Liang used fluorescence microscopy (FM) combined with the phase field model to describe the SBS-phase distribution during storage [28]. Singh reported the rheological property changes with storage time [29]. Typically, each asphalt tank truck can load approximately 40 tons of asphalt. However, using a truck of asphalt in pavement engineering in a short time is difficult, particularly during bad weather. Therefore, before asphalt mixture mixing, SBSPMB is exposed to elevated temperatures (100–180 °C) in closed metal containers with vents for a long time [30], occasionally for over a week. However, only a few studies have focused on the performance decay of SBSPMB during long-term thermal storage.

This study aims to identify the macro-performance and micro-characteristics attenuation of SBSPMB during long-term storage to control the quality of SBSPMB in the field. Segregation test was conducted to simulate the thermal storage procedure. First, macro-performance was used to determine the conventional and rheological index attenuations as storage time increases, which were obtained from softening point, multi-stress creep recovery, and small strain oscillatory rheological tests. Second, micro-characteristics of the SBS degradation during storage were investigated via attenuated total reflection infrared spectroscopy (ATR-FTIR), FM, and atomic force microscopy (AFM). Lastly, a grey relation was used to determine the relationship between macro-performance and micro-characteristics, and the partial least squares (PLS) method was used to predict rheological properties based on conventional physical indexes. This research will be beneficial in markedly improving our understanding of the durability of PMB and controlling its quality in the field.

2. Materials and Methods

2.1. Materials

The grade of base asphalt used is Penetration 70 dmm (marked as A70#). A linear type SBS with molecular weight 110,000 g/mol was used in this research. Commercial sulfur stabilizers were used as asphalt stabilizers. The properties and characteristics of base asphalt and SBS modifier are listed in Table 1.

Table 1. Properties and characteristics of base asphalt and SBS modifiers.

Base Asphalt	**Ductility @ 10 °C/cm**		**Softening Point/°C**	**Penetration @ 25 °C/dmm**	
	38		49	70	
SBS Modifiers	**Type**	**Size**	**Polybutadiene Content/%**	**Polystyrene Content/%**	**Molecular Weight/g·mol^{-1}**
	Linear	20 mesh	70	30	110,000

The SBSPMB was prepared using a high-shear mixer (WeiYu Machine Co., Ltd., Shanghai, China). Base asphalt was heated at 160 °C for improved flow. SBS polymers (4.5 wt.% by asphalt weight) were added to the asphalt at 175 °C for 30 min at a fixed rotation speed for 3000–3500 r/min. Thereafter, asphalt stabilizer (0.25 wt.% by asphalt weight) was added into the mixtures at 180 °C for 5 min. Finally, it was polymer swelling at low temperature and shearing speed to prevent base asphalt aging, and the swelling conditions were 165 °C, 60 min at a fixed rotation speed of 2500 r/min.

2.2. Methods

The entire experimental design is summarized in Figure 1. Long-term thermal storage samples were obtained from segregation test in accordance with ASTM D 5975. The storage time was 0, 2, 4, 6, 8, and 10 days.

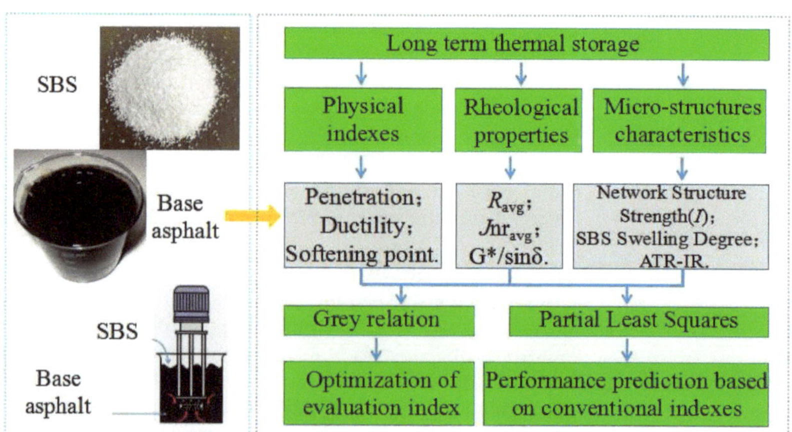

Figure 1. Experimental design flowchart.

This figure shows that storage samples were obtained from segregation tests in accordance with ASTM D 5975. However, the storage time was not only 48 h. Macro-performance testing of SBSPMB included conventional physical and rheological indexes. Conventional physical indexes were obtained by softening point, penetration at 25 °C, and ductility at 5 °C, which were in accordance with ASTM D2398, ASTM D 5, and ASTM D 113, respectively. Softening point difference (ΔSP_{48}) was obtained from ASTM D5976. Rheological indexes were obtained from a Multiple Stress Creep Recovery Test (MSCR) and small strain oscillatory rheological test. MSCR was used to determine irrecoverable compliance (Jnr), recovery rate ($R\%$), and accumulative strain in accordance with ASTM D7405-10a. Small strain oscillatory rheological test was used to obtain the complex modulus (G^*) and phase angle (δ) on the dynamic rheological remoter (DSR) at 70 °C.

Micro-characteristics included experiments conducted using ATR-FTIR, FM, and AFM. ATR-FTIR experiments performed on a TENSOR II instrument (Bruker, German) were used to determine the SBS characteristic peak changes, which occurred during the SBS PMA storage process. FM was used to observe the SBS-rich phase distribution, and AFM was used to identify the particle characteristics of the SBS-rich phase. FM was conducted on a DM 2500 system (Leica, German) and AFM was performed using a Dimension Fast Scan system (Bruker, German). Digital imaging technology was used to find the SBS swelling degree and particle characteristic indexes. For FM, the ultraviolet light type used was blue violet light, the magnification times of the ocular glass was 10×, and that of the object glass was 40×. Asphalt films for FM and AFM were obtained by using the heat-casting method. Hot asphalt binder was cast into glass slides to achieve effective flow with a heating temperature of approximately 160 °C. Bitumen-covered sample holders were left overnight at room temperature before testing.

3. Results

3.1. *Macro Performance Indexes*

3.1.1. Conventional Physical Indexes

Changes in conventional physical indexes as storage time increases are shown in Figure 2. Softening point was typically iso-viscous temperature, penetration was iso-temperature viscosity, and ductility was used to identify the low temperature flexibility of SBSPMB.

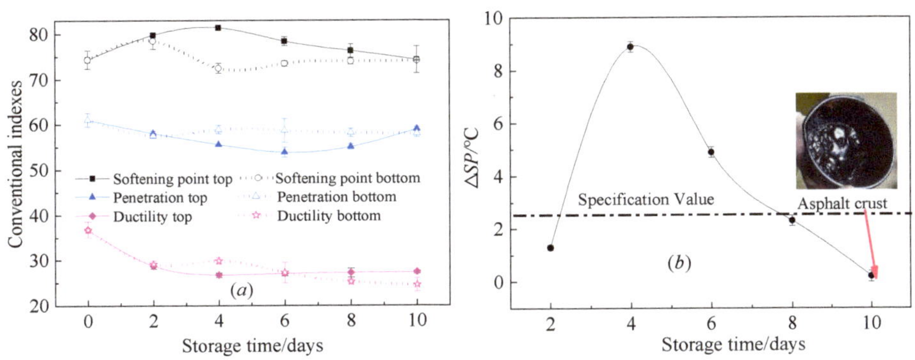

Figure 2. Performance evaluation of SBS modified asphalt based on conventional indexes. The error bar represents the standard deviation of the results of two separate tests. (**a**) Conventional index at top and bottom part during storage. (**b**) Softening point difference within 48 h.

As shown in Figure 2a in the top part of SBSPMB, softening point was increased initially and decreased thereafter, penetration was decreased and increased, and peak values ocurred on the 4th and 6th days. Moreover, Figure 2a shows that the softening point at the bottom of SBSPMB was increased and then decreased, and the peak happened on the 2nd day; and penetration was decreased and then increased, and the valley value was seen on the 2nd day. Ductility was decreased at all times either in the top or bottom part. Thus, long-term storage results in decreased flexibility at low temperature but increased viscidity at high temperature.

In Figure 2b, the max value of ΔSP_{48} was observed on the 4th day, thereby demonstrating the occurrence of the heaviest segregation or phase separation. However, ΔSP_{48} was decreased after a 4-day storage. The reason is that all softening points at the top and bottom decreased, indicating a lack of phase separation improvement. On the 8th and 10th days, ΔSP_{48} satisfied the specification requirement of SBS I-C of China (i.e., below 2.5 °C). Asphalt crusts were extremely issues after a 10-day storage, as shown in Figure 2b. Thus, the attenuation degree of asphalt performance in field storage was not effectively characterized by ΔSP_{48} alone.

3.1.2. Rheological Indexes from MSCR

Accumulated strain, R_{avg}, and J_{nravg} of three modified asphalts were selected to determine the creep characteristic differences of SBSPMB during long-term thermal storage, as shown in Figure 3.

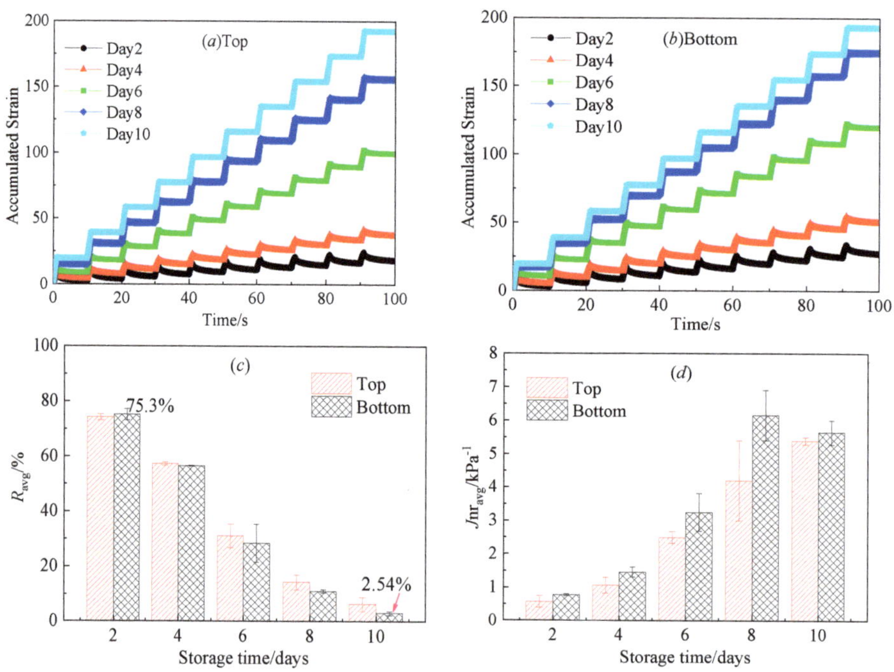

Figure 3. Accumulated strain, R_{avg}, and J_{nravg} of samples with different storage time. The error bar represents the standard deviation of the results of two separate tests. (**a**) Accumulated strain at 3.2 kPa at the top of SBSPMB; (**b**) Accumulated strain at 3.2 kPa at the bottom of SBSPMB; (**c**) Average recovery rate at 0.1 kPa and 3.2 kPa; (**d**) Average irrecoverable compliance at 0.1 kPa and 3.2 kPa.

Equations (1) and (2) were used to obtain the values of R_{avg} and J_{nravg}, respectively:

$$R_{avg} = \frac{R_{0.1kPa} + R_{3.2kPa}}{2} \quad (1)$$

$$Jnr_{avg} = \frac{Jnr_{0.1kPa} + Jnr_{3.2kPa}}{2} \quad (2)$$

where $R_{0.1kPa}$ is the recovery rate at 0.1 kPa, $R_{3.2kPa}$ is the recovery rate at 3.2 kPa, R_{avg} is the average value of $R_{0.1kPa}$ and $R_{3.2kPa}$, $J_{nr0.1kPa}$ is the irrecoverable compliance at 0.1 kPa, $J_{nr3.2kPa}$ is the irrecoverable compliance at 3.2 kPa, and J_{nravg} is the average value of $J_{nr0.1kPa}$ and $J_{nr3.2kPa}$. For accumulated strain, the results in Figure 3a,b show that the accumulated strain increased with an increase in storage time either at the top or bottom of SBSPMB. A larger accumulated strain demonstrated worse elastic recovery ability at high temperature. The results in Figure 3c,d verify this conclusion. The value of R decreased as storage time increased, while the J_{nr} increased at the top or bottom of SBSPMB. High R and low J_{nr} suggested considerable elasticity at high temperature [31]. Thus, higher R and lower J_{nr} would provide SBSPMB with better anti-rutting property. However, on the 10th day of storage, R% reduced to 2.54% and lost 96% elastic ability, resulting in a significantly high J_{nr}. After a two-day storage, J_{nr} was below 0.5 (1/kPa). Thus, long-term hot storage would destroy the three-dimensional network structure, which was proven by the micro-characteristic indexes.

3.1.3. Rheological Indexes from Small Strain Oscillatory Rheological Test

Complex modulus (G^*) is an important index to evaluate the load bearing capacity of PMB. Changes in G^* at 70 °C during long-term storage is summarized in Figure 4a. The results showed that G^* decreased substantially as storage time increased either at the top or bottom part. A smaller G^* provided a lower strength, which was ascribed to the SBS network structure decay and base asphalt aging. In pavement engineering, the ratio of complex modulus to the sine value of the phase angle ($G^*/\sin\delta$) is called the anti-rutting factor, performing the high temperature in performance grade (PG). $G^*/\sin\delta$ variation with storage time is listed in Figure 4b. The result showed that $G^*/\sin\delta$ decreased with an increase in storage time either at the top or bottom part. By contrast, a slightly higher value at the top was observed than the one at the bottom. A larger $G^*/\sin\delta$ indicated that this binder could be used in higher air temperature condition. Thus, SBS particle was implied to float up to enhance the modulus and decrease the phase angle of SBSPMB. By contrast, the bottom was completely the opposite. In general, deformation resistance at high temperature of SBSPMB was weakened after long-term storage.

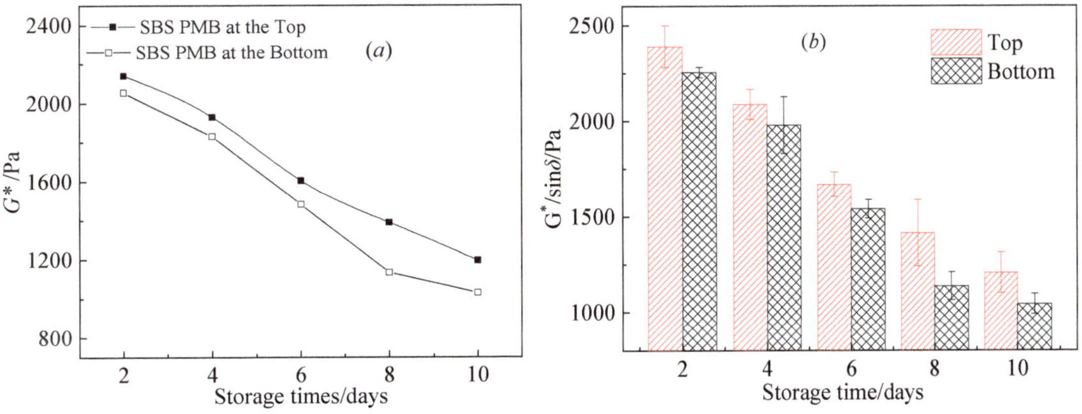

Figure 4. G*/sinδ of samples with different storage time. The error bar represents the standard deviation of the results of two separate tests. (**a**) is complex modulus. (**b**) is anti-rutting factor.

3.2. *Micro Characteristics Indexes*

3.2.1. Network Structure Strength from Stress-Time Curves

A rubbery supporting three-dimensional network formed from the SBS phase is the modification nature of SBSPMB. Wehumbura believed that if a modified asphalt exists in a cross-linked network, then the shear stress–time curve from the continuous application of force at a fixed shear rate in strain control mode would show evident peaks and valleys [32]. Thus, network structure strength (*I*) was proposed to characterize the crosslinking degree of network as follows:

$$I = \frac{S_f}{S_p} \quad (3)$$

where S_f is the peak of the shearing stress, S_p is the flat value of the shearing stress, and all units are in Pa. A large *I* identified a strong network among the polymer phase that is difficult to destroy by external force. The shearing stress–time curves at the top and bottom of SBSPMB are shown in Figure 5, and the value of *I* is presented in Figure 6.

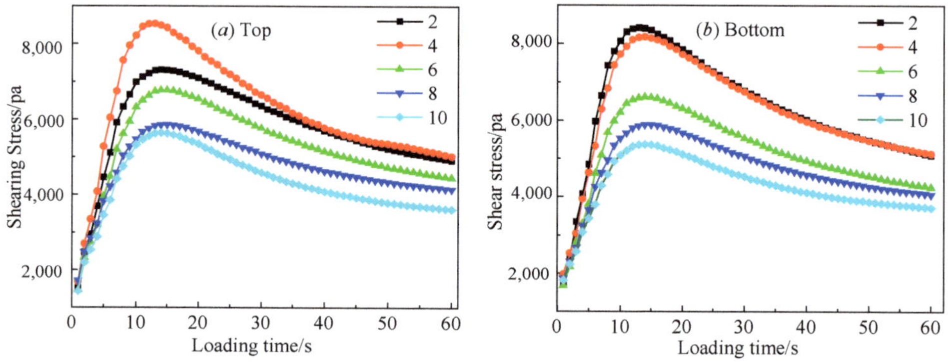

Figure 5. Evaluate of network structure strength. (**a**) Shearing stress at the top. (**b**) Shearing stress at the bottom.

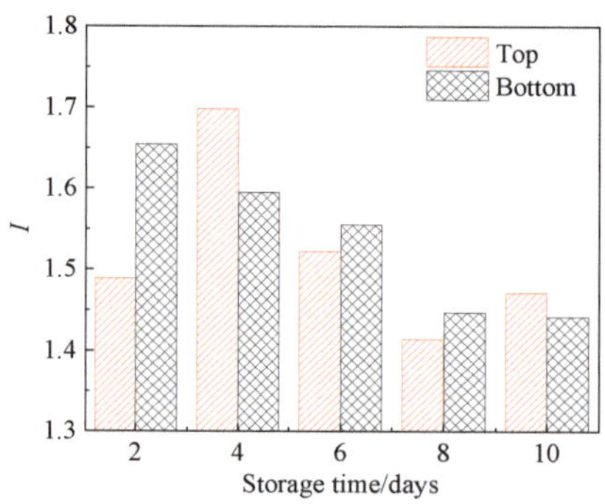

Figure 6. Evaluate of network structure strength.

The results in Figure 5 show that the peak value of the shearing stress–time curve decreased at the bottom with an increase in storage time. For samples at the top, the value at the 2nd storage was smaller than that the 4th day of storage but larger than that at the 6th day of storage. As shown in Figure 6, the value of *I* increased initially and decreased thereafter at the top, but decreased at all times at the bottom. The low *I* value at the bottom was mainly attributed to the thermal degradation of SBS. *I* on the 2nd day was significantly lower than that on the 4th day at the top of the SBSPMB sample. Macro properties showed that heavy segregation occurred on the 4th day. Thus, the SBS value on the 4th day was considerably higher than that on the 2nd day, indicating a strong network. Moreover, *I* on the 2nd day was lower than that at the bottom on the 2nd day because the upper part was easily exposed to oxygen, resulting in SBS oxidation. The flowing up of SBS particle during storage was due to its smaller density than asphalt (i.e., approximately 0.9 g/cm^3). Thus, factors affecting network strength included SBS degradation, loss of light component, and oxidation of the base asphalt and SBS.

3.2.2. SBS Swelling Degree from FM

FM was used to determine the SBS-rich phase distribution characteristics of SBSPMB, as shown in Figure 7. FM images revealed that the SBS swelling degree was calculated on the basis of Equation (4) using digital image technology [33], which indirectly evaluated the compatibility of PMB:

$$SBS\ swelling\ degree = \frac{A_{PRP}}{A_{Total}} \times 100\% \quad (4)$$

where A_{PRP} is the area of the SBS-rich phases and A_{total} is the total area of SBSPMB in the FM images. The original images were processed using MATLAB to convert them into binary images, which only contained two-pixel values (i.e., 0 and 1), and to binary black (asphaltene-rich phases) or white (SBS-rich phases) thereafter. A_{PRP} and A_{total} were eventually obtained. The SBS swelling degree is illustrated in Figure 8.

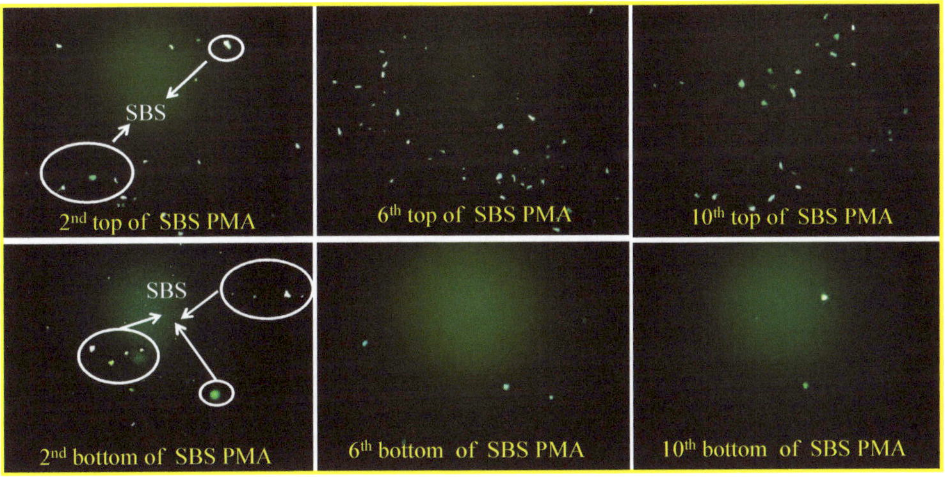

Figure 7. Fluorescence images with different storage time.

The polymer-rich phase of SBSPMB exhibited a higher fluorescence to ultraviolet rays than the bitumen phase; thus, the bright white dots in the FM images were SBS polymers, and the rest was asphalt matrix [34]. As shown in Figure 7, SBS particles increased at the top as storage time increased but decreased at the bottom. The swelling degree of the SBS phase at the top substantially increased on the 4th day and slightly decreased thereafter, as shown in Figure 8. This significant increase was attributed to the upflow of the SBS particles during the first four days. The SBS swelling degree over the range from 4 to 10 days slightly decreased. The main cause of this steady change was the SBS degradation, and a minor consideration may be the SBS phase separation. However, SBS swelling degree at the bottom decreased with an increase in storage time, mainly considering the result of the SBS amount decrease owing to phase separation.

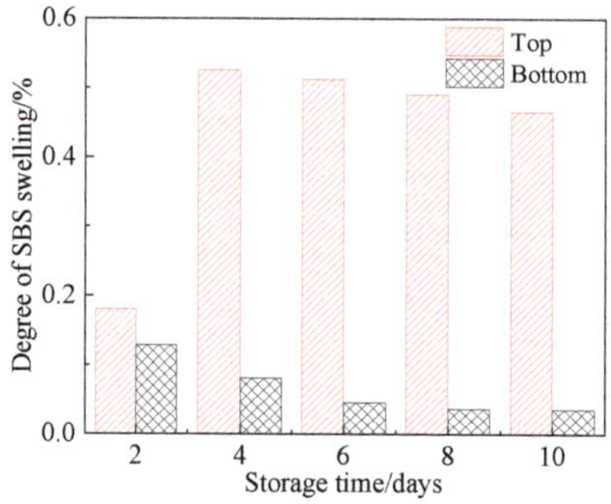

Figure 8. The SBS swelling degree with the increasement of storage time.

3.2.3. SBS Amount from ATR-FTIR

Characteristic peaks in the ATR-FTIR spectrum provided the functional group composition in SBSPMB. ATR is an IR sampling technique, which generally enables qualitative or quantitative analysis of samples with minimal or no sample preparation. Thus, this technique could be used for semi-quantitative analysis in pavement engineering. The ATR-FTIR spectra of SBSPMB with different storage time is summarized in Figure 9.

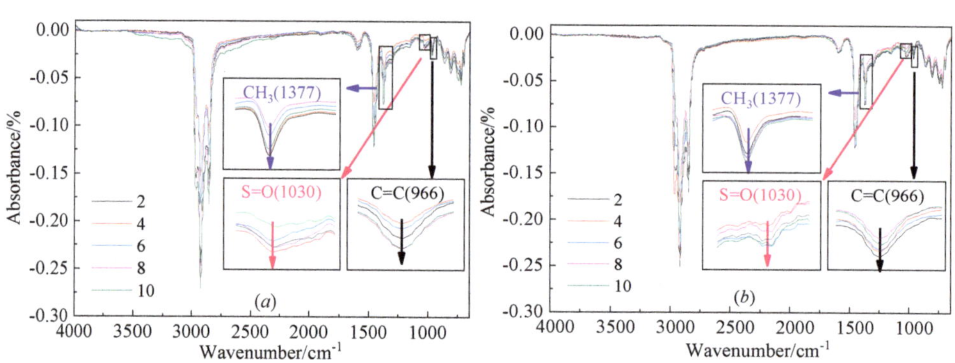

Figure 9. Infrared spectra of different storage time. (**a**) Infrared spectra at the top; (**b**) Infrared spectra at the bottom.

As shown in Figure 9, the observed wavenumbers at 2850, 2920, and 1450–1475 cm^{-1} demonstrate the existence of an aliphatic long chain in saturated hydrocarbon. The characteristic peaks at 1377, 1450–1475, 2850, and 2920 cm^{-1} belonged to the base asphalt. The wavenumber at 1020 cm^{-1} typically represented asphalt aging degree, and 1377 cm^{-1} represented the base asphalt [35]. SBS showed characteristic peaks at wavenumbers 699, 760, 910, and 966 cm^{-1}. Peaks at 699 cm^{-1} and 966 cm^{-1} belonged to the polystyrene phase of SBS [36]. Meanwhile, 966 cm^{-1} was the butadiene stretching vibration of SBS, and showed a relatively stronger peak intensity than 699 cm^{-1}. Thus, the characteristic peak index of SBS polymers (A_{SBS}) (i.e., 966 cm^{-1}) was used to determine the changes in the SBS amount in the asphalt matrix, as shown in Equation (5), where S_{966} and S_{1377} are the peak areas at 966 cm^{-1} and 1377 cm^{-1}, respectively:

$$A_{SBS} = \frac{S_{966}}{S_{1377}} \times 100\% \quad (5)$$

As shown in Figure 10, the characteristic peaks of SBS polymers at the top increased initially and decreased thereafter, and the peak value was obtained on the 4th day. At the bottom, the characteristic peaks of the SBS polymers decreased gradually with storage time. The rule was the same as the macro-performance and other micro-characteristics results. Yan reported that the SBS polymer network could provide a barrier's effect for the oxidation of the actual binder, thereby resulting in improved anti-aging resistant performance when the SBS amount is sufficiently high [32]. Thus, asphalt aging was not considerable apparent as storage time increases, as shown in Figure 9.

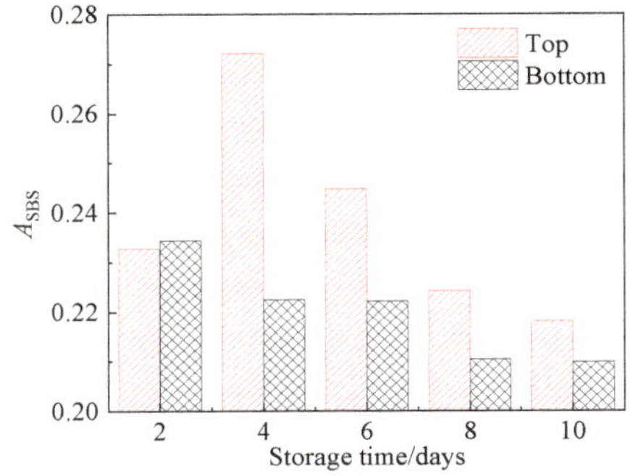

Figure 10. SBS characteristic index $A_{966/1377}$.

The Beer–Lambert law states that absorbance is proportional to the concentration of a substance at a given wavenumber. Thus, A_{SBS} was used to determine the SBS amount from the standard curve of SBSPMB using ATR-FTIR. The standard curve of SBSPMB was obtained from the same prepared process and raw materials but different SBS amounts, as discussed in the SBSPMB preparation part. In this case, the amounts were 3%, 3.5%, 4%, 4.5%, and 5% based on asphalt quality. The calculation of the SBS amount was similar to that in Wang [37]. The standard curve of the SBS amount for SBSPMB is shown in Figure 11a. The correlation index R^2 was 0.9933, showing that the standard curve was effective.

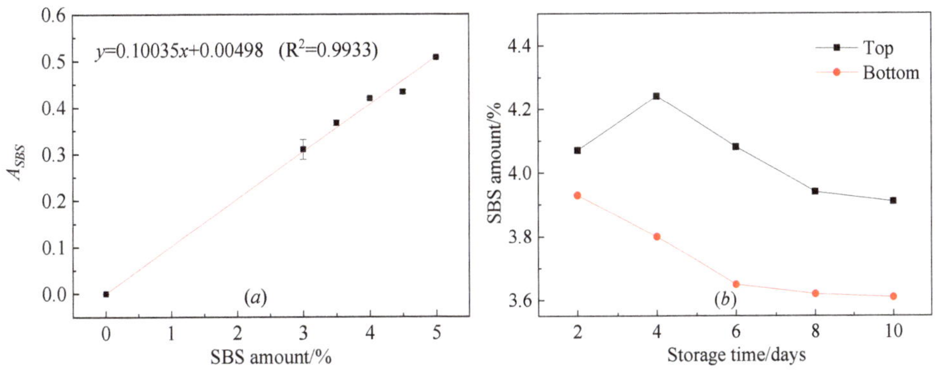

Figure 11. Standard curve of SBS amount for SBSPMB and SBS amount changes as storage time. The error bar represents the standard deviation of the results of five separate tests. (**a**) Standard curve. (**b**) The changes of SBS amount.

As shown in Figure 11b, the SBS amount at the top was significantly larger than that at the bottom, demonstrating that SBS flowed up during long-term storage. This result could explain why the 4th day had a large ΔSP, and the upper part dislayed a better macro-performance than the one at the bottom. Meanwhile, the SBS amount at the top increased initially and decreased thereafter as storage time increased, while the value at the bottom decreased with an increase in storage time. The reduction of the SBS amount was due to SBS degradation in hot condition, while the increase in SBS amount on the 4th day of the top part was attributed to SBS flowing up.

3.3. *Performance Prediction after Long-Term Storage*

3.3.1. Grey Relation

Grey relation is used to explore the relationship between two systems that vary with different objects. Correlation degree is used to measure the degree of similarity or difference of development trend between the two. This study used grey correlation degree to identify the most representative index of the SBS amount. The correlation degree of the macro–micro indexes referenced as SBS amount are shown in Figure 12.

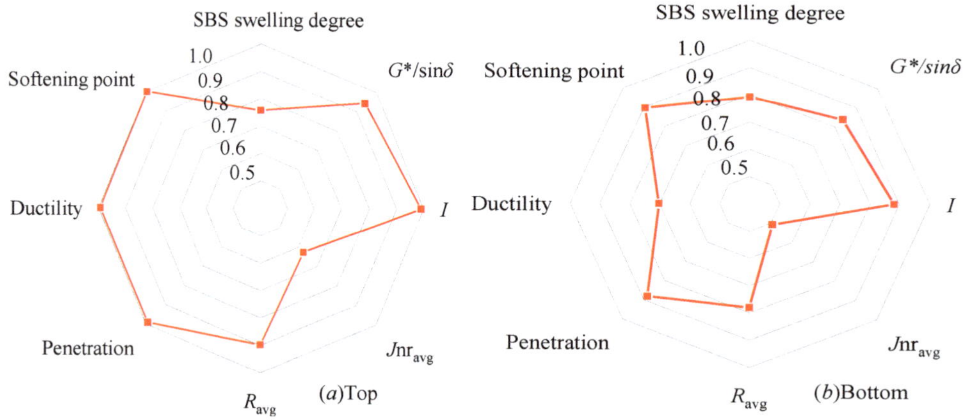

Figure 12. Grey correlation degree between evaluation indexes of top and bottom samples and SBS content. (**a**) Correlation degree of top sample index. (**b**) Correlation degree of bottom sample index.

For the samples at the top part in Figure 12a, the softening point, ductility, penetration, I, R_{avg}, and $G^*/\sin\delta$ showed a grey correlation degree near 1.0, implying they had a closed relationship with SBS amount. However, the grey correlation degree among penetration, softening point, I, R_{avg}, $G^*/\sin\delta$, and SBS amount at the bottom are above 0.8. However, indexes at the bottom provided smaller grey correlation degree than that at the top. SBS swelling degree and J_{nravg} showed slightly smaller correlation with SBS amount. For segregation, the phase separation degrees were determined using Equation (6):

$$\text{Degree of separation}(\Delta_i) = \left|\frac{\text{Top parameters} - \text{Bottom parameters}}{\text{Average of the top and bottom parameters}}\right| \times 100\% \quad (6)$$

The grey correlation degree between the segregation degree based on different indexes and SBS amount is shown in Figure 13. Meanwhile, ΔSP_{48} showed a superior relationship with SBS amount either at the top or bottom part, yet it was not accurate diagnotic of the performance decay during long-term storage though satisfied specification requirement. $G^*/\sin\delta$ has a high grey correlation degree at the top/bottom part and segregation degree. On a micro-scale, I and SBS swelling degree hold a correlation degree above 0.7, although I has a higher correlation degree at the top and bottom with its SBS amount. Thus, the network structure strength I would be a potential indicator to describe the SBS phase separation during storage.

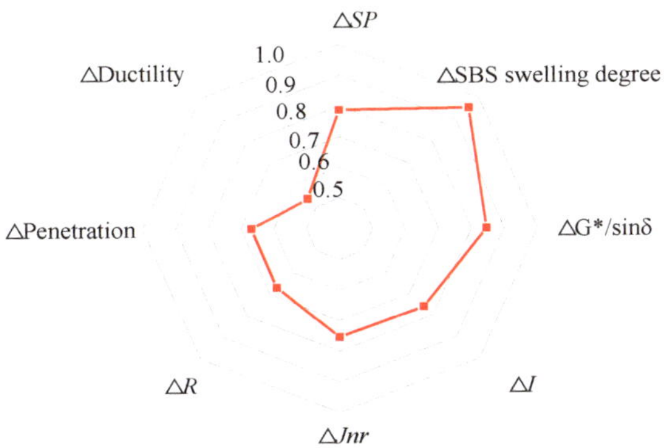

Figure 13. The grey correlation degree between SBS amount and segregation degree based on different indexes.

3.3.2. PLS Regression Model

PLS is a multivariate statistical linear regression technique based on estimated latent variables, which extracts the relationships between dependent and independent variables [37]. In this study, the purpose of PLS regression is to build a linear model to predict the rheological properties and micro-characteristic index based on conventional physical indexes, including penetration, ductility, and softening point during storage. The reason is that conventional physical indexes are easily obtained in all testing organizations.

The prediction models are presented in Figure 14. The highest prediction accuracy was $G^*/\sin\delta$ and its R^2 was 0.9111, as shown in Figure 14b. The second was R_{avg} and its R^2 was 0.8828, as presented in Figure 14c. The prediction accuracy of the SBS amount and J_{nrave} were 0.8540 and 0.8390, respectively, as shown in Figure 14a,d, respectively. However, the prediction accuracy of I and SBS swelling degree was below 0.8, demonstrating a lack of relevance with the conventional physical indexes. Conventional physical indexes were not identifying network strength and the changes of the SBS swelling degree during storage. Thus, segregation based on ΔSP_{48} only was not suitable to determine the macro-performance and micro-characteristics changes. The current research suggested that $G^*/\sin\delta$ and I may be superior indexes to monitor the macro- and micro-changes of SBSPMB with storage time.

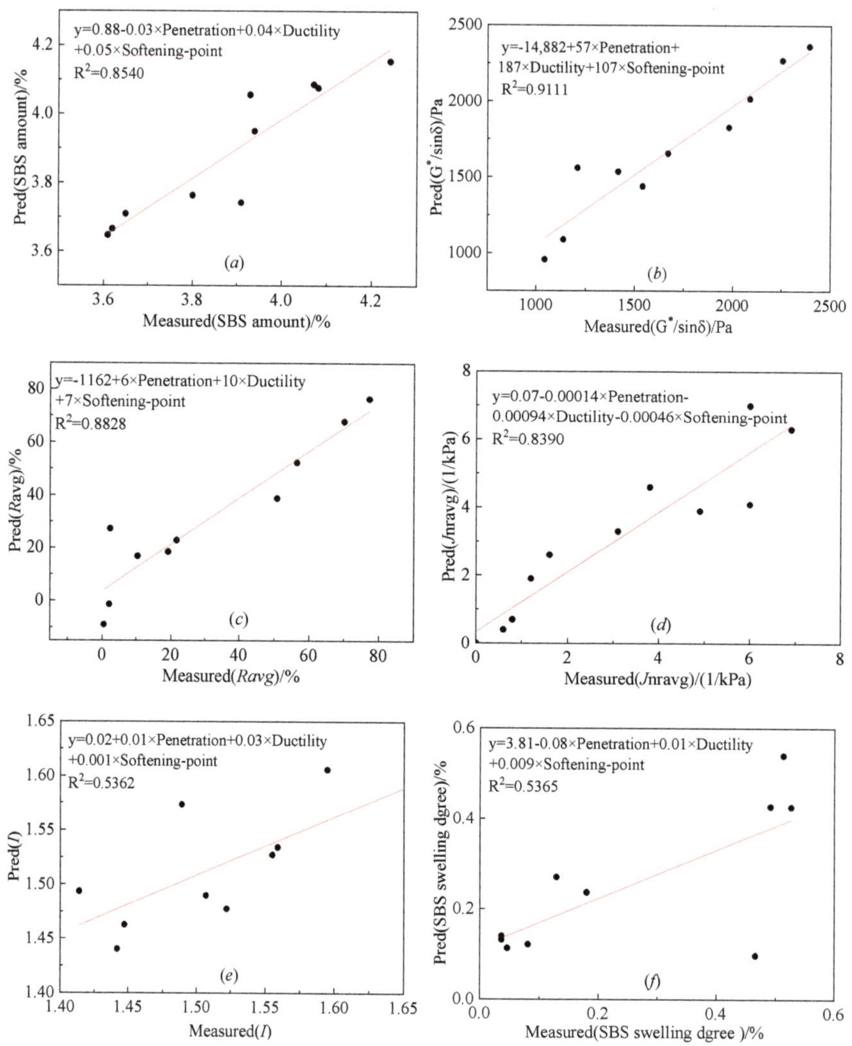

Figure 14. Measured and predicted values of indicators. (**a**) Prediction of SBS amount. (**b**) Prediction of $G^*/\sin\delta$. (**c**) Prediction of R_{avg}. (**d**) Prediction of J_{nravg}. (**e**) Prediction of I. (**f**) Prediction of SBS swelling degree.

4. Conclusions

The elusive thermal storage stability of SBSPMB has attracted wide attention during over half a century of intensive research. However, the existing softening point difference evaluation index within 48 h could not distinguish between phase separation and performance decay during thermal storage. Thus, this research investigated macro performance, micro-characteristic indexes, and their correlation. Lastly, this study provided prediction models based on conventional physical indexes using PLS regression.

The results from the macro performance indexes indicated that ΔSP_{48} increased initially and decreased thereafter to even below 2.5 °C after an 8-day storage. The largest ΔSP was observed on the 4th day of storage. Ductility either at the top or bottom decreased with an increase in storage time. Penetration either at the top or bottom slightly decreased initially and increased thereafter. For rheological indexes, Jnr increased, while R and

$G^*/\sin\delta$ decreased with an increase in storage time. The results implied a worse high-temperature performance after long-term storage, and the values of the bottom samples decayed more substantially than those of the top owing to SBS flowing up.

The results from the micro-characteristic indexes revealed that the network structure strengths were reduced with an increase in storage time either at the top or bottom, except the one at the top on the second day. The top on the first two days can easily contact oxygen, thereby leading SBS to undergo oxidation although the swelling degree and SBS amount was not changed substantially. For the SBS swelling degree, the values of the top were significantly higher than those of the bottom. Meanwhile, the top showed higher SBS amount than that of the bottom, demonstrating SBS flowing up. Phase separation of SBS mainly occurred after a 4-day storage, and the decay of performance for SBSPMB was attributed to SBS degradation and base asphalt aging.

The grey correlation degree was used to find the relationship between SBS amount and other macro- and micro-indexes. The results showed that softening point, penetration, I, R_{avg}, and $G^*/\sin\delta$ either at the top or bottom part, as well as their segregation degree, showed a close relationship with SBS amount. Penetration, ductility, and softening point were used to build a prediction model for rheological properties after long-term storage using PLS. The model can predict I, R, Jnr, $G^*/\sin\delta$, and SBS amount effectively. $G^*/\sin\delta$ and I at the top or bottom storage location of SBSPMB have high coefficients with SBS amount. Thus, they were better in evaluating the phase separation of SBSPMB during thermal storage. This study will be beneficial in obtaining an improved understanding of the durability of PMB and controlling its quality in the field.

Author Contributions: Writing—original draft preparation, P.W. and X.-Y.L.; writing—review and editing, P.W. and H.-R.W.; visualization, R.-B.R.; supervision, L.-Z.W.; project administration, R.-B.R.; funding acquisition, P.W. All authors have read and agreed to the published version of the manuscript.

Funding: The research was funded by National Natural Science Foundation of China (grant no. 51808322), Key Program of Shandong Provincial Natural Science (grant no. ZR2020KE007), and Shandong Provincial Young Scholars Innovative Research Team Development Program in Colleges and Universities (grant no. 2019KJG004).

Institutional Review Board Statement: Ethical review and approval were waived for this study, due to the studies not involving humans or animals.

Informed Consent Statement: Informed consent was obtained from all subjects involved in the study.

Data Availability Statement: The data used to support the findings of this study are included within the article.

Acknowledgments: This study was completed at the School of Transportation Engineering in Shandong Jianzhu University.

Conflicts of Interest: The authors declare no conflict of interest.

References

1. Chen, J.S.; Liao, M.C.; Shiah, M.S. Asphalt modified by styrene-butadiene-styrene triblock copolymer: Morphology and model. *J. Mater. Civ. Eng.* **2002**, *14*, 224–229. [CrossRef]
2. Ali, M.B.; Hamad, A.A.; Abdulrahman, S.A. Comparison of rheological properties for polymer modified asphalt produced in Riyadh. *Int. J. Civ. Environ. Struct. Constr. Archit. Eng.* **2016**, *10*, 197–201.
3. Zhao, Y.J.; Jiang, J.W.; Dai, Y.Q.; Lan, Z.; Ni, F.J. Thermal property evaluation of porous asphalt concrete based on heterogeneous meso-structure finite element simulation. *Appl. Sci.* **2020**, *10*, 1671. [CrossRef]
4. Dai, Y.Q.; Jiang, J.W.; Gu, X.Y.; Zhao, Y.J.; Ni, F.J. Sustainable urban street comprising permeable pavement and bioretention facilities: A practice. *Sustainability* **2020**, *12*, 8288. [CrossRef]
5. Zhu, J.; Birgisson, B.; Kringos, N. Polymer modification of bitumen: Advances and challenges. *Eur. Polym. J.* **2014**, *54*, 18–38. [CrossRef]
6. Xia, T.; Xu, J.; Huang, T. Viscoelastic phase behavior in sbs modified bitumen studied by morphology evolution and viscoelasticity change. *Constr. Build. Mater.* **2016**, *105*, 589–594. [CrossRef]

7. Zhu, J.; Kringos, N. Towards the development of a viscoelastic model for phase separation in polymer-modified bitumen. *Road Mater. Pavement.* **2015**, *16*, 39–49. [CrossRef]
8. Lu, X.; Isacsson, U.; Ekblad, J. Phase separation of SBS polymer modified bitumens. *J. Mater. Civ. Eng.* **1999**, *11*, 51–57. [CrossRef]
9. Liu, Y.; Zhang, J.; Jiang, Y. Investigation of secondary phase separation and mechanical properties of epoxy SBS-modified asphalts. *Constr. Build. Mater.* **2018**, *165*, 163–172. [CrossRef]
10. Leng, Z.; Tan, Z.; Yu, H.; Guo, J. Improvement of storage stability of SBS-modified asphalt with nanoclay using a new mixing method. *Road Mater. Pavement* **2019**, *20*, 1601–1614. [CrossRef]
11. Nciri, N.; Kim, N.; Cho, N. New insights into the effects of styrene-butadiene-styrene polymer modifier on the structure, properties, and performance of asphalt binder: The case of ap-5 asphalt and solvent deasphalting pitch. *Mater. Chem Phys.* **2017**, *193*, 477–495. [CrossRef]
12. Li, Y.; Li, L.; Zhang, Y. Improving the aging resistance of styrene-butadiene-styrene tri-block copolymer and application in polymer-modified asphalt. *J. Appl. Polym. Sci.* **2010**, *116*, 754–761. [CrossRef]
13. Zhu, J.; Lu, X.; Kringos, N. Experimental investigation on storage stability and phase separation behaviour of polymer-modified bitumen. *Int. J. Pavement Eng.* **2016**, *19*, 832–841. [CrossRef]
14. Ali, M.B.; Suhana, K.; Nor, H.R.S.; Mohamed, R.K. Optimization of mixing time for polymer modified asphalt. *Mater. Sci. Eng.* **2019**, *512*, 012030.
15. Fu, H.Y.; Xie, L.D.; Dou, D.Y.; Li, L.F.; Yu, M.; Side, Y. Storage stability and compatibility of asphalt binder modified by SBS graft copolymer. *Constr. Build. Mater.* **2007**, *21*, 1528–1533. [CrossRef]
16. Ren, S.S.; Liu, X.Y.; Fan, W.Y.; Wang, H.P.; Sandra, E. Rheological properties, compatibility, and storage stability of SBS latex-modified asphalt. *Materials* **2019**, *12*, 3683. [CrossRef]
17. Zhang, F.; Yu, J.Y.; Wu, S.P. Effect of ageing on rheological properties of storage-stable SBS/sulfur-modified asphalt. *J. Hazard. Mater.* **2010**, *182*, 507–517. [CrossRef]
18. Liang, M.; Xin, X.; Fan, W. Effects of polymerized sulfur on rheological properties, morphology and stability of SBS modified asphalt. *Constr. Build. Mater.* **2017**, *150*, 860–871. [CrossRef]
19. Carcer, I.A.; De Masegosa, R.M.; Vinas, M.T. Storage stability of SBS/sulfur modified bitumens at high temperature: Influence of bitumen composition and structure. *Constr. Build. Mater.* **2014**, *52*, 245–252. [CrossRef]
20. Larsen, D.O.; Alessandrini, J.L.; Bosch, A.; Cortizo, M.S. Micro-structural and rheological characteristics of SBS-asphalt blends during their manufacturing. *Constr. Build. Mater.* **2009**, *23*, 2769–2774. [CrossRef]
21. Lu, H.; Ye, F.; Yuan, J.; Yin, W. Properties comparison and mechanism analysis of naphthenic oil/SBS and nano-mmt/SBS modified asphalt. *Constr. Build. Mater.* **2018**, *187*, 1147–1157. [CrossRef]
22. Tarsi, G.; Varveri, A.; Lantieril, C.; Scarpas, A. Effects of different aging methods on chemical and rheological properties of bitumen. *J. Mater. Civ. Eng.* **2018**, *30*, 4018009–4018017. [CrossRef]
23. Zhang, D.; Zhang, H.; Zhu, C.; Shi, C. Synergetic effect of multi-dimensional nanomaterials for anti-aging properties of SBS modified bitumen. *Constr. Build. Mater.* **2017**, *144*, 423–431. [CrossRef]
24. Martínez-Anzures, J.D.; Zapién-Castillo, S.; Salazar-Cruz, B.A. Preparation and properties of modified asphalt using branch SBS/nanoclay nanocomposite as a modifier. *Road Mater. Pavement* **2019**, *20*, 1275–1290. [CrossRef]
25. Goli, A.; Ziari, H.; Amini, A. Influence of carbon nanotubes on performance properties and storage stability of SBS modified asphalt binders. *J. Mater. Civ. Eng.* **2017**, *29*, 4017070. [CrossRef]
26. Liang, M.; Liang, P.; Fan, W. Thermo-rheological behavior and compatibility of modified asphalt with various styrene–butadiene structures in SBS copolymers. *Mater. Des.* **2015**, *88*, 177–185. [CrossRef]
27. Zhu, J.; Lu, X.; Balieu, R.; Kringos, N. Modelling and numerical simulation of phase separation in polymer modified bitumen by phase-field method. *Mater. Des.* **2016**, *107*, 322–332. [CrossRef]
28. Liang, M.; Xin, X.; Fan, W.; Wang, H.; Sun, W. Phase field simulation and microscopic observation of phase separation and thermal stability of polymer modified asphalt. *Constr. Build. Mater.* **2019**, *204*, 132–143. [CrossRef]
29. Singh, S.K.; Kumar, Y.; Ravindranath, S.S. Thermal degradation of SBS in bitumen during storage: Influence of temperature, SBS concentration, polymer type and base bitumen. *Polym. Degrad. Stab.* **2018**, *147*, 64–75. [CrossRef]
30. Masson, F.; Collins, P.; Robertson, G.; Woods, J.R.; Margeson, J. Thermodynamics, phase diagrams, and stability of bitumen−polymer blends. *Energy Fuel* **2003**, *17*, 714–724. [CrossRef]
31. Yan, C.; Huang, W.; Lin, P.; Zhang, Y.; Lv, Q. Chemical and rheological evaluation of aging properties of high content SBS polymer modified asphalt. *Fuel* **2019**, *252*, 417–426. [CrossRef]
32. Wekumbura, C.; Stastna, J.; Zanzotto, L. Destruction and recovery of internal structure in polymer-modified asphalts. *J. Mater. Civ. Eng.* **2007**, *19*, 227–232. [CrossRef]
33. Sengoz, B.; Isikyakar, G. Analysis of styrene-butadiene-styrene polymer modified bitumen using fluorescent microscopy and conventional test methods. *J. Hazard. Mater.* **2008**, *150*, 424–432. [CrossRef]
34. Sugano, M.; Iwabuchi, Y.; Watanabe, T. Relations between thermal degradations of SBS copolymer and asphalt substrate in polymer modified asphalt. *Clean Technol. Environ.* **2010**, *12*, 653–659. [CrossRef]
35. Wang, P.; Dong, Z.; Tan, Y.; Liu, Z. Anti-ageing properties of styrene–butadiene–styrene copolymer-modified asphalt combined with multi-walled carbon nanotubes. *Road Mater. Pavement Design.* **2016**, *18*, 533–549. [CrossRef]

36. Wang, P.; Shil, F.; Liu, X. Role of aliphatic chain characteristics on the anti-cracking properties of polymer-modified asphalt at low temperatures. *Polymers* **2019**, *11*, 31817766. [CrossRef] [PubMed]
37. Wang, K.; Yuan, Y.; Han, S.; Yang, Y. Application of FTIR spectroscopy with solvent-cast film and PLS regression for the quantification of SBS content in modified asphalt. *Int. J. Pavement Eng.* **2019**, *20*, 1336–1341. [CrossRef]

sustainability

Article

Comprehensive Life Cycle Environmental Assessment of Preventive Maintenance Techniques for Asphalt Pavement

Mulian Zheng [1,*], Wang Chen [1], Xiaoyan Ding [2], Wenwu Zhang [2] and Sixin Yu [2]

[1] Key Laboratory for Special Region Highway Engineering of Ministry of Education, Chang'an University, Mid-South Erhuan Road, Xi'an 710064, China; chris_chenwang@chd.edu.cn
[2] Shandong Hi-Speed Group, Jinan 250098, China; dingxiaoyan_sdgs@outlook.com (X.D.); zhangwenwu_sdgs@outlook.com (W.Z.); yusixin_sdgu@outlook.com (S.Y.)
* Correspondence: zhengml@chd.edu.cn; Tel.: +86-029-8233-4846

Abstract: Preventive maintenance (PM) is regarded as the most economical maintenance strategy for asphalt pavement, but the life cycle environmental impacts (LCEI) of different PM techniques have not yet been comprehensively assessed and compared, thus hindering sustainable PM planning. This study aims to comprehensively estimate and compared the LCEI of five PM techniques then propose measures to reduce environmental impacts in PM design by using life cycle assessment (LCA), including fog seal with sand, micro-surfacing, composite seal, ultra-thin asphalt overlay, and thin asphalt overlay. Afterwards, ten kinds of LCEI categories and energy consumption of PM techniques were compared from the LCA phases, and inventory inputs perspectives, respectively. Results show that fog seal with sand and micro-surfacing can lower all LCEI scores by more than 50%. The environmental performance of five PM techniques provided by sensitivity analysis indicated that service life may not create significant impact on LCA results to some extent. Moreover, four PM combination plans were developed and compared for environmental performance, and results show that the PM plan only includes seal coat techniques that can reduce the total LCEI by 7–29% in pavement life. Increasing the frequency of seal coat techniques can make the PM plans more sustainable.

Keywords: life cycle assessment; asphalt pavement; preventive maintenance; sustainable; environmental impacts; sensitivity analysis

Citation: Zheng, M.; Chen, W.; Ding, X.; Zhang, W.; Yu, S. Comprehensive Life Cycle Environmental Assessment of Preventive Maintenance Techniques for Asphalt Pavement. *Sustainability* **2021**, *13*, 4887. https://doi.org/10.3390/su13094887

Academic Editors: Joel R. M. Oliveira, Hugo Silva, R. Christopher Williams, Zejiao Dong and Adelino Jorge Lopes Ferreira

Received: 21 March 2021
Accepted: 22 April 2021
Published: 27 April 2021

1. Introduction

With the ever-increasing service time, different types and degrees of distress appear on the surface of asphalt pavement, such as rutting and cracking [1]. In the past ten to twenty years, milling and resurfacing have been used as the primary structural rehabilitation techniques [2]. However, considering the life cycle cost-effectiveness of asphalt pavement, preventive maintenance (PM) is widely used in the early distress treatment of asphalt pavement to restore pavement performance in time [3,4]. The PM techniques mainly include fog seal, micro-surfacing, gravel seal, composite seal, ultra-thin asphalt overlay, and thin asphalt overlay [5]. Among these techniques, asphalt overlay has been widely used in PM activities for high-grade asphalt pavements. Furthermore, these techniques are required to heat up the asphalt and aggregate, which are not conducive to energy saving and environmental protection. In contrast, seal coat techniques (fog seal, micro-surfacing, gravel seal, composite seal) are different from asphalt overlay in that they may not require heating of raw materials, reducing energy consumption and emissions.

To extend the service life of pavement and achieve sustainable maintenance management, PM techniques usually need to be implemented serval times in pavement life cycle. Currently, micro-surfacing and asphalt overlay are still considered as the primary PM techniques in designing maintenance plans because they offer significant advantages in restoring pavement surface conditions. However, the fog seal has obvious advantages in suppressing the deterioration of early cracks, and the composite seal also has better economic benefits than asphalt overlay, giving the seal coat techniques great potential application in future PM planning. In contrast, although asphalt overlay possesses better performance, their implementation will consume more natural resources, especially bitumen and modified bitumen, which are harmful to the environment. The production of bitumen will emit large amounts of carbon oxides, nitrogen oxides, and sulfides, which are important components of global warming, acidification, and photochemical pollution [6–9].

In the field of eco-friendly pavement research, it is necessary to conduct a life cycle environmental assessment (LCEA) of different maintenance techniques to characterize their sustainability. Life cycle assessment (LCA) is a commonly used method to assess the environmental impacts of a complete product throughout its life cycle stages [9]. Häkkinen and Mäkelä [10] first introduced the LCA method to the field of pavement engineering, aiming to evaluate the differences in energy consumption between asphalt pavement and cement concrete pavement. Since then, the LCA method has attracted much academic attention for quantifying the life cycle environmental impacts (LCEI) of road infrastructure. The main application areas include LCEI evaluation of road projects or alternative designs [11–14], environmental assessment of maintenance and rehabilitation alternatives [15–18], and materials sustainability evaluation and comparison [19–22], etc. With the accelerated development of pavement maintenance techniques and the increasing public and governmental concern for road environment, the LCA method has been gradually applied [23]. Simões, et al. [24] conducted a sustainability analysis for four different micro-surfacing structures by using LCA method. Ma, et al. [25] adopted the LCA method to compare the LCEI and energy consumption of warm-mixed asphalt mixture (WMA) and hot-mixed asphalt mixture (HMA). Cong, et al. [26] evaluated the environmental emissions and energy consumption of polyurethane pavements over the life cycle. Furthermore, LCA was also usually combined with life cycle costs (LCC); for example, Cao, et al. [27] applied the Eco-efficiency analysis (EEA) framework to evaluate the eco-efficiency of hot-in-place recycling and milling-and-filling under variable lifespan in conjunction with LCA. Santos, et al. [28] developed a highly integrated LCC-LCA model to evaluate the sustainability of five pavement maintenance techniques, and proposed the most suitable maintenance strategy for decision-makers. It is worth noting that the seal coat techniques are different from the traditional asphalt overlay techniques in that it requires fewer raw materials, but some polymers are necessary to be added to enhance its performance. As Santos, et al. [29] pointed out, the raw material and use phases of pavement are the main phases of LCEI. Wang, et al. [30] evaluated the LCEI of various polymer mixtures (improve the anti-rutting performance), and found that polymers may cause intensive greenhouse gas emissions and energy consumption. Samieadel and Fini [31] pointed out that the production of bio-adhesives would reduce five times for carbon dioxide emissions and three times for methane to the environment.

Until now, the focus points of existing pavement LCA studies are mainly restorative maintenance techniques and their environmental assessments, with the evaluation categories mostly limited to energy consumption or carbon dioxide emissions [25,32–35]. In addition, as for pavement PM techniques, Ma, et al. [36] estimated greenhouse gas emissions of 16 pavement maintenance techniques with the system boundary limited to cradle-to-gate, and indicated that increasing the frequency of PM techniques could achieve low-emissions maintenance planning. In addition, similar work was carried out by Qi [37] and Han, et al. [38]. Furthermore, Wang, et al. [39] pointed out that PM techniques are essential to reduce energy consumption by analyzing the energy consumption of pavement projects in Inner Mongolia, China. Overall, it is well known that PM plays an important role in mitigating environmental emissions throughout the whole life cycle of asphalt pavement. Although LCA of pavement PM techniques has been carried out by some researchers, previous studies have failed to assess the comprehensive environmental impacts of commonly used PM techniques, leading to a lack of LCEI inventories of PM techniques, which are a critical database for sustainable maintenance planning in the pavement life cycle [40,41].

Given this, five typical PM techniques were selected and a life cycle environmental impact assessment was conducted. The main objectives of this work are (1) to quantify the LCEI of five typical pavement PM techniques, covering global warming potential, acidification, human health, and energy consumption, and (2) to identify the PM techniques that have the lowest environmental impacts, and propose measures to reduce environmental impacts in designing PM plans throughout the pavement life cycle. Results will help highway management agencies to achieve more sustainable pavement PM management.

2. Materials and Methods

LCA was used to evaluate and compare the environmental performance of different PM techniques (List of the abbreviations used in this work is presented in Appendix A). In this study, LCA was carried out according to the ISO 14,040 series of standards [42], and ISO14044 [42] defines a typical framework for LCA studies, including goals and scope definition, life cycle inventory (LCI) analysis, life cycle impact assessment (LCIA) and life cycle interpretation.

In this study, a life cycle environmental assessment of typical PM techniques was conducted; the overall framework for LCEA of PM techniques is shown in Figure 1. The system boundaries can be determined by investigating pavement PM history, which allowed access to commonly-used PM techniques and their life cycle phases. The inventory inputs and outputs can be calculated by using the quota methods. The quota methods are based on the current Chinese Specifications for *Budget Quota of Highway Engineering (JTG/T 3832-2018)* [43] and *Budget Quota of Highway Engineering Machinery Shift (JTG/T 3833-2018)* [44]. Raw materials can be obtained by collecting maintenance history, and the fuel consumption of each life cycle phase in a given unit can be determined according to the construction process specified in the budget quota. By combining the LCA database of pavement construction materials and fuels published on a global scale, the LCEI of different PM techniques in each phase can be calculated. Based on that, the LCEI of PM techniques were compared from the LCA phases, inventory inputs, and service life attenuation perspectives, respectively. Finally, methods to reduce the LCEI in the pavement life cycle were proposed by comparing the environmental performance of different PM technique combination plans.

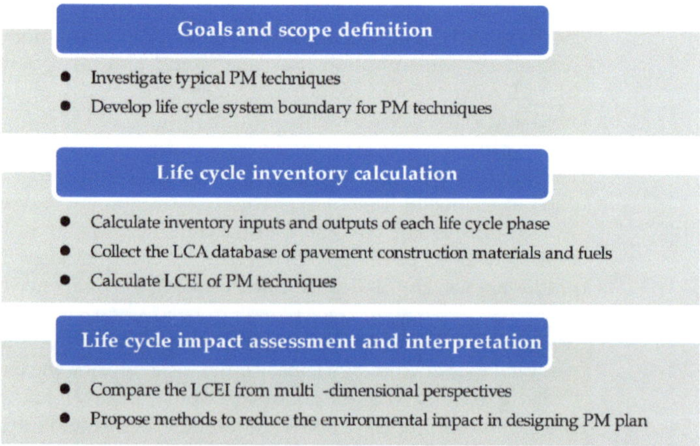

Figure 1. Overall framework for LCEA of PM techniques in this study.

2.1. PM Techniques and Materials

PM techniques involved in this study are mainly commonly applied to high-grade asphalt pavement. Generally, these PM techniques can be classified into the following three categories according to maintenance requirements and their thickness [45,46], as shown in Figure 2.

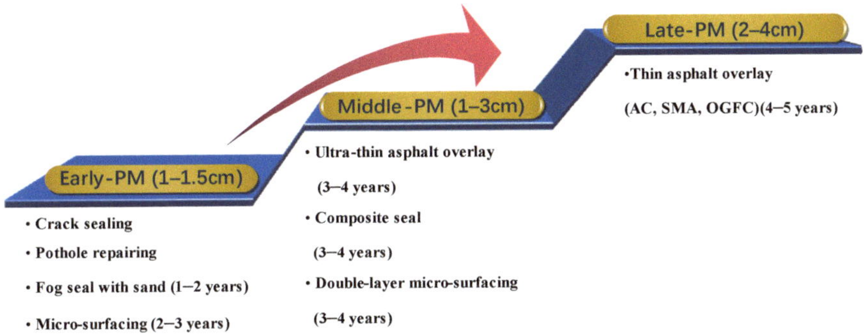

Figure 2. Main PM techniques for high-grade highways.

- The early-type (early-PM)

The early-PM refers to light intensity maintenance measure, it ranges from daily maintenance technique to seal coat technique (1–1.5 cm of the thickness), such as crack sealing, fog seal with sand, micro-surfacing, which are mainly suitable for early distress treatment. The main raw materials include asphalt, crushed stone, modifiers, and adhesives (polymers), which are added to ensure their performance [47,48].

- The middle-type (middle-PM)

The middle-PM refers to moderate intensity maintenance measure with a thickness of about 1~3 cm, including ultra-thin asphalt overlay, composite seal, and double-layer micro-surfacing, which are important for national and provincial highways due to its effectiveness in slowing down the deterioration of pavement cracks, and restoring flatness and roughness for asphalt pavement. There are two kinds of structures in composite seal, including gravel seal + micro-surfacing, gravel seal + slurry seal. Studies have shown that the combination of rubber asphalt gravel seal and micro-surfacing performs better water stability and bonding ability, as well as fast construction speed [49].

- The late-type (late-PM)

The late-PM refers to strong intensity maintenance measure, aiming to manage serious diseases on the upper layer of high-grade asphalt pavement, but the performance of the rest pavement structure is still relatively intact. In this case, part of the surface layer can be milled and sprinkled with a thin layer of asphalt overlay (thickness is generally 2~4 cm). The late-PM technique includes AC, SMA, and OGFC, which usually require heating and mixing of asphalt and stone in the plant [50].

In this study, based on the experience of preventive maintenance for high-grade asphalt pavement in China, five commonly-used PM techniques were selected from the above three categories, including fog seal with sand, micro-surfacing (1 cm), composite seal (2 cm), ultra-thin asphalt overlay (2.5 cm), and thin asphalt overlay (3 cm). The materials and their compositions of each PM technique were obtained by investigating the maintenance history of Shandong Province from 2015 to 2020 (Table 1).

Table 1. Compositions of individual PM techniques (unit: ton/1000 m^2). (Note: the materials, and corresponding compositions in this table were obtained from maintenance history).

Raw Material	Fog Seal with Sand	Micro-Surfacing	Composite Seal	Ultra-Thin Asphalt Overlay	Thin Asphalt Overlay
Emulsified Bitumen	0.53	1.91	1.91	-	-
Bitumen	-	-	1.24	-	4.22
Modified Bitumen	-	-	-	3.858	-
Crushed stone	-	16.24	25.24	31.848	49.84
Sand	0.80	-	-	-	-
Cement	-	0.24	0.24	-	-
Epoxy resin	0.02	-	-	-	-
Dummy rubber	-	-	0.36	-	-

2.2. *Goal and Scope Definition*

2.2.1. Goal

The main purpose of this study was to evaluate the LCEI of five commonly used PM techniques for high-grade asphalt pavement based on past historical maintenance data and standards, which can be summarized as following three aspects:

- Quantify the LCEI of five typical pavement PM techniques, covering global warming potential, acidification, human health, and energy consumption, etc., which can be used to establish comprehensive LCA inventory for sustainable pavement management.
- Conduct sensitivity analysis to explore the impact of PM techniques performance variations on LCA results.
- Identify the PM techniques that have the lowest environmental impacts, and propose measures to reduce environmental impacts in designing PM plans throughout the pavement life cycle.

2.2.2. System Boundary

The LCA in this study was conducted according to the method from cradle-to-grave, for pavement projects, the life cycle starts from the extraction of raw materials and ends at the end-of-life [51]. Figure 3 shows the system boundary, six stages were included: (i) raw materials extraction; (ii) the mixture production at the asphalt mixing plant; (iii) the mixture transportation from the mixing plant to the construction site; (iv) construction; (v) maintenance; (vi) solid waste demolition and transportation at the end-of-life. In addition, the boundary of the pavement structure is limited to the PM paving structure layer.

2.2.3. Functional Unit

In this study, the functional unit was defined as the provision of the typical expressway section used for pavement maintenance quota calculation in China, different PM techniques were considered for a one-way, two-lane with a hard shoulder expressway pavement. Single carriageway width is 3.75 m, hard shoulder width is 2.5 m, and the length of this area is 100 m for typical PM activity. That is, a total pavement area is constant at 1000 m^2. Besides, the thickness of micro-surfacing, ultra-thin asphalt, composite seal, thin asphalt overlay was 1 cm, 2 cm, 2.5 cm, 3 cm, respectively.

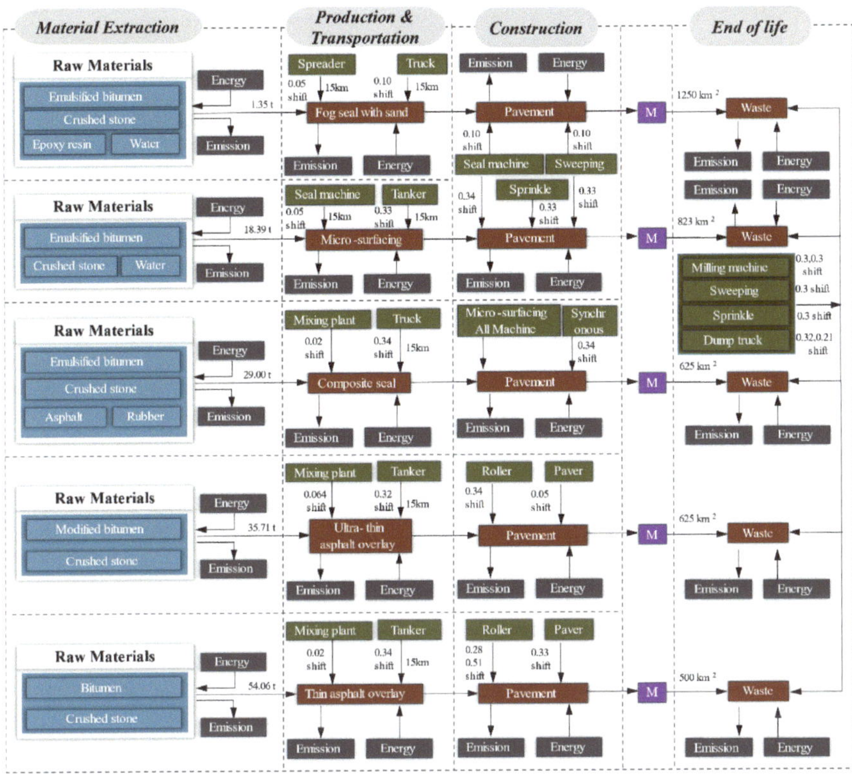

Figure 3. The system boundary involved in five PM techniques. (Note: the materials, equipment, and corresponding parameters in this figure were determined based on the method in Section 2.3).

2.3. Life Cycle Inventory Calculation Model

The parameters of equipment for each pavement construction or maintenance project can be determined by the quota in China [43,44], including the type of equipment, the number of mechanical shifts, and the fuel consumption per shift, which can minimize the variability of construction processes due to geography as possible. This provides great convenience for the development of LCA, and once sufficient field observation data are not available, the LCI can be calculated based on the relevant parameters recorded in the quota. Thus, in this study, a calculation model was used to quantify the energy consumption and emissions per functional unit for each PM technique (List of the notations used in this work is presented in Appendix B), as shown in Equations (1) and (2). Furthermore, Equation (2) shows the calculation method of the emission value for any kind of emissions.

$$\text{Energy consumption (MJ/Funtion)} = \sum_{p=1}^{P}\sum_{j=1}^{J}\sum_{i=1}^{I}\left(ME_i \times Q_{i,p} + \sum_{f=1}^{F}\left(FE_f \times FC_{f,j,p} \times C_{j,p}\right)\right) \quad (1)$$

$$\text{One type of embodied emmission (g/Funtion)} = \sum_{p=1}^{P}\sum_{j=1}^{J}\sum_{i=1}^{I}\left(MG_i \times Q_{i,p} + \sum_{f=1}^{F}\left(FG_f \times FC_{f,j,p} \times C_{j,p}\right)\right) \quad (2)$$

It is worth noting that the material consumption of each PM technique with different corresponding thicknesses under the same functional unit can be acquired in maintenance history data, and the mechanical shifts can be found in the quota [43,44,52]. Therefore, combining the LCI of different materials and fuels, all LCEI categories in each life cycle phase can be calculated.

2.3.1. Phase 1: Raw Materials Extraction

This stage involves the LCEI generated by raw materials. All LCEI of PM techniques were considered in this study, the LCI of each raw material was collected mainly according to the LCA database from public reports published by research institutions. According to the trade-off rules in the current life cycle analysis, the upstream production data of the material can be ignored as the mass of ordinary materials is less than 1% of the whole product [53]. Thus, based on the identified system boundary, the authors collected four types of environmental profiles for the different elements considered in this study, as follows:

- Bitumen, emulsified bitumen, and modified bitumen [54];
- Crushed stone [55];
- Cement, epoxy resin, and sand [56];
- Dummy rubber [57].

Environmental impacts from phases such as transportation and construction are mainly a conversion of energy consumed by vehicles and equipment, and the types of energy consumed in these processes include diesel, petrol, electricity, and heavy oil. Based on the emission factors provided by the National Climate Center of China, the calorific values per unit of main energy types can be derived [26], as illustrated in Table 2. In addition, the environmental impacts per calorific value of each energy type can be obtained from Ecoinvent [56].

Table 2. Calorific value of main energy types.

Energy Type	Calorific Values [26]	Unit	LCA Database
Diesel	42.7	MJ/Kg	Ecoinvent [56]
Petrol	43.1	MJ/Kg	Ecoinvent [56]
Electricity	3.6	MJ/KWh	Ecoinvent [56]
Heavy oil	41.8	MJ/Kg	Ecoinvent [56]

2.3.2. Phase 2: Mixture Production

The mixture production may vary significantly due to the various construction process requirements for each PM technique. More specifically, fog seal with sand and micro-surfacing do not require heating the aggregate and bitumen. However, ultra-thin asphalt overlay, thin asphalt overlay, and composite seal (the rubber asphalt gravel used for the lower seal coat layer) require heating of raw materials at this phase. In this case, mechanical shifts and fuel consumption can be determined by using the quota methods.

2.3.3. Phase 3: Mixture Transportation

This study focuses on transportation from the mixing plant to the construction site in the transportation phase. The environmental impacts mainly come from the fuel consumption of vehicles. It is worth noting that the transportation of composite seal needs to be divided into two steps, which belong to the construction process of micro-surfacing and gravel seal, respectively. Transportation equipment includes spreaders, sealers, cargo trucks, liquid bitumen tankers, dump trucks, while other facilities with lower energy consumption were ignored. Through the investigation of a PM construction project in Shandong, China, it was found that in most projects, the average transportation distance between the cement stabilized material mixing plant and the construction site was 10 km, and the distance between the asphalt mixture mixing plant and the construction site was 15 km. Thus, it is assumed that the transportation distance for each PM technique is 15 km.

2.3.4. Phase 4: Construction

For the sake of minimizing the differences in construction processes due to construction policy, climate environment, regional economy, etc., the authors selected representative construction techniques and machinery used for pavement maintenance projects. Thus, the construction equipment categories, fuel types, and consumption values, as well as machinery shifts for each technique were collected by querying quotas. Consequently, the mass of each kind of fuel for all PM techniques during the construction can be obtained.

2.3.5. Phase 5: Maintenance

The performance of asphalt pavement will be gradually decreased under repeated traffic loads, so maintenance activities should be implemented on the old pavement. The Chinese *Specifications for maintenance design of highway asphalt pavement* (JTG 5421-2018) [58] summarized the service life of typical maintenance techniques from long term practice in China; in particular, the service life of fog seal with sand, micro-surfacing, composite seal, ultra-thin asphalt overlay, and thin asphalt overlay were fixed to 1–2 years, 2–3 years, 3–4 years, 3–4 years, and 4–5 years. In this study, the maximum service life of each technique was chosen to calculate the environmental impacts generated during the maintenance phase. More specifically, the service life of fog seal with sand, micro-surfacing, composite seal, ultra-thin asphalt overlay, and thin asphalt overlay are 2 years, 3 years, 4 years, 4 years, and 5 years, respectively. Furthermore, the PM techniques with the longest service life were selected as the benchmark, so the analysis period was 5 years. It is assumed that the pavement after the implementation of PM technique is only maintained once during the analysis period, and the maintenance times for the remaining techniques is the ratio of the corresponding service life to the benchmark.

In the choice of maintenance techniques, only repeated overlay was selected—that is, it is consistent with the previous maintenance technique. To simplify the subsequent calculation of LCA, the ratio of maintenance times was regarded as the ratio of the maintenance area since the former may not be an integer value. Due to the lack of sufficient historical data on re-maintenance after the implementation of PM techniques, the definition method of maintenance area for the maintenance phase in previous studies was adopted in this study [26,59]. Hence, it is assumed that the basic maintenance area is 50% of the pavement unit (1000 m^2). Through gathering the LCI of individual techniques in the whole construction phase, including raw material extraction, mixture production, transportation, and construction. The LCEI of each PM technique in the maintenance phase can be calculated according to the maintenance area.

2.3.6. Phase 6: End of Life

The pavement is required to be milled as its performance decreases to the limit value. In this study, the application of the waste materials was not considered, and only the milling and solid waste transportation were considered to calculate the LCEI. In addition, through the investigation of the maintenance projects in Shandong, China, it was found that the transportation distance from the road demolition sites to recycling plants was about 20 km–40 km. In this instance, the intermediate value (30 km) was used to represent the transportation distance. It should be mentioned that not all pavements were required to be milled at this phase; for example, the fog seal with sand was paved on the asphalt pavement then it will form a super-thin layer, and the thickness can be ignored at the end-of-life phase since it is subjected to ever-increasing traffic loads, which means that there was no milling phase for fog seal with sand. As for the other techniques, the milling area is the sum of the area in the original pavement area (construction phase) and additional pavement area (maintenance phase), and the milling thickness is the weighted average thickness.

Based on the above, the construction equipment required by different PM techniques at each phase of the life cycle can be obtained. Then, the corresponding calorific consumption value also can be calculated by Equation (3), and results were presented in Table 3.

$$\text{Calorific consumption value of each machine (MJ)} = FE_f \times FC_{f,j,p} \times C_{j,p} \quad (3)$$

Table 3. The Calorific consumption value of each machine in the life cycle stage of PM (unit: MJ). (Note: the basic data used to calculate this table were from the Budget quota of highway engineering machinery shift (JTG/T 3833-2018)).

Phase	Items	Fog Seal with Sand	Micro-Surfacing	Composite Seal	Ultra-Thin Asphalt Overlay	Thin Asphalt Overlay	Type of Fuels
Mixture production	Asphalt mixing plant (240 t/h)				2.8×10^4		Heavy oil
	Asphalt mixing plant (240 t/h)				9.0×10^2		Electricity
	Asphalt mixing plant (30 t/h)					2.7×10^4	Heavy oil
	Asphalt mixing plant (30 t/h)					1.6×10^3	Electricity
	Asphalt mixing plant (320 t/h)			1.2×10^4			Heavy oil
	Asphalt mixing plant (320 t/h)			3.7×10^2			Electricity
	Wheeled loader (2.0 m^3) ZL-40				5.7×10^2		Diesel
	Wheeled loader (3.0 m^3) ZL-50			2.3×10^3		2.7×10^3	Diesel
Materials transportation	Dump truck (18–20 t) BJ374			1.1×10^3			Diesel
	Dump truck (2–3 t)				7.2×10^1		Petrol
	Dump truck (20–30 t)				9.9×10^2	1.5×10^3	Diesel
	Fog seal spreader 500 L	7.8×10^1					Diesel
	Liquid bitumen tanker (7000 L)		6.4×10^2	6.4×10^2			Diesel
	Micro-surfacing seal machine		2.2×10^2	2.2×10^2			Diesel
	Truck (2 t)	8.7×10^1					Petrol
Construction	Asphalt mixture paver (3.6–4.5 m)					8.6×10^2	Diesel
	Dump truck (2–3 t)				1.5×10^2		Diesel
	Flatbed trailer group (30 t)				2.9×10^2		Diesel
	Fog seal spreader 500 L	1.6×10^3					Diesel
	Micro-surfacing seal machine		1.5×10^3	1.5×10^3			Diesel
	Motorized air compressor (12 m^3/min)			6.3×10^2			Diesel
	Pavement milling machine (500 mm)					3.5×10^2	Diesel
	Pavement sweeping machine (DF)					7.2×10^2	Diesel
	Asphalt mixture paver (S1800-2SJ)				7.1×10^2		Diesel
	Pavement sweeping machine (YD80Q-1)	7.0×10^1	2.3×10^2	5.7×10^2			Diesel
	Rubber-tire roller (16–20 t)			1.3×10^3			Diesel
	Smooth-wheel roller (10–12 t)					1.4×10^3	Diesel
	Smooth-wheel roller (6–8 t)					4.9×10^2	Diesel
	Sprinkler (4000–6000 L)				3.8×10^1	4.1×10^2	Petrol
	Sprinkler (6000–8000 L)	2.1×10^3	6.7×10^2	6.7×10^2			Diesel
	Sprinkler (8000–10,000 L)			1.1×10^3			Diesel
	Synchronous stone machine			1.9×10^3			Diesel
	Vibratory roller with double steel wheel (10–15 t)				1.2×10^3		Diesel
End of Life	Pavement milling machine (2000 mm)		4.1×10^3	3.8×10^3	3.5×10^3	3.5×10^3	Diesel
	Pavement milling machine (500 mm)		6.2×10^2	5.7×10^2	5.3×10^2	5.3×10^2	Diesel
	Pavement sweeping machine (DF)		1.3×10^3	1.2×10^3	1.1×10^3	1.1×10^3	Diesel
	Sprinkler (4000–6000 L) YGJ5102GSSEQ		7.4×10^2	6.8×10^2	6.3×10^2	6.3×10^2	Diesel
	Dump truck (12 t)		4.7×10^2	4.3×10^2	3.9×10^2	4.0×10^2	Diesel
	Dump truck (6 t)		6.7×10^2	6.1×10^2	5.6×10^2	5.7×10^2	Diesel

2.4. *Life Cycle Impact Assessment (LCIA)*

LCIA aims to comprehensively assess the potential LCEI of different PM techniques and their differences using the results of LCI, which shows the harmful emissions from various raw material production processes and energy consumption processes, including emissions to air, soil, and water. Therefore, the comprehensive LCEI results of different PM techniques should be classified and compared by using characterization factors. In this study, the impact assessment categories defined in TRACI v.2.1 were adopted [60], including Ozone depletion (OD), Global warming (GW), photochemical smog formation

(PSF), Acidification (AC), Eutrophication (EU), Human health carcinogenic (HHC), Human health noncarcinogenic (HHN), Human health particulate (HHP), Ecotoxicity (EC), Fossil fuel depletion (FFD). Besides, the total embodied energy (TEE) was also incorporated in this LCA assessment.

3. Results and Discussion

3.1. Total Life Cycle Assessment Results

Based on the LCI calculation method described in Section 2.3, the potential LCEI were calculated. Figure 4 shows the composition ratio of each PM technique to the LCEI categories. It can be found that the LCEI categories generated by PM techniques were mainly TEE, FFD, EC, GW, and PSF, with proportions of 23.7–43.1%, 20.3–28.6%, 31.5–44.2%, 4.0–5.0%, and 0.4–0.7%, respectively. However, FFD and EC were often neglected by previous research [25,26]. It is noteworthy that the composition ratios of the ultra-thin asphalt overlay and thin asphalt overlay were almost similar to those of LCEI, which stems from their similar construction technology processes and raw materials. However, the differences in the proportion of various indicators between the three seal coat techniques were significant.

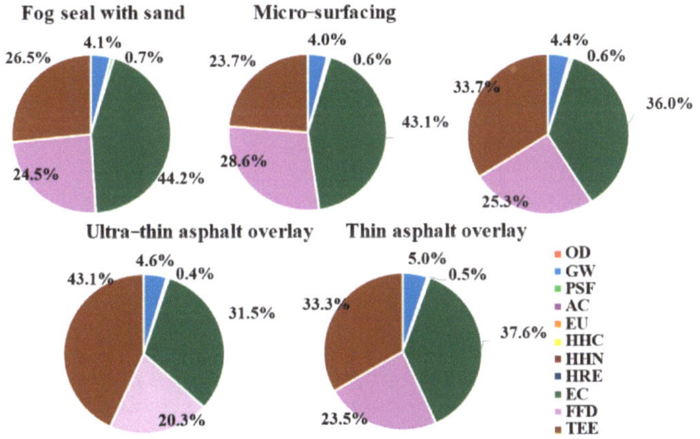

Figure 4. The LCEI categories composition of different PM techniques. (Note: the basic data used to calculate the results of this figure comes from Eurobitume [54], European Commission [55], Ecoinvent [56], National Renewable Energy Laboratory [57], National Climate Center of China [26], which were introduced in Section 2.3).

Figure 5 shows the relative LCEI of each PM technique, the principle of interpretation of the results was based on the impact scores of thin asphalt overlay, with a positive number indicating a benefit to environmental protection and the opposite being harmful to the environment. The results show that the fog seal with sand and the micro-surfacing have a great effect on the environment protection, with an improvement of more than 50% in all impact scores, followed by composite seal, and ultra-thin asphalt overlay. Indeed, the seal coat techniques can be considered as an eco-friendly alternative technique for PM because these three PM techniques can reduce the potential score in all impact categories. Among the ten LCEI categories, the most improvements are EU (57.5–87.6%), HRE (48.8–86.5%), GW (36.4–85.7%), OD (30.5–85.6%), AC (33.2–84.8%), and TEE (26.3–86.0%) also can be saved. Moreover, ultra-thin asphalt overlay exhibits a negative impact on energy saving (15.1% increase in TEE) due to the use of modified asphalt. However, although these five impact categories accounted for a large proportion of the total, there were also significant differences between each PM technique, such as the most significant reduction of FDD by fog seal with sand, which was about 30.2% and 60.1% higher than that of micro-surfacing and composite seal, respectively. This was due to the variability in the construction process and raw material composition of the different PM techniques. Thus, it is necessary to analyze the effects of different life cycle phases and inventory inputs in detail.

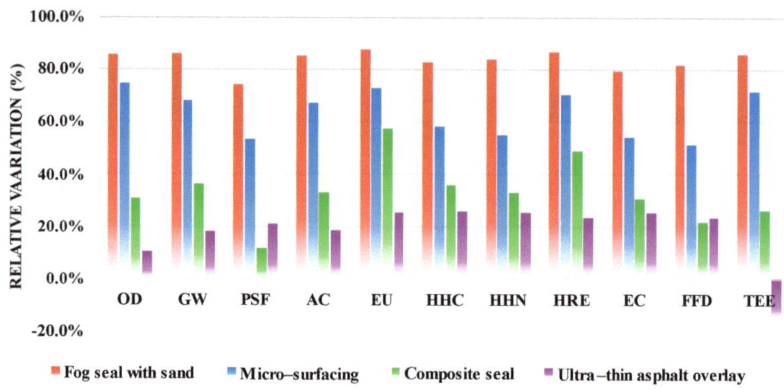

Figure 5. The LCEI of different PM techniques relative to thin asphalt overlay. (Note: the basic data used to calculate the results of this figure come from Eurobitume [54], European Commission [55], Ecoinvent [56], National Renewable Energy Laboratory [57], National Climate Center of China [26], which were introduced in Section 2.3).

3.2. Contribution Analysis

Figure 6 shows the relative contribution of the six phases to LCEI. It can be seen that the LCEI characteristics of different PM techniques are mainly generated by the raw material phase and maintenance phase, although the exact order varies depending on the impact category. Regardless of the PM techniques, the raw materials phase has a significance impact on EU, HHC, HHN, HRE, EC and FFD, with relative contributions of 35.6–59.3%, 36.0–59.2%, 42.0–62.2%, 24.9–50.6%, 40.3–59.8%, and 37.5–52.9%, respectively. While in the production phase, OD, GW, and AC were the main contributors with relative contributions of 26.2–44.1%, 16.3–25.8%, and 15.5–24.6%, respectively. In particular, TEE still accounts for a large proportion in this phase due to the mixture production. The construction phase and end-of-life phase mainly affect PSF and OD. Due to the differing performance (equivalent to service life), the maintenance phase of fog seal with sand has the largest contribution (55.6%), and the raw material phase was still the main contributor in terms of the other four main impact categories.

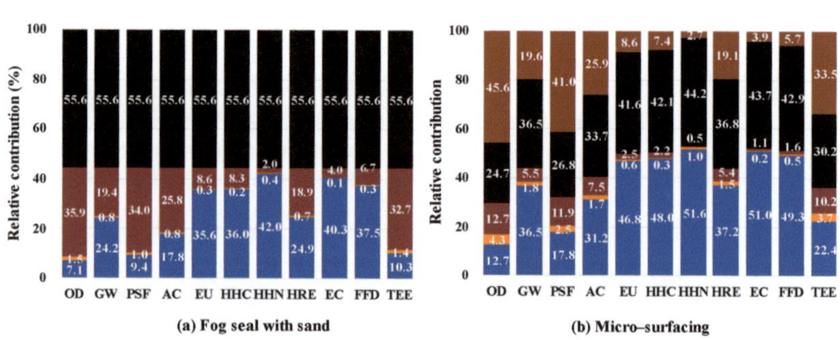

Figure 6. *Cont.*

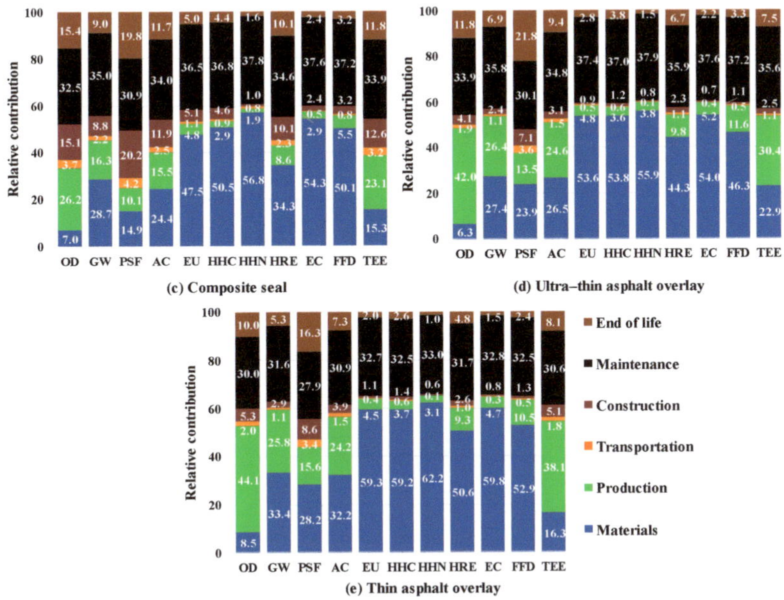

Figure 6. The contribution of different stages to the LCEI score. (Note: the basic data used to calculate the results of this figure come from Eurobitume [54], European Commission [55], Ecoinvent [56], National Renewable Energy Laboratory [57], National Climate Center of China [26], which were introduced in Section 2.3).

Figure 6 also shows that the LCEI of thin asphalt overlayer and ultra-thin asphalt overlay were extremely closed. Combined with Figure 5, it can be seen that the LCEI of the five PM techniques is as follows: fog seal with sand < micro-surfacing < composite seal < ultra-thin asphalt overlay < thin asphalt overlay. The main reasons can be concluded as follows:

1. Fog seal with sand requires the least amounts of raw materials and eliminates the production and end-of-life phase.
2. The raw material required by micro-surfacing is only higher than fog seal with sand, while the former can reduce the frequency of maintenance during the maintenance phase due to its longer service life.
3. For ultra-thin asphalt overlay and composite seal, the difference is mainly in the construction phase as the latter integrates the construction process of the gravel seal and micro-surfacing. However, the former consumes 1.22 times more mass of asphalt than the latter.

Therefore, in terms of environmental performance, compared to ultra-thin asphalt overlay, composite seal can be regarded as a sustainable alternative for middle-PM due to its lower price and similar service life. However, considering the social benefits, the sustainability of both techniques needs to be further evaluated.

In terms of the pavement life cycle from the raw materials extraction to the construction completion, the raw material phase plays an extremely important role in the LCEI results, and the production phase was also the main source of energy consumption generated by fuels. In order to analyze the LCEI of different PM techniques deeply, the contribution of each inventory inputs to the LCEI was calculated, as shown in Figure 7.

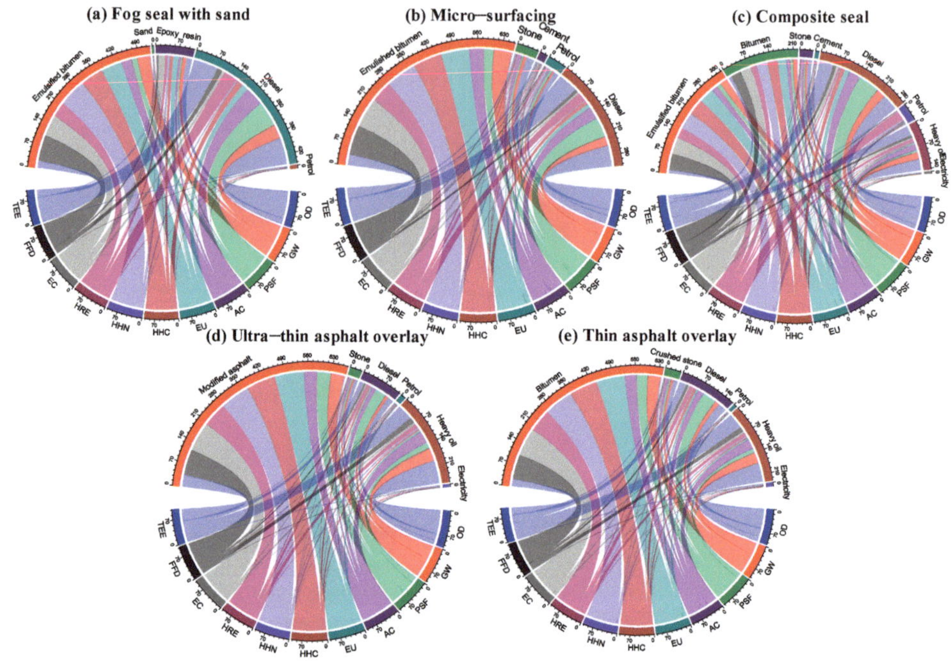

Figure 7. The contribution of different inventory inputs to the LCEI score. (Note: the basic data used to calculate the results of this figure come from Eurobitume [54], European Commission [55], Ecoinvent [56], National Renewable Energy Laboratory [57], National Climate Center of China [26], which were introduced in Section 2.3).

The enrichment string chart was originally used in bioinformatics engineering to characterize which proteins are involved in "Go term" synthesis. In this study, it was used to characterize the percentage contribution of different inventory inputs to ten LCEI categories; that is, all LCEI categories (lower semicircle part) are presented at 100% of the length. The inventory inputs (upper semicircle part) emit lines of different intensities for each LCEI category, where a line of greater intensity indicates a greater contribution to that LCEI category and vice versa, when no line is emitted, no contribution is generated. Regardless of the PM techniques, bitumen materials inputs took up the vast majority of the proportion. Nevertheless, the remaining inputs also play an important role in the LCEI. In terms of fog seal with sand, diesel has the second highest consumption with 205.6 kg, contributing 82.4%, 78.1%, and 75.2%to OD, PSF, and TEE, respectively. It is noteworthy that epoxy resin, with a mass of only 1.5% of raw materials, became the third largest contributor to LCEI, especially EC (22.2%), since the production of polymers releases toxic gases such as aromatic hydrocarbons. Similar to fog seal with sand, diesel was also the second largest contributor to micro-surfacing with consumption of 325.4 kg. Unlike that, the other three techniques (composite seal, ultra-thin asphalt overlay, thin asphalt overlay) consume heavy oil and electricity in addition to diesel, it can be found to consume 448.1 kg, 315.7 kg, and 437.0 kg of diesel and 448.1 kg, 1075.4 kg, and 982.9 kg of heavy oil, respectively, indicating that the fundamental difference in environmental impacts between composite seal and the ultra-thin asphalt overlay lies in the production phase. Although the former requires more machinery and consumes more diesel in construction phase, the latter consumes much more heavy oil in the production phase than that. Therefore, the environmental performance of composite seal is better than ultra-thin asphalt overlay in terms of middle-PM.

3.3. Sensitivity Analysis

Life attenuation is generally regarded as one of the important parameters to evaluate the effectiveness of long-term maintenance [3]. In order to compare the variability of LCEI results generated by different PM techniques under variable lifespan, a sensitivity analysis was conducted in this study. Due to the lack of statistical parameters such as mean value and standard deviation of the actual service life distribution characteristics of these five preventive maintenance techniques, it is difficult to set up different scenarios by means of probability distributions. Given this, in the setting of scenarios and the calculation of LCEI, the authors proposed the following four criteria and principles to simplify the sensitivity analysis process.

1. Unlike hot in-place recycling, its long service life allows researchers to form scenarios by increasing years [27]. In contrast, PM techniques have a short service life and the change in their performance over a short service time can be approximated as a linear decay [61,62], thus, the life descending rate was used for the setting of five scenarios.
2. The thin asphalt overlay was used as the baseline; that is, its lifetime was maintained at 5 years, while the remaining four PM techniques decayed at 0% (Scenario 1), 10%, 20%, 30%, and 40% of the pre-defined lifetime (Section 2.3.5), as shown in Figure 8. For example, the service life in Scenario 2 was decreased by 10%, so the service life of fog seal with sand, micro-surfacing, composite seal, and ultra-thin asphalt overlay were descending from the 4, 4, 3, and 2 years to 3.6, 3.6, 2.7, and 1.8 years, respectively.
3. During the entire analysis period (5 years), the impacts brought by the life decaying only considered as the environmental burden resulting from the repeated maintenance times of the corresponding PM technique.
4. For those old pavements where PM techniques have been implemented, the same PM techniques would be applied continuously at the end of their life.

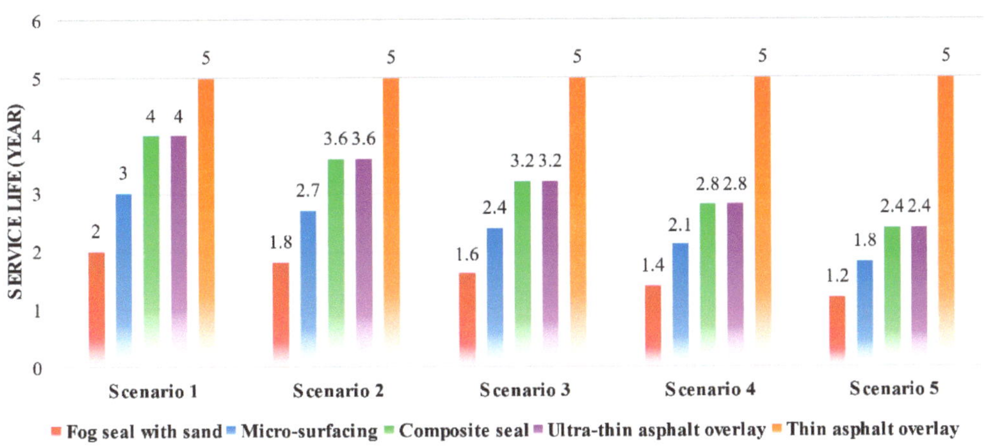

Figure 8. The five analysis scenarios.

Based on the results in Section 3.1, the LCEI under different scenarios can be calculated, and the relative values were used to compare the LCEI under five scenarios from fog seal with sand, micro-surfacing, composite seal, to ultra-thin asphalt. Figure 9 shows the relative environmental performance and trends of the four techniques at different service lives. In the scenario of life decreasing from 0% to 40%, the scale changes in LCEI occur between 5% and 10% due to the smaller impact score of the fog seal compared to thin asphalt overlay, although the frequency of maintenance increases. The PSF, HHC, HHN, EC, FFD generated by micro-surfacing increased with the decay of life, with an average rate of change was 11%. For composite seal and ultra-thin asphalt overlay, the average rate of change was

17% and 21%, respectively. It can be seen that the LCEI results of different PM techniques can be affected by the service life performance and their own LCEI score, because the life attenuation can only affect the maintenance and end-of-life phase. However, the end-of-life phase itself generates less LCEI scores, while the maintenance phase is equivalent to include the raw material phase, the production phase, and the transportation phase. Therefore, when comparing the LCEI of the two maintenance techniques, it is extremely important to use service life performance as a source of sensitivity analysis once the LCEI score of both are close, which may lead to the opposite result. For example, the LCEI scores of ultra-thin asphalt overlay were smaller than those of thin asphalt overlay at 20% life decreasing, but the scores of OD, GW, PSF, AC, etc. were higher than those of thin asphalt overlay with 20% or more.

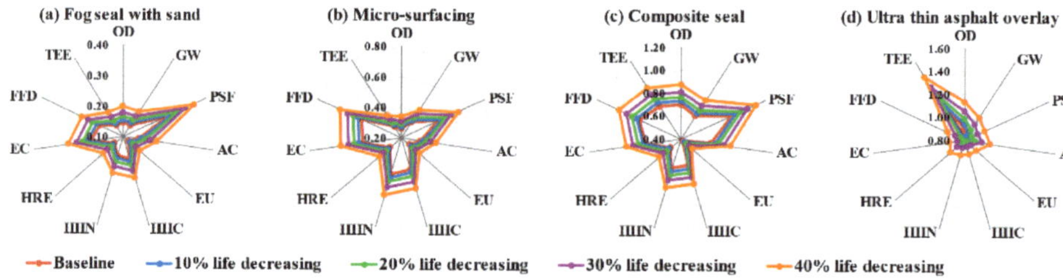

Figure 9. The relative environmental performance and trends under different life decreasing. (Note: the basic data used to calculate the results of this figure come from Eurobitume [54], European Commission [55], Ecoinvent [56], National Renewable Energy Laboratory [57], National Climate Center of China [26], which were introduced in Section 2.3).

3.4. Scenario Analysis

After the LCEI of each PM technique was evaluated, the LCEI of PM planning for the entire life cycle of asphalt pavement can be assessed. Generally, the reduction of pavement service life to less than 75% is the implementation period of PM, while the life of expressway asphalt pavement is 15 years in China. Zheng, et al. [63] stated that 60% of highways in China will undergo major and medium maintenance after 10–12 years. Therefore, the PM planning period for asphalt pavement is 12 years in this study, and four PM plans have been formulated, as shown in Figure 10. The design principles are as follows:

1. The PM techniques from early-PM to later-PM were selected in turn.
2. The performance of PM techniques is not considered to reach the maximum service life at the later PM planning period due to the performance of other structural layers in the pavement may be damaged.
3. PM Plan#1 represents the combination of seal coat and asphalt overlay techniques, PM Plan#2 represents a combination of seal coat techniques. PM Plan#3 and PM Plan#4 has the smallest, largest implementation interval, respectively.
4. The maintenance phase will no longer be included in the life cycle system boundary, but the remaining phases will keep unchanged.

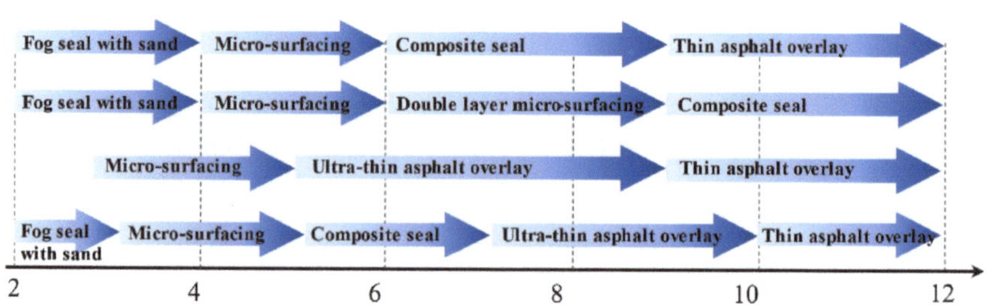

Figure 10. PM plans for the life cycle of the asphalt pavement.

Figure 11 shows the relative LCEI of each PM plan (based on PM Plan#1). The results show that the LCEI generated by PM Plan#3 and PM Plan#1 was relatively small, as the total increment was on average 1.06% higher. Nevertheless, there are still some differences in the individual LCEI categories; specifically, the EU generated by PM Plan#3 was 12% higher than PM Plan#1. The fact that PM Plan#3 requires only three maintenance times, along with the reduction in raw materials and mechanical equipment, making it possible for PM Plan#3 to generate better economic benefits, which will require further consideration of social and economic indicators in the future. Therefore, PM Plan#3 may be better than PM Plan#1.

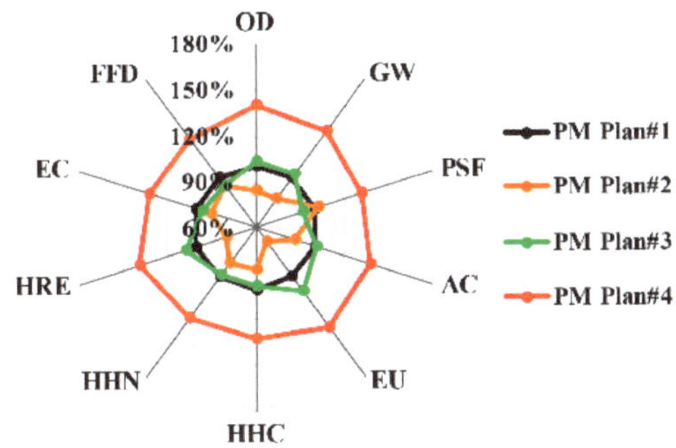

Figure 11. Relative variation of the LCIA results in relation to the baseline PM Plan#1. (Note: the basic data used to calculate the results of this figure come from Eurobitume [54], European Commission [55], Ecoinvent [56], National Renewable Energy Laboratory [57], National Climate Center of China [26], which were introduced in Section 2.3).

PM Plan#2 only included seal coat techniques; in order to maintain the same level of service life performance over the analysis period, more frequent maintenance activities should be required. Strictly, though PM Plan#2 implemented micro-surfacing three times, it still produced the smallest LCEI. In particular, it could reduce the total LCEI by 7–29% compared to PM Plan#1, and all LCEI categories were lower than the other PM Plan, indicating that the implementation of seal coat techniques can allow for asphalt pavement to be eco-friendly and energy-saving. PM Plan#4 had the largest maintenance interval and could achieve the highest performance throughout the analysis period, but it also meant that more agency costs would invested. Compared to PM Plan#1, the implementation of PM Plan#4 would result in a 31~41% improvement in scores for each LCEI category.

As can be seen in Figure 12, thin asphalt overlay caused a large proportion of the LCEI, accounting for 39.3–57.4% in PM Plan#1 and 42.9–51.3% in PM Plan#3. In order to reduce the maintenance frequency, the best techniques from the early-PM, mid-PM, and late-PM were selected for PM Plan#3. It can be found that the LCEI generated by micro-surfacing was the smallest. Therefore, in PM Plan#2, the implementation times of micro-surfacing was increased to avoid the use of a thin asphalt layer, which made it possible to reduce all LCEI categories. Although the seal coat technique was added in PM Plan#4, the asphalt overlay technique was also dominant in the LCEI, which demonstrates the important role of seal coat technique in environmental protection. Despite the increased maintenance, interval is not conducive to the environment, but is beneficial for maintaining performance.

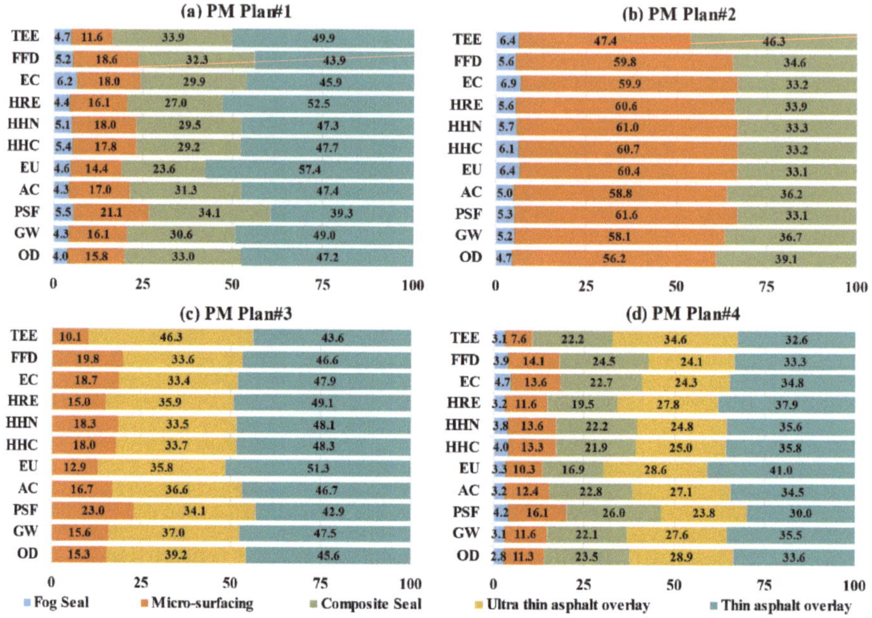

Figure 12. Proportion of LCEI in PM Plan#1–4. (Note: The basic data used to calculate the results of this figure come from Eurobitume [54], European Commission [55], Ecoinvent [56], National Renewable Energy Laboratory [57], National Climate Center of China [26], which were introduced in Section 2.3).

In the planning for the entire life cycle of asphalt pavement, a plan with more maintenance frequency may be selected once only minimizing the LCEI is considered. As the maintenance frequency increases, more user costs will be incurred, and more raw materials will be consumed, thus increasing agency costs. This is not in line with the concept of sustainable pavement management. In future research, it is necessary to comprehensively consider social and economic development indicators to obtain more sustainable PM plans.

4. Conclusions

In this study, the LCEI of five typical PM techniques was evaluated comprehensively by using the LCA method, involving the entire life cycle of asphalt pavement. Then, the sensitivity analysis was also conducted to investigate the LCA results at different decreasing service life rates (0%, 10%, 20%, 30%, and 40%). Finally, four preventive maintenance plans were designed for the 12-year PM period of asphalt pavement, and the LCEI of all PM plans were compared within the given system boundary. The following conclusions are drawn:

1. The seal coat techniques can be seen as an eco-friendly alternative to conventional PM because they can reduce scores in all LCEI categories. Both fog seal with sand and micro-surfacing exhibit great significant environmental performance. Moreover, the composite seal as a technique for middle-PM has better environmental performance than the ultra-thin asphalt overlay.
2. The LCEI categories of the five PM techniques are mainly FFD, EC, GW, and PSF. However, FFD and EC are often neglected by previous researches. Besides, the seal coat technique can reduce TEE by 26.3–86.0% compared to thin asphalt overlay due to the omission or reduction of the mixture production phase.
3. In the entire LCEA of PM techniques, the perceptiveness in performance may not necessarily have a significant impact on the assessment results. It is mainly related to the LCEI categories scores from the raw materials to the construction phase, and once

the scores of both are close, it is highly necessary to consider performance, because it may have the opposite result. In turn, it would not.

4. As the frequency of seal coat techniques increases, the LCEI results show that PM plans can towards more eco-friendly under the same maintenance interval, and the PM plan that includes only seal coat techniques can reduce the total LCEI score by 7–29% compared to the baseline plan.

In contrast to previous studies, this study systematically quantifies all LCEI of the typical PM techniques for asphalt pavement, and discusses the environmental performance involving the entire life cycle of PM planning. However, this is limited to a discussion of environmental performance. Future research can trade-off the entire life cycle agency cost and life cycle user cost to more thoroughly analyze the sustainability of PM techniques and PM planning, which is critical to the development of sustainable pavement management.

Author Contributions: M.Z., W.C.: Conceptualization. W.C.: Methodology. X.D., W.Z., S.Y.: Investigation. M.Z.: Supervision. M.Z.: Project administration. W.C.: Writing—original draft. W.C., M.Z.: Writing—review & editing. X.D.: Funding acquisition. All authors have read and agreed to the published version of the manuscript.

Funding: This research was funded by the National Natural Science Foundation of China (Grant No. 52078051), the Transportation Department of Shandong Province (Lujiaokeji [2017] 28), the Technique Innovation Project of Shandong Department of Industry and Information (Grant No. Lugongxinji [2020] 8).

Institutional Review Board Statement: Not applicable.

Informed Consent Statement: Not applicable.

Data Availability Statement: The data of this study are available from the authors upon request.

Conflicts of Interest: The authors declare no conflict of interest.

Appendix A

List of abbreviations.

AC	Acidification
EC	Ecotoxicity
EEA	Eco-efficiency analysis
EU	Eutrophication
FFD	Fossil fuel depletion
GW	Global warming
HHC	Human health carcinogenic
HHN	Human health noncarcinogenic
HHP	Human health particulate
HMA	Hot-mixed asphalt mixture
LCA	Life cycle assessment
LCC	Life cycle costs
LCEA	Life cycle environmental assessment
LCEI	Life cycle environmental impacts
LCI	Life cycle inventory
LCIA	Life cycle impact assessment
OD	Ozone depletion
PM	Preventive maintenance
PSF	Photochemical smog formation
TEE	Total embodied energy
WMA	Warm-mixed asphalt mixture

Appendix B

List of the notations used in the equations.

P	Total phase of the life cycle.
J	Total category of equipment in the life cycle.
I	Total category of materials in the life cycle.
F	Total category of fuels in the life cycle.
ME_i	Calorific values per mass in one functional unit of i material, MJ/t.
$Q_{i,p}$	Consumption mass of i material in p phase, t.
FE_j	Calorific values per mass of f fuels, MJ/t.
$FC_{f,j,p}$	The f fuels consumption per shift used for j equipment in p phase, t/shift.
$C_{j,p}$	The shifts per functional unit of j equipment in p phase, shift.
MG_i	Emission values per mass per functional unit of i material, g/F.
FG_j	Emission values per mass per functional unit of f fuel, MJ/t.

References

1. Shen, S.; Zhang, W.; Shen, L.; Huang, H. A statistical based framework for predicting field cracking performance of asphalt pavements: Application to top-down cracking prediction. *Constr. Build. Mater.* **2016**, *116*, 226–234. [CrossRef]
2. Chen, J.; Huang, X.-M. Overlay design considering remnant life of old asphalt pavement. *J. Traffic Transp. Eng.* **2009**, *9*, 17–21.
3. Mamlouk, M.S.; Dosa, M. Verification of effectiveness of chip seal as a pavement preventive maintenance treatment using ltpp data. *Int. J. Pavement Eng.* **2014**, *15*, 879–888. [CrossRef]
4. Wei, C.J.; Tighe, S.; Transportat Res, B. Development of preventive maintenance decision trees based on cost-effectiveness analysis—An ontario case study. In *Maintenance and Management of Pavement and Structures*; Transportation Research Board Natl Research Council: Washington, DC, USA, 2004; pp. 9–19.
5. Liu, Z. Asphalt pavement preventive maintenance technology overview. In *Progress in Industrial and Civil Engineering iii, Pt 1*; Liang, J., Wu, X., Yang, W., Chen, W., Eds.; Trans Tech Publications: Zürich, Switzerland, 2014; Volume 638–640, pp. 1135–1138.
6. Cass, D.; Mukherjee, A. Calculation of greenhouse gas emissions for highway construction operations by using a hybrid life-cycle assessment approach: Case study for pavement operations. *J. Constr. Eng. Manag.* **2011**, *137*, 1015–1025. [CrossRef]
7. Chowdhury, R.; Apul, D.; Fry, T. A life cycle based environmental impacts assessment of construction materials used in road construction. *Resour. Conserv. Recycl.* **2010**, *54*, 250–255. [CrossRef]
8. Huang, Y.; Bird, R.; Heidrich, O. Development of a life cycle assessment tool for construction and maintenance of asphalt pavements. *J. Clean. Prod.* **2009**, *17*, 283–296. [CrossRef]
9. Santero, N.J.; Masanet, E.; Horvath, A. Life-cycle assessment of pavements. Part i: Critical review. *Resour. Conserv. Recycl.* **2011**, *55*, 801–809. [CrossRef]
10. Häkkinen, T.; Mäkelä, K. *Environmental Adaption of Concrete: Environmental Impact of Concrete and Asphalt Pavements*; VTT Technical Research Centre of Finland: Espoo, Finland, 1996.
11. Wang, R.R.; Eckelman, M.J.; Zimmerman, J.B. Consequential environmental and economic life cycle assessment of green and gray stormwater infrastructures for combined sewer systems. *Environ. Sci. Technol.* **2013**, *47*, 11189–11198. [CrossRef] [PubMed]
12. Biswas, W.K. Carbon footprint and embodied energy assessment of a civil works program in a residential estate of western australia. *Int. J. Life Cycle Assess.* **2014**, *19*, 732–744. [CrossRef]
13. Haslett, K.E.; Dave, E.V.; Mo, W. Realistic traffic condition informed life cycle assessment: Interstate 495 maintenance and rehabilitation case study. *Sustainability* **2019**, *11*, 3245. [CrossRef]
14. Butt, A.A.; Harvey, J.; Saboori, A.; Ostovar, M.; Bejarano, M.; Garg, N. Decision support in selecting airfield pavement design alternatives using life cycle assessment: Case study of nashville airport. *Sustainability* **2021**, *13*, 299. [CrossRef]
15. Giani, M.I.; Dotelli, G.; Brandini, N.; Zampori, L. Comparative life cycle assessment of asphalt pavements using reclaimed asphalt, warm mix technology and cold in-place recycling. *Resour. Conserv. Recycl.* **2015**, *104*, 224–238. [CrossRef]
16. Turk, J.; Pranjic, A.M.; Mladenovic, A.; Cotic, Z.; Jurjavcic, P. Environmental comparison of two alternative road pavement rehabilitation techniques: Cold-in-place-recycling versus traditional reconstruction. *J. Clean. Prod.* **2016**, *121*, 45–55. [CrossRef]
17. Sollazzo, G.; Longo, S.; Cellura, M.; Celauro, C. Impact analysis using life cycle assessment of asphalt production from primary data. *Sustainability* **2020**, *12*, 10171. [CrossRef]
18. Wang, H. 25-life-cycle analysis of repair of concrete pavements. In *Eco-Efficient Repair and Rehabilitation of Concrete infrastructures*; Pacheco-Torgal, F., Melchers, R.E., Shi, X., Belie, N.D., Tittelboom, K.V., Sáez, A., Eds.; Woodhead Publishing: Cambridge, UK, 2018; pp. 723–738.
19. Choudhary, J.; Kumar, B.; Gupta, A. Utilization of solid waste materials as alternative fillers in asphalt mixes: A review. *Constr. Build. Mater.* **2020**, *234*, 117271. [CrossRef]
20. Anastasiou, E.K.; Liapis, A.; Papayianni, I. Comparative life cycle assessment of concrete road pavements using industrial by-products as alternative materials. *Resour. Conserv. Recycl.* **2015**, *101*, 1–8. [CrossRef]
21. Pratico, F.G.; Giunta, M.; Mistretta, M.; Gulotta, T.M. Energy and environmental life cycle assessment of sustainable pavement materials and technologies for urban roads. *Sustainability* **2020**, *12*, 704. [CrossRef]

22. Tokede, O.O.; Whittaker, A.; Mankaa, R.; Traverso, M. Life cycle assessment of asphalt variants in infrastructures: The case of lignin in australian road pavements. *Structures* **2020**, *25*, 190–199. [CrossRef]
23. Jiang, R.; Wu, P. Estimation of environmental impacts of roads through life cycle assessment: A critical review and future directions. *Transp. Res. Part D* **2019**, *77*, 148–163. [CrossRef]
24. Simões, D.; Almeida-Costa, A.; Benta, A. Preventive maintenance of road pavement with microsurfacing—an economic and sustainable strategy. *Int. J. Sustain. Transp.* **2017**, *11*, 670–680. [CrossRef]
25. Ma, H.; Zhang, Z.G.; Zhao, X.; Wu, S. A comparative life cycle assessment (lca) of warm mix asphalt (wma) and hot mix asphalt (hma) pavement: A case study in china. *Adv. Civ. Eng.* **2019**, *2019*, 9391857. [CrossRef]
26. Cong, L.; Guo, G.; Yu, M.; Yang, F.; Tan, L. The energy consumption and emission of polyurethane pavement construction based on life cycle assessment. *J. Clean. Prod.* **2020**, *256*, 120395. [CrossRef]
27. Cao, R.; Leng, Z.; Hsu, S.-C. Comparative eco-efficiency analysis on asphalt pavement rehabilitation alternatives: Hot in-place recycling and milling-and-filling. *J. Clean. Prod.* **2019**, *210*, 1385–1395. [CrossRef]
28. Santos, J.; Flintsch, G.; Ferreira, A. Environmental and economic assessment of pavement construction and management practices for enhancing pavement sustainability. *Resour. Conserv. Recycl.* **2017**, *116*, 15–31. [CrossRef]
29. Santos, J.; Ferreira, A.; Flintsch, G. A life cycle assessment model for pavement management: Methodology and computational framework. *Int. J. Pavement Eng.* **2015**, *16*, 268–286. [CrossRef]
30. Wang, J.Y.; Yuan, J.; Xiao, F.P.; Li, Z.Z.; Wang, J.; Xu, Z.Z. Performance investigation and sustainability evaluation of multiple polymer asphalt mixtures in airfield pavement. *J. Clean. Prod.* **2018**, *189*, 67–77. [CrossRef]
31. Samieadel, A.; Fini, E.H. Interplay between wax and polyphosphoric acid and its effect on bitumen thermomechanical properties. *Constr. Build. Mater.* **2020**, *243*, 118194. [CrossRef]
32. Thenoux, G.; González, Á.; Dowling, R. Energy consumption comparison for different asphalt pavements rehabilitation techniques used in chile. *Resour. Conserv. Recycl.* **2007**, *49*, 325–339. [CrossRef]
33. Wang, T.; Lee, I.-S.; Kendall, A.; Harvey, J.; Lee, E.-B.; Kim, C. Life cycle energy consumption and ghg emission from pavement rehabilitation with different rolling resistance. *J. Clean. Prod.* **2012**, *33*, 86–96. [CrossRef]
34. Rubio, M.d.C.; Moreno, F.; Martínez-Echevarría, M.J.; Martínez, G.; Vázquez, J.M. Comparative analysis of emissions from the manufacture and use of hot and half-warm mix asphalt. *J. Clean. Prod.* **2013**, *41*, 1–6. [CrossRef]
35. Choi, J.-h. Strategy for reducing carbon dioxide emissions from maintenance and rehabilitation of highway pavement. *J. Clean. Prod.* **2019**, *209*, 88–100. [CrossRef]
36. Ma, F.; Dong, W.; Fu, Z.; Wang, R.; Huang, Y.; Liu, J. Life cycle assessment of greenhouse gas emissions from asphalt pavement maintenance: A case study in china. *J. Clean. Prod.* **2021**, *288*, 125595. [CrossRef]
37. Qi, X.f. Quantization and analysis of carbon emission of asphalt pavement preventive maintenance technology. *Highway* **2017**, *62*, 227–232.
38. Han, B.; Ling, J.; Zhao, H. *Environmental Impacts of Different Maintenance and Rehabilitation Strategies for Asphalt Pavement*; Transportation Research Congress: Beijing, China, 2018; pp. 312–322.
39. Wang, F.; Xie, J.; Wu, S.; Li, J.; Barbieri, D.M.; Zhang, L. Life cycle energy consumption by roads and associated interpretative analysis of sustainable policies. *Renew. Sustain. Energy Rev.* **2021**, *141*, 110823. [CrossRef]
40. Liu, C.L.; Du, Y.C.; Wong, S.C.; Chang, G.Z.; Jiang, S.C. Eco-based pavement lifecycle maintenance scheduling optimization for equilibrated networks. *Transp. Res. Part D* **2020**, *86*, 18. [CrossRef]
41. Umer, A.; Hewage, K.; Haider, H.; Sadiq, R. Sustainability evaluation framework for pavement technologies: An integrated life cycle economic and environmental trade-off analysis. *Transp. Res. Part D* **2017**, *53*, 88–101. [CrossRef]
42. ISO14044. *Environmental Management: Life Cycle Assessment*; Requirements and Guidelines; International Organization for Standardization: Geneva, Switzerland, 2006.
43. MOT. *Jtg/t 3832-2018: Budget Quota of Highway Engineering*; Ministry of Transport of China: Beijing, China, 2018.
44. MOT. *Jtg/t 3833-2018: Budget Quota of Highway Engineering Machinery Shift*; Ministry of Transport of China: Beijing, China, 2018.
45. Zhang, H. Effectiveness Evaluation of Asphalt Pavement Preventive Maintenance Treatments in Freeway. Master's Thesis, South China University of Technology, Guangzhou, China, 2015.
46. Editorial Department of China Journal of Highuay and Transport. Review on China's pavement engineering research·2020. *China J. Highw. Transp.* **2020**, *33*, 1–66.
47. Liu, M.M.; Han, S.; Wang, Z.Y.; Ren, W.Y.; Li, W. Performance evaluation of new waterborne epoxy resin modified emulsified asphalt micro-surfacing. *Constr. Build. Mater.* **2019**, *214*, 93–100. [CrossRef]
48. Sun, X.L.; Liu, Z.S.; Qin, X.; Zeng, D.F.; Yin, Y.M. Purifying effect evaluation of pavement surfacing materials modified by novel modifying agent. *Front. Mater.* **2020**, *7*, 180. [CrossRef]
49. Bai, R.; Lyu, C. Application of cape seal technology in the preventive maintenance engineering of national and provincial trunk road. *Highway* **2019**, *64*, 252–258.
50. Im, S.; You, T.; Kim, Y.-R.; Nsengiyumva, G.; Rea, R.; Haghshenas, H. Evaluation of thin-lift overlay pavement preservation practice: Mixture testing, pavement performance, and lifecycle cost analysis. *J. Transp. Eng. Part B Pavements* **2018**, *144*, 04018037. [CrossRef]
51. Li, J.; Xiao, F.; Zhang, L.; Amirkhanian, S.N. Life cycle assessment and life cycle cost analysis of recycled solid waste materials in highway pavement: A review. *J. Clean. Prod.* **2019**, *233*, 1182–1206. [CrossRef]

52. BRE. *Jlzj-y-gl-003-2020: Budget Quota for Major and Medium Repair Projects of Beijing Ordinary Highways*; Beijing Road Engineering Cost Quota Management Station: Beijing, China, 2020.
53. European Commission—Joint Research Centre—Institute for Environment and Sustainability. General Guide for Life Cycle Assessment—Detailed Guidance. 2010. Available online: https://eplca.jrc.ec.europa.eu/uploads/ILCD-Handbook-General-guide-for-LCA-DETAILED-GUIDANCE-12March2010-ISBN-fin-v1.0-EN.pdf (accessed on 10 October 2020).
54. Eurobitume. *Life Cycle Inventory: Bitumen*; European Bitumen Association: Brussels, Belgium, 2012.
55. European Commission. Elcd European Life Cycle Database. 2013. Available online: Http://eplca.Jrc.Ec.Europa.Eu/?Page_id=126 (accessed on 22 November 2014).
56. Ecoinvent. *Ecoinvent Database v3.0. St-Gallen: Swiss Centre for Life Cycle Inventories*; Ecoinvent: St. Gallen, Switzerland, 2013.
57. National Renewable Energy Laboratory. U.S. Life Cycle Inventory Database. 2012. Available online: Https://www.Lcacommons.Gov/nrel/search (accessed on 19 November 2012).
58. MOT. *Jtg 5421-2018: Specifications for Maintenance Design of Highway Asphalt Pavement*; Ministry of Transport of China: Beijing, China, 2019.
59. Wu, S. Research on the Invironmental Impact of HMA Pavement and WMA Pavement Based on LCA. Master's Thesis, School of Transportation Southeast University, Nanjing, China, 2015.
60. Bare, J.C. Tools for the Reduction and Assessment of Chemical and Other Environmental Impacts (Traci)-User's Manual. EPA/600/R-12/554: 2012. Available online: Https://www.Epa.Gov/chemical-research/tool-reduction-and-assessment-chemicals-and-other-environmental-impacts-traci (accessed on 11 December 2014).
61. Dong, Q.; Huang, B. Cost-effectiveness evaluation of pavement maintenance treatments by optime. In Proceedings of the GeoShanghai 2010 International Conference, Shanghai, China, 3–5 June 2010; pp. 435–440.
62. Dong, Q.; Huang, B.S.; Richards, S.H.; Yan, X.D. Cost-effectiveness analyses of maintenance treatments for low- and moderate-traffic asphalt pavements in tennessee. *J. Transp. Eng.* **2013**, *139*, 797–803. [CrossRef]
63. Zheng, J.; Lü, S.; Liu, C. Technical system, key scientific problems and technical frontier of long-life pavement. *Chin. Sci. Bull.* **2020**, *65*, 3219–3229.

MDPI AG
Grosspeteranlage 5
4052 Basel
Switzerland
Tel.: +41 61 683 77 34

Sustainability Editorial Office
E-mail: sustainability@mdpi.com
www.mdpi.com/journal/sustainability

www.ingramcontent.com/pod-product-compliance
Lightning Source LLC
Chambersburg PA
CBHW041418210926
51850CB00013B/3305
* 9 7 8 3 7 2 5 8 2 9 0 2 6 *